Ulrich Schubert (ed.)

Silicon Chemistry

SpringerWienNewYork

Univ.-Prof. Dr. Ulrich Schubert
Institut für Anorganische Chemie, Technische Universität, Wien, Österreich

Typesetting: Thomson Press (India) Ltd., New Delhi
Printing: Druckerei Ketterl Ges.m.b.H., A-1180 Wien
Printed on acid-free and chlorine-free bleached paper

With 89 partly coloured Figures

CIP data applied for

Special Edition of "Monatshefte für Chemie / Chemical Monthly", Vol. 130, No. 1, 1999

ISBN 3-211-83278-5 Springer-Verlag Wien New York

Contents

Editorial

The central position of silicon in the periodic system of the elements, at the borderline between organic and inorganic chemistry on one hand and between metallic and nonmetallic elements on the other hand, results in a manifold and unique chemical behavior of its compounds, hardly paralleled by any other element. Furthermore, silicon and silicon compounds have contributed decisively to the technical progress. Technical applications range from mass commodities to highly sophisticated special materials, from ceramics to polymers, from medicine to microelectronics.

To keep pace with scientific and technical developments in other industrial countries in this important area, the *Deutsche Forschungsgemeinschaft* (DFG) and the *Fonds zur Förderung der wissenschaftlichen Forschung in Österreich* (FWF) decided to establish national priority programs (called *Schwerpunktprogramm* in Germany and *Forschungsschwerpunkt* in Austria). The programs are strongly linked to each other as well as to some Swiss groups and help to strengthen the scientific "silicon community" in the German-speaking countries.

The German *Schwerpunktprogramm* "Specific phenomena in silicon chemistry: compounds for the construction and understanding of extended systems with novel properties" was established by the DFG in October 1994 and started its work in July 1995 with 34 projects. The program was evaluated in 1997 and then extended into a second funding period with 35 projects. The Austrian *Forschungsschwerpunkt* "Novel approaches to the formation and reactivity of compounds with silicon-silicon bonds" was established in October 1996 and started its work in January 1997.

At mid-term of the German program and the end of the first funding period of the Austrian program it appears appropriate to summarize part of the results in a special publication. I am very glad that the Editors of *Monatshefte für Chemie* and *Springer-Verlag* providing the opportunity on that occasion to publish a special issue of "Monatshefte für Chemie / Chemical Monthly", of which this volume is a special edition, and that most of the chemical groups participating in the projects agreed to contribute research papers. Although most of the papers are rather short due to page restrictions, this volume gives a very good overview on the current chemical (and partly physical) activities in the joint Austrian/German/Swiss program.

My special thanks go to *Isabella Krcmaricic* and Dr. *Hermann Kalchhauser* for their technical assistance in publishing this volume.

Ulrich Schubert

Invited Review

Carbosilane Dendrimers – Synthesis, Functionalization, Application

Christian Schlenk and **Holger Frey***

Freiburger Materialforschungszentrum und Institut für Makromolekulare Chemie, Albert Ludwigs Universität Freiburg, D-79104 Freiburg, Germany

Summary. This micro-review summarizes historical and recent developments in synthesis, functionalization, and characterization of carbosilane dendrimers. These compounds are the most important among the heteroatom-based dendrimers at present due to the flexibility of the synthetic route and to their chemical stability which allows versatile further functionalization. Synthesis as well as application potential of carbosilane dendrimers for catalysis, host-guest chemistry, liquid crystals, and novel polymer topologies is discussed.

Keywords. Carbosilane; Dendrimer; Catalysis; Liquid crystals; Hyperbranched.

Carbosilandendrimere – Synthese, Funktionalisierung und Anwendung

Zusammenfassung. Der vorliegende Mikroreview faßt historische und aktuelle Entwicklungen bezüglich Synthese, Funktionalisierung und Charakterisierung von Carbosilandendrimeren zusammen. Wegen der Variabilität der Syntheseroute und ihrer chemischen Stabilität, die weitere Funktionalisierung erlaubt, zählen diese Verbindungen zu den wichtigsten unter den Heteroatomdendrimeren. Synthese und Potential der Carbosilandendrimere auf den Gebieten der Katalyse, Wirt-Gast-Chemie, Flüssigkristalle und neuer Polymertopologien werden diskutiert.

1. Introduction

Dendrimers are three-dimensional, highly branched, monodisperse (in the ideal case) macromolecules with perfectly defined branching structure. Together with the hyperbranched polymers they represent the class of the cascade molecules. However, in contrast to the randomly branched hyperbranched polymers, dendrimers possess defined branches emanating from a central core (Fig. 1).

The dendrimer scaffold can be subdivided into three regions: (*i*) the core from which the branching units emanate, (*ii*) the region of the inner repeat units, and (*iii*) the outer region with the end groups. The individual layers around the core are designated "generations". Dendrimers are built up by iterative stepwise reaction

* Corresponding author

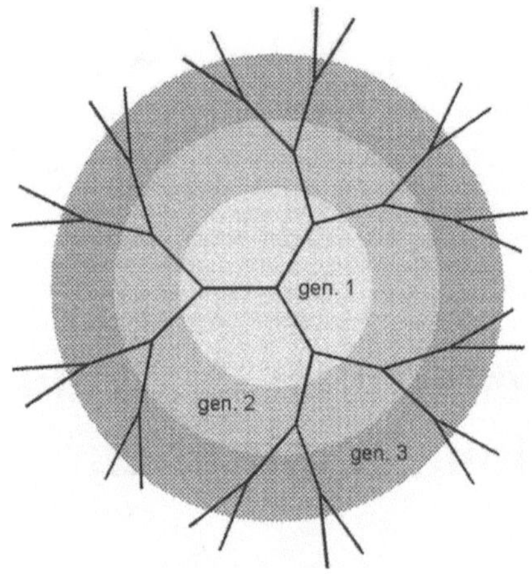

Fig. 1. Schematic representation of the dendrimer structure

sequences, either in the convergent [1, 2] or divergent [3] approach. In the divergent strategy, dendrimers are built from the central core outwards to the periphery. The convergent approach constructs the dendrimer from the periphery towards the central core.

Some of the developments in the area of the heteroatom-based dendrimers have been summarized in recent reviews [4, 5]. Silicon chemistry offers several quantitative reactions suitable for the construction of perfect dendrimers, such as transformation of chlorosilanes with organometallic reagents, *e.g. Grignard* reagents, as well as Pt-catalyzed hydrosilylation. To date carbosilane, carbosiloxane, and small silane dendrimers have been described. At present, carbosilane dendrimers represent the most important class of Si-based dendrimers. They are kinetically as well as thermodynamically very stable molecules owing to the dissociation energy of the Si–C bond (306 kJ/mol), which is comparable to that of C–C bonds (345 kJ/mol), and the low polarity of the Si–C bond. This is an important prerequisite for further functionalization. This micro-review aims at a summary of the developments in the synthesis, functionalization, and characterization of carbosilane dendrimers as well as some of their potential applications.

2. Synthesis and Characterization

2.1. General Synthetic Strategy

To date, all reported carbosilane dendrimers have been synthesized *via* the divergent approach (Fig. 2). *Fetters et al.* reported the use of a G1 carbosilane dendrimer with 12 end groups for the synthesis of a star polymer as early as 1978 [6]. However, *van der Made et al.* [7, 8], *Roovers et al.* [9, 10], and *Muzafarov et al.* [11] independently reported the first syntheses aiming at carbosilane dendrimers of various generations. Starting from the central core (*G0*) possessing *f* alkenyl groups, the dendrimer is constructed using repeating sequences of alternating hydrosilylations with chlorosilanes and ω-alkenylations with *Grignard* reagents. *Van der Made et al.* used tetraallylsilane as core, $HSiCl_3$ as hydrosilylation reagent,

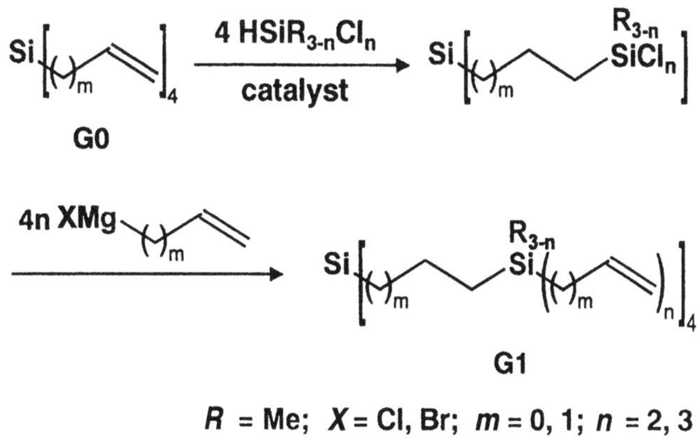

Fig. 2. Divergent synthesis of a typical carbosilane dendrimer

and allylmagnesium bromide as ω-alkenylation reagent to obtain dendrimers up to the fifth generation. However, it should be emphasized that the molecular weight and the structural perfection of these dendrimers were not substantiated by appropriate analytical methods. *Roovers et al.* started from a tetravinylsilane core and used $HMeSiCl_2$ and vinylmagnesium bromide to obtain dendrimers possessing a somewhat open structure. Dendrimers with an even more open structure were obtained by *Muzafarov* and coworkers who chose $HMeSiCl_2$ as hydrosilylation reagent, allylmagnesium chloride as ω-alkenylation reagent, and triallylmethylsilane as a core.

Obviously, the synthetic route to carbosilane dendrimers offers high flexibility and versatility. Not only the hydrosilylation reagent and the ω-alkenylation reagent (to some extent) but also the core molecule can be varied without drastic changes in the reaction conditions. Table 1 summarizes the carbosilane dendrimers reported to date (core molecules, segment length, and branching multiplicity).

The most popular core molecules are tetraallylsilane and tetravinylsilane, which lead to dendrimers of spherosymmetrical topology. To date, the branching multiplicity has been varied from two to three and the length of the alkyl spacer from two to three methylene units (cf. Table 1). The use of a long alkylmagnesium bromide as *Grignard* reagent has been reported in one case [8]. However, reaction between long alkylmagnesium bromide and tetrahedral silicon is problematic [12].

Numerous reports on the synthesis of carbosilane dendrimers with allyl end groups have been published by *Kim et al.* [42–44, 46] who used various core molecules (2,4,6,8-tetramethyl-2,4,6,8-tetravinyltetrasiloxane, diallylphenylmethylsilane, and triallylmethylsilane) and constructed the dendrimers with allylmagnesium bromide as *Grignard* reagent and either $HSiCl_3$ or $HMeSiCl_2$ as hydrosilylation reagent. In further publications, these authors report the synthesis of carbosilane dendrimers terminated by phenylethynyl and *p*-bromophenoxy and biphenyl groups, respectively [47, 48]. In one case, the obtained products were characterized by MALDI-TOF mass spectrometry [48].

The synthesis of carbosilane dendrimers bearing 12 or 36 phenyl rings at the periphery has been reported by *Friedmann* and coworkers [37]. The obtained dendrimers were characterized by X-ray diffraction analysis and NMR spectroscopy. NMR showed the retention of solvent with the second generation.

Gossage et al. described the synthesis and characterization of a carbosilane dendrimer with a functionalizable core [52, 53]: 4-triallylsilylphenol, whose phenolic hydroxy group was protected by a *tert*-butyldimethylsilyl group while constructing the dendrimer scaffold. After deprotection, this core permits facile attachment of transition metal complexes or molecular probes.

Table 1. Carbosilane dendrimers reported to date: core molecules, segment structure, and branching multiplicity (b.m.)

Core (*G0*)	Spacer	b.m.	Ref.	Core (*G0*)	Spacer	b.m.	Ref.
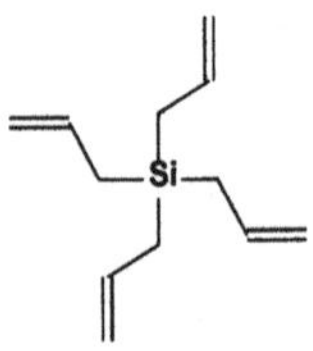	C3	3 2	[7, 8, 13–23} [24–34]	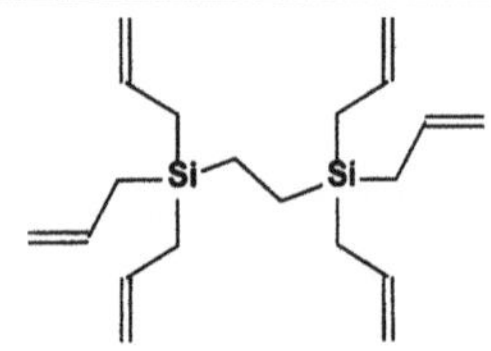	C3	2	[48]
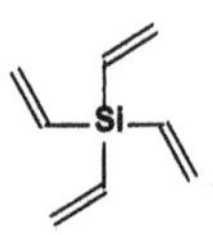	C2	3 2	[35–37] [9, 10, 38–41]	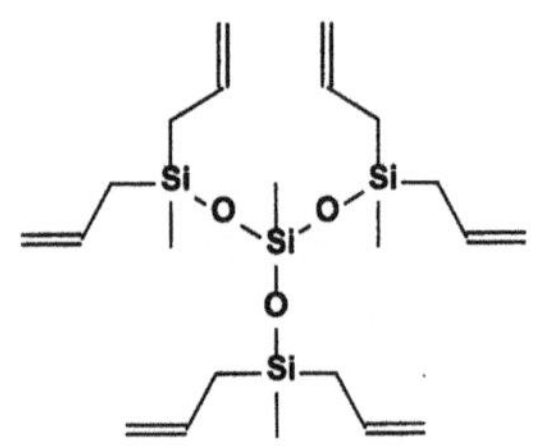	C3	2	[49]
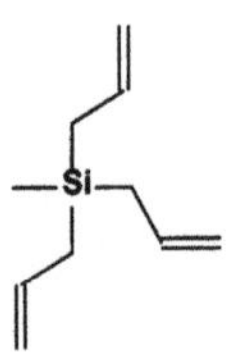	C3	2	[11, 42]	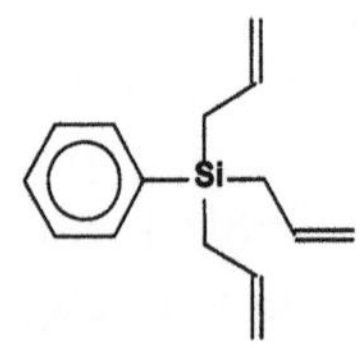	C3	2	[50]
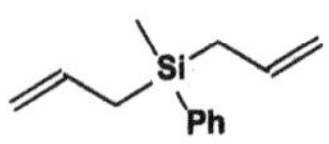	C3	3 2	[43] [44]	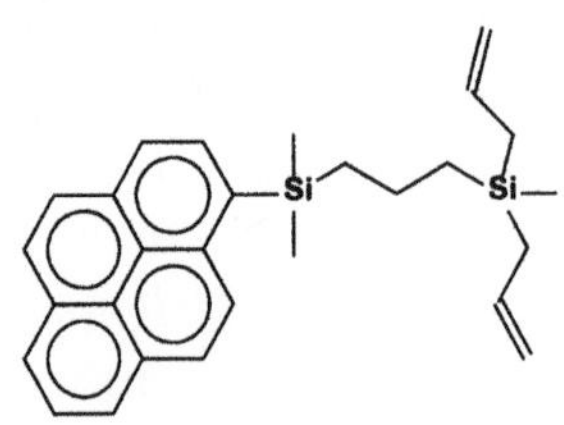	C3	2	[51]
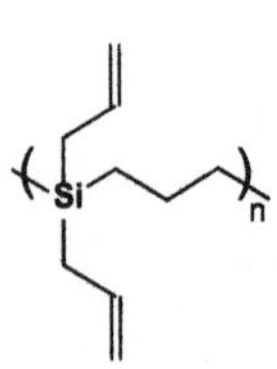	C3	3	[45]	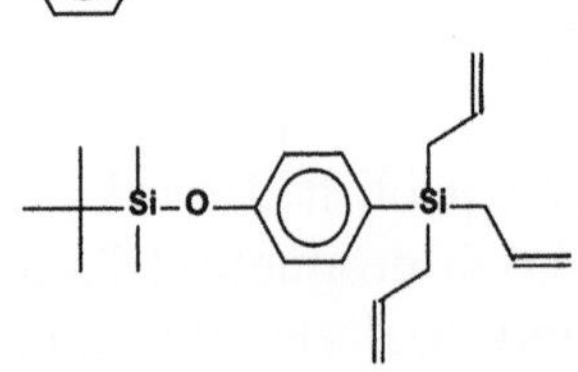	C3	3	[52]
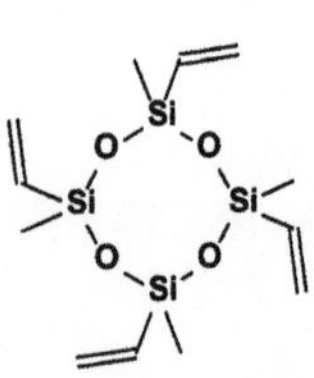	C3	3/2 [2]	[46] [47]				

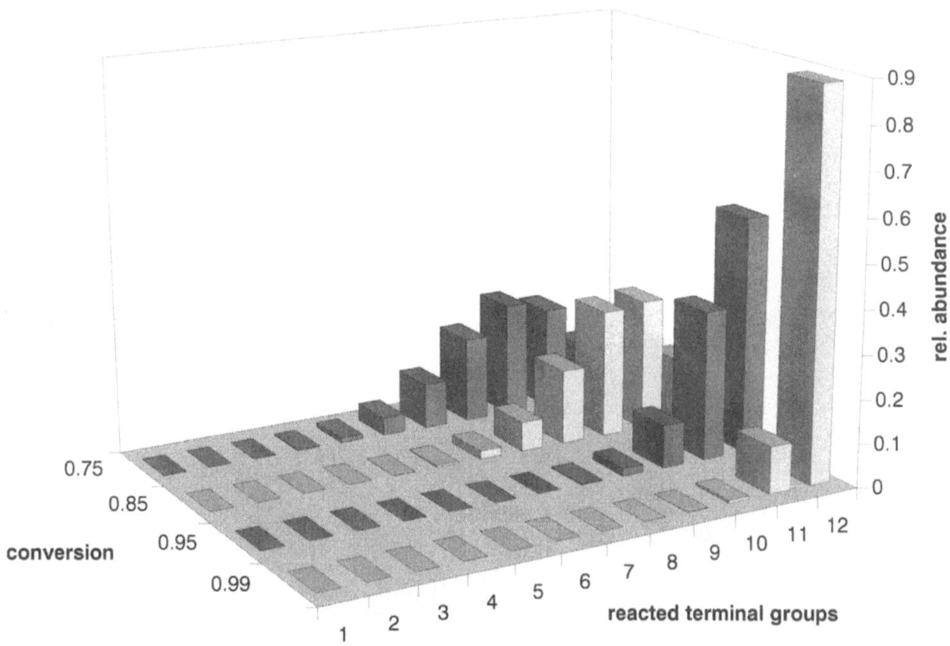

Fig. 3. Relative abundance of dendrimer molecules *vs.* conversion and number of reacted terminal groups, demonstrating the importance of full conversion in dendrimer synthesis

2.2. Other Systems

Besides the carbosilane dendrimers whose synthesis is based on the repeating sequences of alternating hydrosilylations with chlorosilanes and ω-alkenylations with *Grignard* reagents, only a few other systems have been developed: *Nakayama* and *Lin* synthesized the first generation of an organosilicon dendrimer composed of thiophene rings [54]. Tetralithiation of tetra-2-thienylsilane followed by reaction with methoxy-tri-2-thienylsilane gave the desired first generation, 5,5′,5″,5‴-*tetrakis*(tri-2-thienylsilyl(tetra-2-thienyl))silane, which forms inclusion complexes with CCl_4, CH_2Cl_2, benzene, and acetone when crystallized from these solvents or mixtures of these solvents and hexane. *Kim et al.* started from *tetrakis*(phenyl-ethynyl)silane and obtained a carbosilane dendrimer of the third generation *via* a repeated sequence of alternating hydrosilylations with dichloromethylsilane and ω-alkynylations with lithium phenylacetylide. Figure 4 shows generation 2 of this interesting new class of carbosilane dendrimers.

2.3. Molecular Characterization

In the early days of dendrimer chemistry, molecular characterization was carried out in most cases by ^1H, ^{13}C, and ^{29}Si NMR spectroscopy as well as size exclusion chromatography (SEC). However, in higher generations the differences between the relative intensities of the signals of the perfect structure and those of the imperfect structure are too marginal to be detected. Thus, NMR spectra can only be

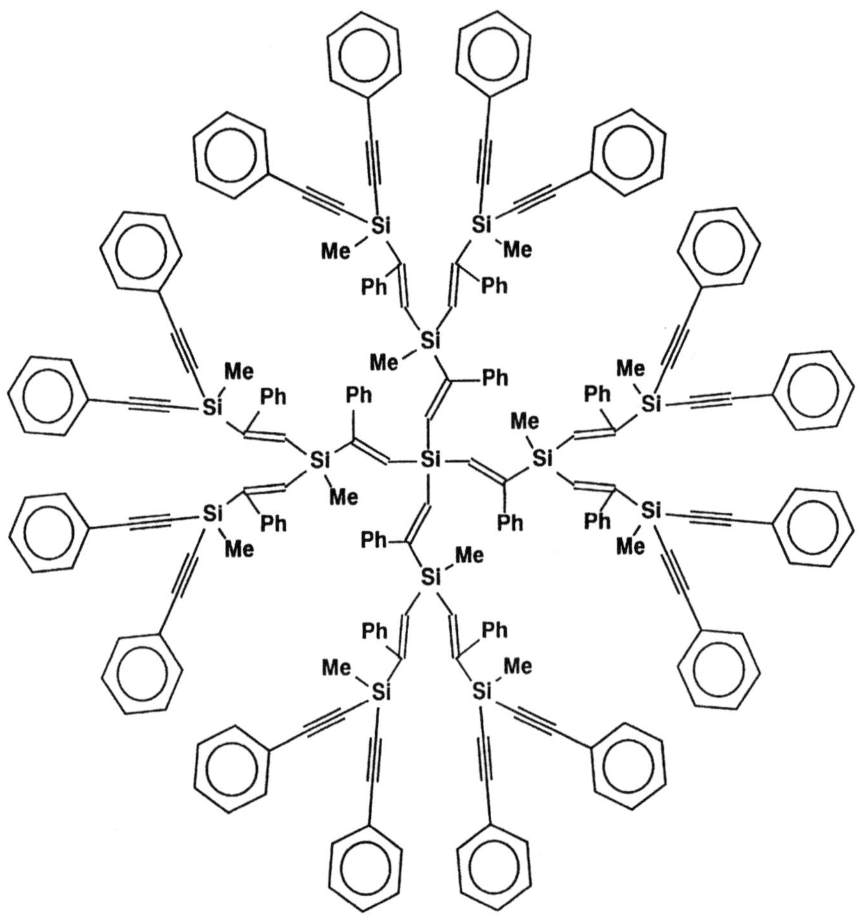

Fig. 4. Generation 2 of carbosilane dendrimers based on phenylethynyl segments

taken as an indication of the perfection of the dendrimer structure. The usefulness of SEC to determine molecular weights or molecular weight distribution of dendrimers is also limited, because dendrimers do not fit the calibration curves based on linear standards. Generally, one observes an overestimation of the molecular weight for dendrimers of lower generations and an underestimation for dendrimers of higher generations. Owing to the broadening of the SEC traces by diffusion, correct molecular weight distributions are not obtained.

With the introduction of MALDI-TOF mass spectrometry (matrix assisted laser desorption ionization and time of flight detection) in dendrimer chemistry, it became possible to determine not only molecular weights but also the molecular weight distribution of dendrimers, as in the MALDI-TOF mass spectrometry fragmentation of the molecular ions does not occur [15, 17, 26, 48, 55].

3. Properties and Application Potential

3.1. Metal Complexes and Catalysis

One of the most promising applications of carbosilane dendrimers is the use as scaffolds for catalytically active metal complexes. Dendrimer-bound catalysts combine the advantages of heterogenous and homogenous catalysis; they allow the accurate control of the number and structure of the active sites, as homogenous catalysts do, and they are easily removed from a product containing solution using ultrafiltration as known from heterogenous catalysts. The first example of a homogenous catalyst based on a dendritic carbosilane scaffold was reported by *van Koten et al.* [13]. They connected 4-amino substituted 2,6-*bis*((dimethylamino)-methyl)-1-bromobenzene, a precursor entity for the organometallic site, to the periphery of generation 0 with 4 end groups and generation 1 with 12 end groups, respectively, by a 1,4-butanediol linker. The obtained dendritic precursor was then reacted with an excess of the zerovalent nickel complex $Ni(PPh_3)_4$ to produce the desired nickel complex containing dendrimers. The prepared dendrimers were successfully employed as homogenous catalysts for the *Kharasch* addition reaction.

A carbosilane dendrimer with 12 peripheral iodoarene groups was prepared from polyol precursors by *Hoare et al.* [21]. The iodoarene groups were attached to the polyols by esterification with 4-iodobenzoyl chloride. The obtained compound was reacted with $Pd(dibenzylideneacetone)_2$ in presence of N,N,N′,N′-tetramethyl-ethylenediamine to yield periphery-palladated complexes. The prepared dendrimer represented the first example of an exclusively σ-bonded completely periphery-palladated dendrimer (Fig. 5).

The strategies developed to introduce metals at the peripheral sites of carbosilane dendrimers were summarized recently by *van Koten* and *Jastrzebski* [14]. In further elegant work the incorporation of reactive Ru(II) *PCP′* complexes into carbosilane dendrimers was described [56].

Alonso et al. reported the first carbosilane dendrimers bearing a controlled number of redox active centres, *i.e.* ferrocenyl units [24]. They attached the corresponding ferrocenyl containing organometallic units to *G0* and *G1 via* two different synthetic routes: the first included the reaction between ferrocenyllithium with chlorosilane terminated dendrimers, the second used also the Si–Cl bonds of the chlorosilane terminated dendrimers, but converted them into Si–N bonds by reaction with β-aminoethylferrocene. The dendrimers were isolated as red-orange materials. The CVs of these compounds exhibited a single reversible oxidation wave. The number of electrons exchanged was found to correspond directly to the number of ferrocenyl units. In addition, differential pulse voltammetry measurements showed only one wave. Thus, the metal centres act independently in redox processes, although they are coupled by the dendrimer scaffold. Such materials have been used to modify redox electrodes [57].

Carbosilane dendrimers have been used furthermore as scaffolds for chromium tricarbonyl moieties [25]. *Lobete* and coworkers prepared a carbosilane dendrimer functionalized with phenyl rings on the periphery by termination of the dendrimer synthesis with a hydrosilylation step with dimethylphenylsilane. The π-coordinating ability of these surface-located arene rings has been utilized to attach the chromium tricarbonyl moieties. It was found that the $Cr(CO)_3$ moieties are bonded by η^6-coordination at the phenyl ring. Cyclopentadienyl functionalized dendrimers used to coordinate $Co(CO)_2$ as well as carbosilane dendrimers containing Si–Co and Si–Fe σ-bonds have also been reported [58]. However, it has to be mentioned that in this case only *G0* dendrimers were used.

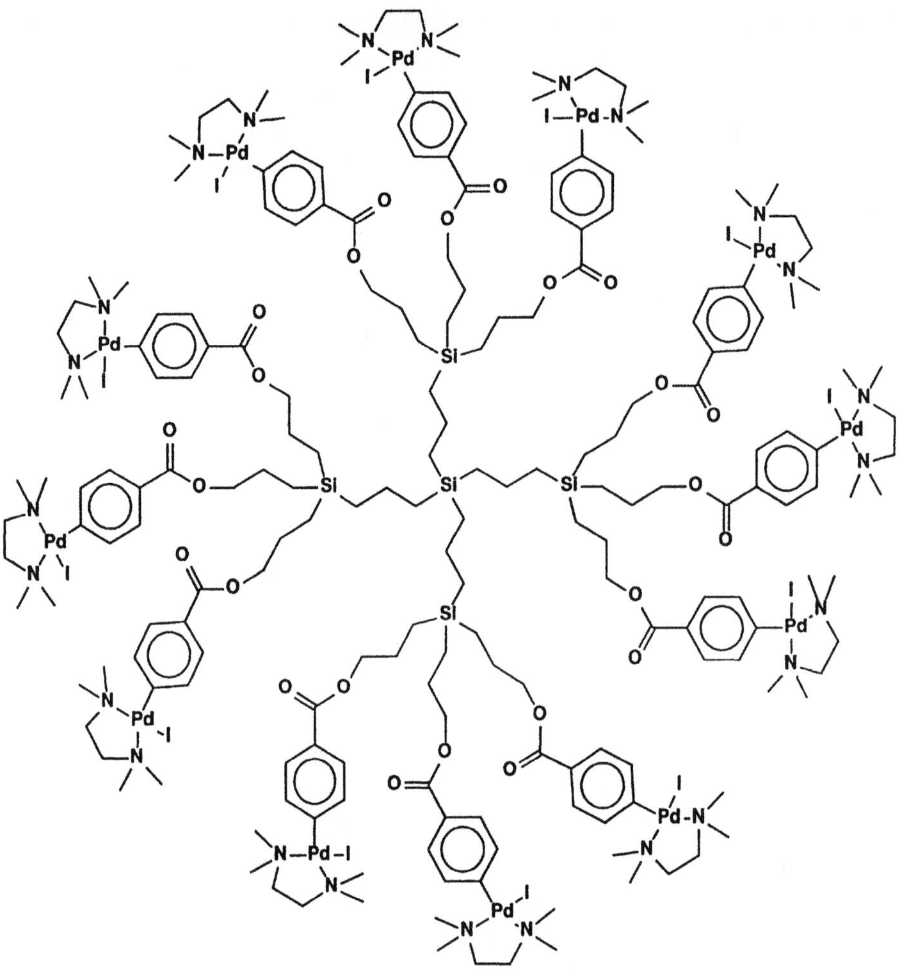

Fig. 5. Organopalladium functionalized carbosilane dendrimer [21]

Carbosilane dendrimers bearing peripheral acetylene-dicobalt hexacarbonyl substituents have been reported by *Seyferth et al.* [36] who prepared dendrimers of the first and second generation that contain peripheral ethynyl groups. These compounds were reacted with dicobalt octacarbonyl to give the desired complexes on the surface of the dendrimers.

3.2. Host-Guest-Chemistry

Due to their inner cavities and the fixed spherical topology, dendrimers with an amphiphilic structure may be regarded as unimolecular analogs of micelles. The first dendrimer that acts like a micelle of common tensides was synthesized by *Newkome et al.* [59] who prepared a carboxylate terminated hydrocarbon dendrimer which – in contrast to usual amphiphiles – shows no critical micelle concentration.

Carbosilane chemistry permits to control the size of the inner cavities. This has been shown by *Lach et al.* in a molecular force field study concerning the host properties of carbosilane dendrimers [60]. *Lach et al.* found that the dimensions of the inner cavities can be varied from 5–15 Å through variation of branching multiplicity and/or spacer length.

Only recently, *Krska* and *Seyferth* reported the synthesis of water soluble carbosilane dendrimers [41]. Nucleophilic reactions between mercapto substituted amphiphiles and carbosilane dendrimers bearing (chloromethyl)silyl groups on their terminal branches led to amphiphilic dendrimers with hydrophobic carbosilane cores and hydroxyl, dimethylamino, or sodium sulfonate amphiphilic groups at the periphery.

3.3. Dendritic Liquid Crystalline Polymers (DLCP)

Currently, there is growing research interest in the combination of branched structures and mesogenic units, motivated by the fact that a globular shape might reduce the bulk viscosity and hence the switching times of the materials. For example, hyperbranched topologies have been combined with mesogenic units by *Bauer et al.* [61] and *Percec et al.* [62]. Furthermore, *Percec et al.* reported on a perfect dendrimer with mesogenic branching units showing thermotropic liquid crystalline phases [63]. A more recent approach is concerned with the coupling of flexible, dendritic carbosilane scaffolds with rigid mesogenic units resulting in dendritic liquid crystalline polymers (DLCPs). The attachment of mesogenic units to the flexible carbosilane dendrimer scaffold leads to a structural conflict between preferential anisotropic order of the mesogenic units and the spherosymmetry of the dendrimer. We have prepared carbosilane dendrimers with 12, 36, and 108 cholesteryl end groups *via* esterification of dendritic carbosilane polyols with cholesteryl chloroformiate [16, 17]. In further studies we have investigated the influence of *i)* generation, *ii)* spacer length, and *iii)* type of mesogen attached on the phase behaviour of the dendritic liquid crystalline polymers [19, 20, 23]. All obtained DLCPs have been fully characterized by polarizing microscopy, DSC, WAXS, and SAXS with respect to their phase behaviour in bulk. It was found that most of the DLCPs investigated develop layered (smectic) structures which are rationalized by separate ordering of the calamitic surface groups and the core which, however, leads to a deformation of the dendritic scaffolds. Reducing the spacer length and/or increasing the number of end groups (*i.e.* the generation number) complicates the formation of well-developped smectic phases.

Similar results were obtained by *Ponomarenko et al.* [27] and *Terunuma et al.* [50]. *Ponomarenko et al.* used cyanobiphenyl, methoxyphenyl benzoate, and cholesteryl groups as mesogenic units. These units were coupled to the carbosilane dendrimers bearing eight allyl groups by hydrosilylation. Smectic phases were also found for carbosilane dendrimers substituted with mesogenic units based on azobenzene at their surface [64].

Besides classical calamitic mesogens, perfluoralkyl groups ($-C_6F_{13}$) have been attached to carbosilane dendrimers [22]. The attachment of the perfluorinated alkyl groups to the allyl end groups of the dendrimers was performed *via* free radical addition of 3,3,4,4,5,5,6,6,7,7,8,8,8-tridecafluoro-*n*-octyl mercaptan, which affords the corresponding thioether functionalized end groups. Perfluorinated dendrimers of generation 0 to generation 3 with 4, 12, 36, and 108 perfluoralkyl end groups, respectively, have been obtained. The dendrimer periphery of these dendrimers is immiscible with the carbosilane interior. Such core-shell type dendrimers also exhibit generation dependent mesophase formation.

3.4. Polymer Architectures Based on Carbosilane Dendrimers

Due to their spherical topology and the large number of end groups, carbosilane dendrimers have been used as core molecules for star shaped polymers. Two elegant approaches have been reported to date. *Roovers et al.* obtained multiarm star polymers with 32, 64, and 128 arms, respectively, by coupling Si–Cl terminated dendrimers with living poly(butadienyl)lithium [9, 39, 65]. The arm molecular weight was varied between 6400 and 72000 g/mol.

Vasilenko et al. reported another way to star shaped polymers based on carbosilane dendrimers [31, 34] using polylithium derivatives of carbosilane dendrimers as initiators for the anionic polymerization of different monomers such as styrene, hexamethylcyclotrisiloxane, and ethylene oxide. The polylithium derivatives were obtained by hydrosilylation of allyl terminated dendrimers with the sterically demanding bidecylmethylsilane. This led to the reaction of only one half of the end groups leaving allyl groups unreacted in the interior of the dendrimer. The reaction of these allyl groups with *sec*-butyllithium yielded the desired polylithium derivative of the carbosilane dendrimer. Due to the location of the lithium atoms in the inner space of the dendrimer, a main problem of polylithium compounds, *i.e.* the high tendency of aggregation, was solved.

4. Conclusions and Outlook

In this short review we have summarized historical and recent developments in the field of carbosilane dendrimers. Due to their chemical stability and the possibility to tailor branching multiplicity, segment length, and core molecule, these are useful scaffolds for supramolecular dendrimer chemistry in general. Their flexibility renders them an important synthetic platform for dendrimer based liquid crystalline polymers that show generation dependent liquid crystalline phases. The versatile construction strategy permits design of dendrimer structures with varying geometry, which is necessary to establish structure-property correlations valid for dendrimers in general. On the other hand, such dendrimers can be used to create novel polymer structures, *e.g.* multiarm star polymers.

Concerning the potential application of such materials, on the long run hyperbranched carbosilane polymers (not discussed in the current article) might become strong competitors for dendrimers, particularly in applications where perfection is not a necessary prerequisite. These polymers are prepared in a one-step procedure by polymerization of AB_2 or AB_3 monomers, *i.e.* dialkenyl-alkylsilanes or trialkenylsilanes, and possess a certain polydispersity and randomly branched structure as expected from the statistics of an AB_m polymerization [4, 66].

References

[1] Hawker CJ, Fréchet JMJ (1990) J Am Chem Soc **112**: 7638
[2] Miller TM, Neenan TX (1990) Chem Mater **2**: 346
[3] Tomalia DA, Baker H, Dewald J, Hall M, Kallos G, Martin S, Roeck J, Ryder J, Smith P (1986) Macromolecules **19**: 2466
[4] Frey H, Lach C, Lorenz K (1998) Adv Mater **10**: 279
[5] Mathias LJ, Carothers TW (1995) Advances in Dendritic Macromolecules, vol 2. JAI Press, p 101
[6] Hadjichristidis N, Guyot A, Fetters LJ (1978) Macromolecules **11**: 668
[7] van der Made AW, van Leeuwen PWNM (1992) J Chem Soc Chem Commun 1400
[8] van der Made AW, van Leeuwen PWNM, de Wilde JC, Brandes RAC (1993) Adv Mater **5**: 466
[9] Roovers J, Toporowski PM, Zhou LL (1992) Polym Prepr (Am Chem Soc Div Polym Chem) **33**: 182
[10] Zhou L-L, Roovers J (1993) Macromolecules **26**: 963
[11] Muzafarov AM, Gorbatsevich OB, Rebrov EA, Ignat'eva GM, Chenskaya TB, Myakushev VD, Bulkin AF, Papkov VS (1993) Polym Sci Ser A **35**: 1575
[12] Lennon PJ, Mack DP, Thompson QE (1989) Organometallics **8**: 1121
[13] Knapen JWJ, van der Made AW, de Wilde JC, van Leeuwen PWNM, Wijkens P, Grove DM, van Koten G (1994) Nature **372**: 659
[14] van Koten G, Jastrzebski JTBH (1997) Polym Mater Sci Eng **77**: 75
[15] Lorenz K, Mülhaupt R, Frey H, Rapp U, Mayer-Posner FJ (1995) Macromolecules **28**: 6657
[16] Frey H, Mülhaupt R, Lorenz K, Rapp U, Mayer-Posner FJ (1995) Polym Mater Sci Eng **73**: 127
[17] Frey H, Lorenz K, Mülhaupt R (1996) Macromol Symp **102**: 19
[18] Collaud Coen M, Lorenz K, Kressler J, Frey H, Mülhaupt R (1996) Macromolecules **29**: 8069
[19] Frey H, Lorenz K, Hölter D, Mülhaupt R (1996) Polym Prep (Am Chem Soc Div Polym Chem) **37**: 758
[20] Lorenz K, Hölter D, Stühn B, Mülhaupt R, Frey H (1996) Adv Mater **8**: 414
[21] Hoare JL, Lorenz K, Hovestad NJ, Smeets WJJ, Spek AL, Canty AJ, Frey H, van Koten G (1997) Organometallics **16**: 4167
[22] Lorenz K, Frey H, Stühn B, Mülhaupt R (1997) Macromolecules **30**: 6860
[23] Lorenz K, Hölter D, Frey H, Stühn B (1997) Polym Mater Sci Eng **77**: 168
[24] Alonso B, Cuadrado I, Moran M, Losada J (1994) J Chem Soc Chem Commun 2575
[25] Lobete F, Cuadrado I, Casado CM, Alonso B, Moran M, Losada J (1996) J Organomet Chem **509**: 109
[26] Sheiko SS, Eckert G, Ignat'eva GM, Muzafarov AM, Spickermann J, Räder HJ, Möller M (1996) Macromol Rapid Commun **17**: 283
[27] Ponomarenko SA, Rebrov EA, Bobrovsky AY, Boiko NI, Muzafarov AM, Shibaev VP (1996) Liq Cryst **21**: 1
[28] Sheiko SS, Muzafarov AM, Winkler RG, Getmanova EV, Eckert G, Reineker P (1997) Langmuir **13**: 4172
[29] Getmanova EV, Chenskaya TB, Gorbatsevich OB, Rebrov EA, Vasilenko NG, Muzafarov MA (1997) React Func Polym **33**: 289
[30] Ignat'eva GM, Rebrov EA, Myakushev VD, Chenskaya TB, Muzafarov AM (1997) Polym Sci Ser A **39**: 843
[31] Vasilenko NG, Getmanova EV, Myakushev VD, Rebrov EA, Möller M, Muzafarov AM (1997) Polym Sci Ser A **39**: 977
[32] Sheiko SS, Buzin AI, Muzafarov AM, Rebrov EA, Getmanova EG (1998) Polym Prepr (Am Chem Soc Div Polym Chem) **39**: 481
[33] Getmanova EV, Rebrov EA, Vasilenko NG, Muzafarov AM (1998) Polym Prepr (Am Chem Soc Div Polym Chem) **39**: 581

[34] Vasilenko NG, Rebrov EA, Muzafarov AM, Sheiko SS, Möller M (1998) Polym Prepr (Am Chem Soc Div Polym Chem) **39**: 479
[35] Seyferth D, Son DY, Rheingold AL, Osrander RL (1994) Organometallics **13**: 2682
[36] Seyferth D, Kugita T, Rheingold AL, Yap GPA (1995) Organometallics **14**: 5362
[37] Friedmann G, Guilbert Y, Wittmann JC (1997) Eur Polym J **33**: 419
[38] Zhou LL, Hadjichristides N, Toporowski PM, Roovers J (1992) Rubber Chem Technol **65**: 303
[39] Roovers J (1997) Macromol Symp **121**: 89
[40] Wu Z, Biemann K (1997) Int J Mass Spectrom Ion Processes **165/166**: 349
[41] Krska SW, Seyferth D (1998) J Am Chem Soc **120**: 3604
[42] Kim C, Park E, Kang E (1995) J Korean Chem Soc **39**: 799
[43] Kim C, Park E, Kang E (1996) Bull Korean Chem Soc **17**: 419
[44] Kim C, Sung D-D, Chung D, Park E, Kang E (1995) J Korean Chem Soc **39**: 789
[45] Kim C, Park E, Jung I (1996) J Korean Chem Soc **40**: 347
[46] Kim C, An K (1997) Bull Korean Chem Soc **18**: 164
[47] Kim C, An K (1997) J Organomet Chem **547**: 55
[48] Kim C, Jeong Y (1997) Bull Korean Chem Soc **18**: 347
[49] Ignat'eva GM, Rebrov EA, Myakushev VD, Muzafarov AM, Il'ina MN, Dubovik II, Papkov VS (1997) Polym Sci Ser A **39**: 874
[50] Terunuma D, Kato T, Nishio R, Matsuoka K, Kuzuhara H, Aoki Y, Nohira H (1998) Chem Lett 59
[51] Sluch MI, Vitukhnovsky AG, Petty MC (1996) Thin Solid Films **284/285**: 284
[52] Gossage RA, Muñoz-Martínez E, Frey H, Burgath A, van Koten G, J Org Chem (to be published)
[53] Gossage RA, Muños-Martínez E, van Koten G (1998) Tetrahedron Lett **39**: 2397
[54] Nakayama J, Lin JS (1997) Tetrahedron Lett **34**: 6043
[55] Schlenk C, Frey H (manuscript in preparation)
[56] Dani P, Karlen T, Gossage RA, Smeets WJJ, Spek AL, van Koten G (1997) J Am Chem Soc **119**: 11317
[57] Alonso B, Morán M, Casado CM, Lobete F, Losada J, Cuadrado I (1995) Chem Mater **7**: 1440
[58] Cuadrado I, Moran M, Moya A, Casado CM, Barranco M, Alonso B (1996) Inorg Chim Acta **251**: 5
[59] Newkome G, Moorefield C, Baker C, Saunders M, Grossman S (1990) Angew Chem **103**: 1205
[60] Lach C, Brizzolara D, Frey H (1997) Macromol Theory Simul **6**; 371
[61] Bauer S, Fischer H, Ringsdorf H (1993) Angew Chem Int Ed Engl **32**: 1589
[62] Percec V, Chu P, Kawasumi M (1994) Macromolecules **27**: 4441
[63] Percec V, Chu P, Ungar G, Zhou J (1995) J Am Chem Soc **117**: 11441
[64] Zhang QZ, Sun JR, Cui L, Pan JX (1997) Chem J Chin Univ **18**: 158
[65] Roovers J, Zhou LL, Toporowski PM, van der Zwan M, Iatrou H, Hadjichristidis N (1993) Macromolecules **26**: 4324
[66] Lach C, Müller P, Frey H, Mülhaupt R (1997) Macromol Rapid Commun **18**: 253

Received June 12, 1998. Accepted (revised) September 7, 1998

Invited Review

Synthesestrategien zur Darstellung von unterschiedlich funktionalisierten Cyclosiloxanen sowie Additions- und Substitutionsreaktionen eines $Al_4(OH)_4$-Achtringes in einem molekularen Oligoalumosiloxan

Michael Veith*, **Andreas Rammo**, **Maria Jarczyk** und **Volker Huch**

Institut für Anorganische Chemie, Universität des Saarlandes, D-66041 Saarbrücken, Germany

Strategies for the Synthesis of Functionalized Cyclosiloxanes and the Addition and Substitution Reactions of an $Al_4(OH)_4$ Eight Membered Cycle within a Molecular Oligoalumosiloxane

Summary. The synthesis of cyclic polysiloxanes of the general formula $(OSiX_2)_n$ with X = halogen, nitrogen, or oxygen centered ligand is achieved using a new strategy of assembling difunctional entities. The actual progress in this field is reviewed. Furthermore, polysiloxanes are described in which at distinct locations the silicon atoms are replaced by aluminum to which hydroxide groups are attached. The central molecule of interest has the formula $[(Ph_2Si)_2O_3Al(OH)]_4$, and its reactions with bases like diethylether, triethylamine, or pyridine are discussed following ideas of host-guest chemistry. Reaction of $[(Ph_2Si)_2O_3Al(OH)]_4$ with lithium organyls are also reviewed.

Keywords. Cyclosiloxanes; Functionalized cyclosiloxanes; Molecular alumosiloxanes; Host-guest chemistry in molecular alumosiloxanes.

Synthesestrategien zur Darstellung von unterschiedlich funktionalisierten Cyclosiloxanen

Cyclosiloxane sind in der Literatur vielfach beschrieben worden, wobei die Siliciumatome meist durch Organylgruppen abgesättigt sind. Dies ist nicht zuletzt auf deren einfache Darstellung durch Hydrolyse von Dihalogendiorganylsilanen zurückzuführen [1]. Im Gegensatz hierzu ist die Synthese von definierten am Silicium unterschiedlich substituierten Siloxanen $R'(R_2SiO)_nSiR_3$ und Cyclo-

* Corresponding author

Syntheseschema 1: Syntheseschema 2:

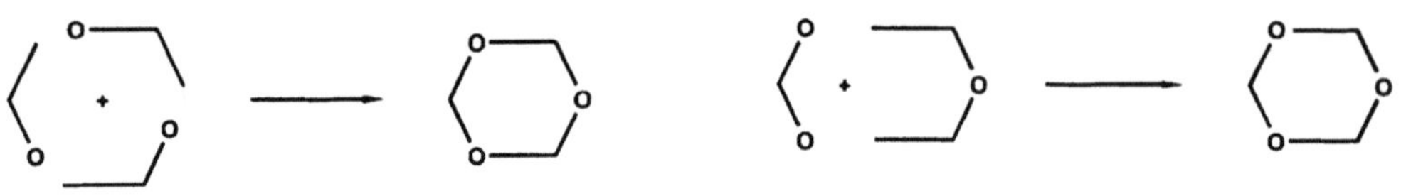

Abb. 1. Allgemeine Syntheseschemata zur Darstellung von funktionalisierten Cyclosiloxanen; die Ecken repräsentieren Siliciumatome, die freien Enden funktionelle Gruppen

siloxanen $(R_2SiO)_n$ bestimmter Ringgröße bis heute problematisch. Das klassische Verfahren zur Herstellung von Cyclosiloxanen durch Hydrolyse von Halogensilanen führt in den meisten Fällen zu zahlreichen ketten- und ringförmigen Si–O-Verbindungen [1, 2]. Die Abtrennung und Isolierung der einzelnen Spezies erweist sich oft als sehr schwierig bzw. ist in vielen Fällen nicht möglich. Deshalb sind die Ausbeuten auch meist sehr niedrig [1].

Anfang der achtziger Jahre publizierten *Klingebiel et al.* die gezielte Synthese von linearen und cyclischen Organosiloxanen [3–10]. Die bis dahin nur als Zwischenstufen bei der Hydrolyse von Chlorsilanen angenommenen, aber nicht nachgewiesenen Organochlorsilanole wurden über gezielte Reaktionen dargestellt und röntgenographisch charakterisiert [6, 11, 12]. Darüber hinaus gelang es auch, die entsprechenden Silanolate zu synthetisieren. Damit standen Bausteine zur Verfügung, die gezielt zur Synthese und zum schrittweisen Aufbau von acyclischen und cyclischen Siloxanen mit überwiegend organischen Resten eingesetzt werden konnten [3, 4, 8].

Zur Darstellung von Perchlorcyclosiloxanen wurden bereits in den sechziger Jahren Anstrengungen in verschiedenen Arbeitskreisen unternommen. In einer Gasphasenreaktion von Siliciumtetrachlorid mit Sauerstoff bei *ca.* 950°C wurden eine große Anzahl an linearen und cyclischen Halogensiloxanen erhalten, was zu niedrigen Ausbeuten und einer präparativ aufwendigen Aufarbeitung führte [13, 14].

Ziel unserer Arbeit ist es, über eine schrittweise Vorgehensweise gezielt Cyclotrisiloxane zu synthetisieren, deren Siliciumatome durch unterschiedliche und weiter substituierbare Gruppen, wie z.B. Amine oder Halogene, abgesättigt sind. Wir verfolgen hierbei zwei Syntheserouten (Abb. 1).

Nach Syntheseschema 1 lassen sich durch Salzeliminierungsreaktionen gezielt aminfunktionalisierte Cyclotrisiloxane in hohen Ausbeuten synthetisieren [15]. Ein noch größeres Synthesepotential eröffnet Syntheseroute 2, wobei unter milden Bedingungen in homogener Phase mit Hilfe von Organodistannoxan als Sauerstofflieferant Si–O–Si-Gerüste mit unterschiedlichen Substituenten aufgebaut werden [16].

Darstellung von aminfunktionalisierten Cyclotrisiloxanen nach Syntheseroute 1

Die Synthese der aminfunktionalisierten Si–O-Ringe verläuft über folgende präparative Stufen, die alle röntgenstrukturanalytisch abgesichert sind (Gl. (1)):

$$Me_2Si(NR)_2SiCl_2 \xrightarrow[-Et_3N\cdot HCl]{H_2O/Et_3N} Me_2Si(NR)_2Si(Cl)OH$$

$R = SiMe_3$ **1a** **2a**

$R = {}^tBu$ **1b** **2a**

$$\xrightarrow[-BuH]{+BuLi} Me_2Si(NR)_2Si(Cl)OLi \xrightarrow[-3LiCl]{3x} [Me_2Si(NR)_2SiO]_3$$

3a **4a**

3b **4b** (1)

Stabile aminsubstituierte Halogensilanole waren bisher unbekannt. Von *U. Klingebiel* wurde nur ein in Lösung existentes aminsubstituiertes Halogensilanol beschrieben [11]. Durch die kontrollierte Hydrolyse der Dichlorocyclodisilazane **2a** [17] bzw. **2b** [18] lassen sich die aminsubstituierten Chlorsilanole **2a** bzw. **2b** herstellen und in Substanz isolieren (Gl. (2)).

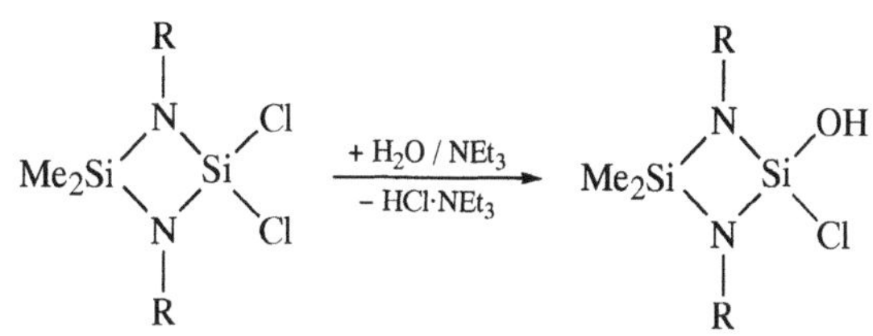

(2)

$R = SiMe_3$ **1a** **2a**

$R = {}^tBu$ **1b** **2b**

Um eine homogene Reaktionsführung zu erzielen, wird eine Dioxanlösung mit bekanntem Wassergehalt zur Hydrolyse verwendet. Zur Gleichgewichtsverlagerung zugunsten der Produkte und zur Unterdrückung einer Spaltung der Silicium-Stickstoff-Bindung durch den frei werdenden Chlorwasserstoff wird der Hydrochloridfänger Triethylamin eingesetzt.

Die in Dioxan durchgeführte Hydrolyse verläuft selektiv, d.h. sie führt nur zur Monosubstitution eines Chloratoms durch eine Hydroxylgruppe, was sich leicht NMR-spektroskopisch nachweisen läßt (Abb. 2). Weitere Produkte, wie z.B. das

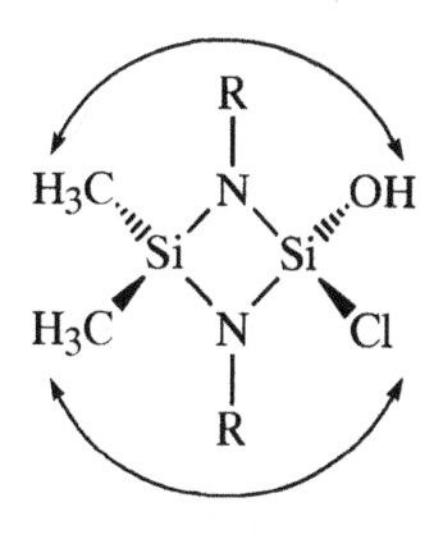

Abb. 2. Die beiden Methylgruppen am Silicium haben unterschiedliche chemische Umgebung und führen daher zu zwei getrennten NMR-Signalen (^1H, ^{13}C)

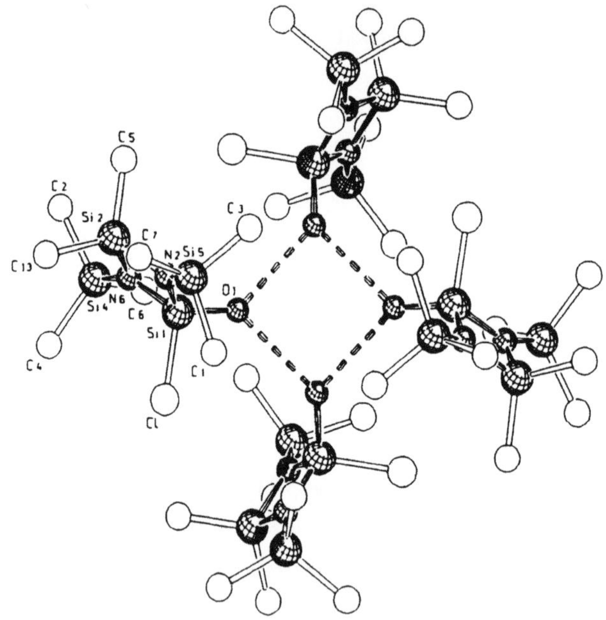

Abb. 3. Durch Ausbildung von Wasserstoff-Sauerstoff-Brückenbindungen lagern sich vier Moleküle der Verbindung **2a** zu einem Tetramer zusammen

Silandiol, werden nicht gebildet. Auch eine Hydrolyse der Si–N-Bindung wird nicht beobachtet.

Aufgrund der unterschiedlichen Subtituenten ergeben die beiden Methylgruppen der Dimethylsilylgruppe zwei unterschiedliche Signale im ^{1}H-NMR-Spektrum; im Gegensatz dazu läßt sich bei einer Disubstitution (Silandiol) nur ein Signal beobachten [19].

Im Unterschied zu den von *Klingebiel et al.* dargestellten aminosubstituierten Chlorsilanolen, die nur in Lösung stabil sind, können die Verbindungen **2a** und **2b** bei niedrigen Temperaturen durch Sublimation in Ausbeuten zwischen 55 und 65% als farblose Feststoffe rein isoliert werden. **2a** und **2b** sind somit die ersten stabilen und röntgenographisch charakterisierten aminsubstituierten Chlorsilanole. Damit ist es erstmals gelungen, ein Chlorsilanol als Zwischenstufe bei einer direkten Hydrolyse von Chlorsilanen nachzuweisen und zu isolieren. *Klingebiel et al.* erhielten die entsprechenden Diorganylhalogensilanole durch Salzeliminierungsreaktionen bzw. indirekt durch Umsetzen von Silandiolen mit Phoshorhalogeniden [12].

Im Festkörper liegen die Verbindungen **2a** und **2b** nicht monomer vor, sondern bilden Tetramere durch Ausbildung von O–H···O-Brückenbindungen mit einem O···O-Abstand von 2.843 Å [20–23] (Abb 3).

In *n*-Hexan lassen sich die Chlorsilanole **2a** und **2b** mit *n*-Butyllithium in einer spontanen und exothermen Reaktion in die *Bis* (amino)chlorsilanolate **3a** bzw. **3b** überführen, die beim Erkalten der Reaktionslösung in großen Kuben auskristallisieren (Gl. (3)).

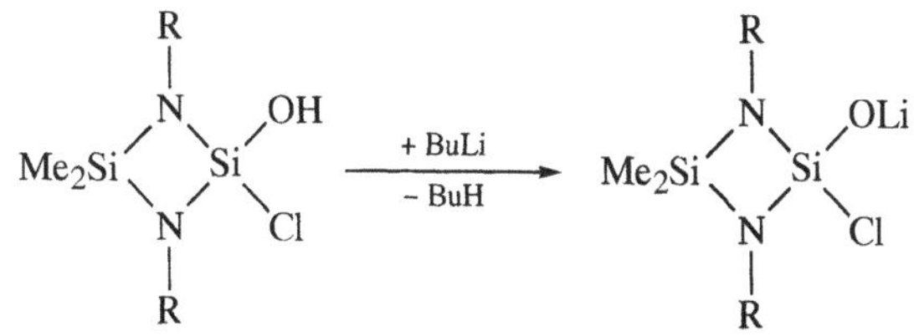

$R = SiMe_3$ **2a** **3a**

$R = {}^tBu$ **2b** **3b**

(3)

Die Ausbeuten der Lithiumsilanolate **3a**, **b** liegen hierbei über 90%. Im Kristall bilden **3a**, **b** Tetramere mit einer durch Lithium- und Sauerstoffatome gebildeten annähernden Cubanstruktur. Das Chloratom nähert sich hierbei intramolekular dem nächstgelegenen Lithiumatom, wodurch das Lithiumatom elektronisch abgesättigt wird und ein Elektronenoktett erlangt (Abb. 4).

3a und **3b** sind außerodentlich beständig gegenüber einer LiCl-Eliminierung, was auch bereits von *U. Klingebiel* bei organosubstituierten Silanolaten beobachtet wurde [6]. Aus heißen, unpolaren Lösungsmitteln, wie z.B. *n*-Hexan oder Toluol, könnnen **3a**, **b** umkristallisiert werden, ohne daß es zu einer Salzeliminierung kommt. Erst fünfzigstündiges Erhitzen von **3a** oder **3b** in Tetrahydrofuran führt unter Lithiumchlorideliminierung zu den Cyclotrisiloxanen **4a** bzw. **4b** mit Ausbeuten von über 90% (Gl. (4)).

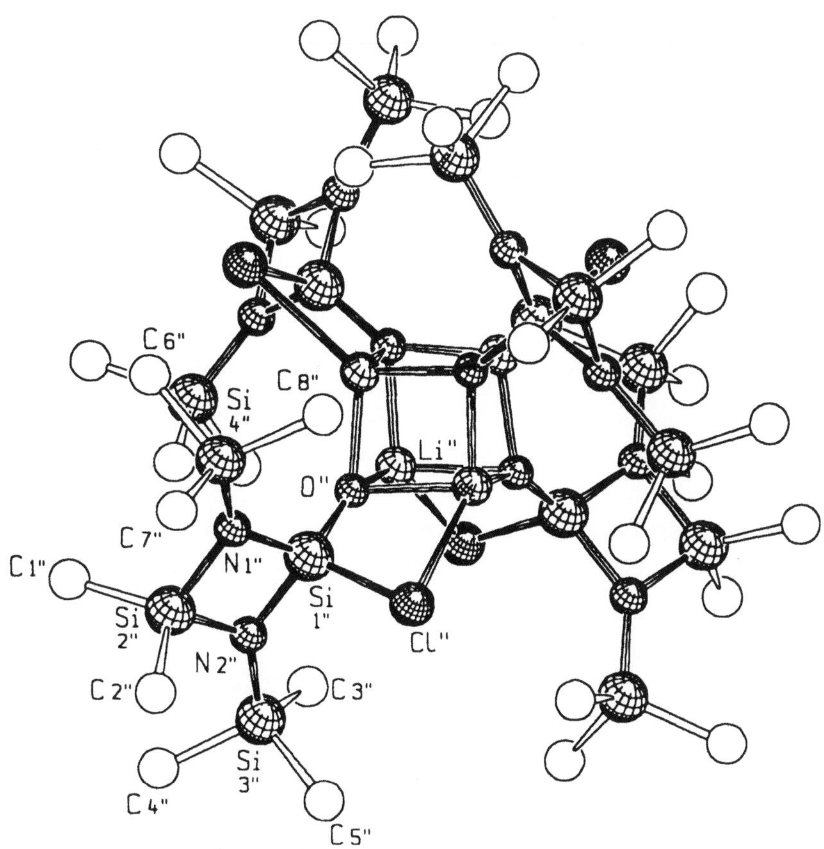

Abb. 4. Struktur des Tetrameren von **3a** [15]

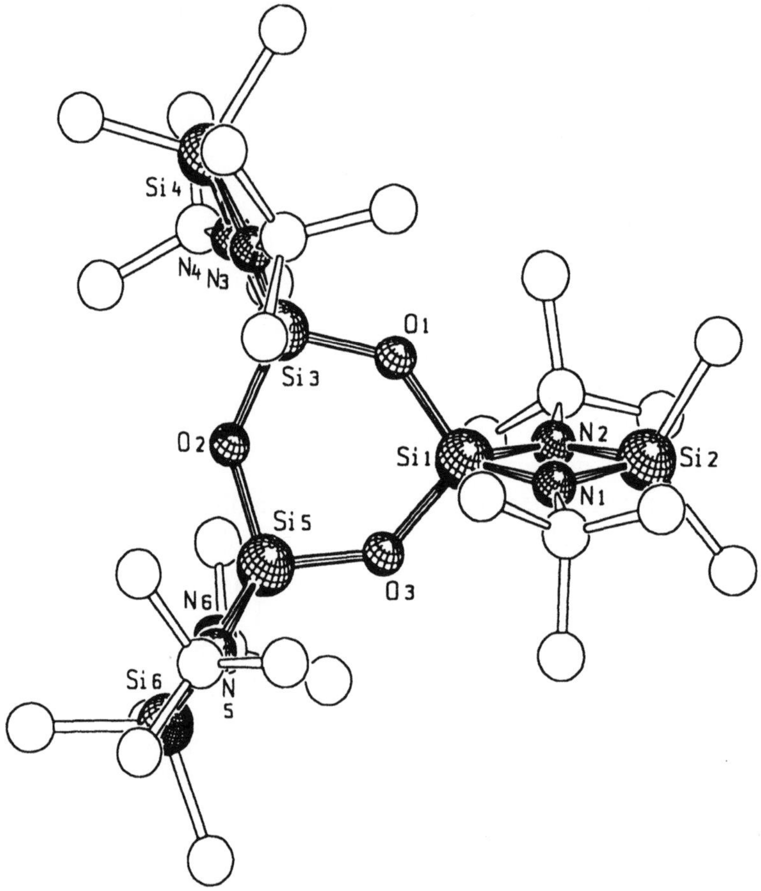

Abb. 5. Struktur des Cyclotrisiloxans **4b** [15]

$$3Me_2Si(NR)_2Si(Cl)OLi \xrightarrow[-3LiCl]{\Delta} [Me_2Si(NR)_2SiO]_3 \qquad (4)$$

$R = SiMe_3$ **3a** **4a**

$R = {}^tBu$ **3b** **4b**

Die cyclischen aminsubstituierten Siloxane **4a, b** sind farblose Feststoffe, die leicht zu kristallisieren sind und sich über Sublimation reinigen und isolieren lassen. Wie erwartet, besitzen die aminsubstituierten Cyclotrisiloxane mit ihren drei spirocyclisch gebundenen Siliciumatomen planare Si–O-Sechsringe [25, 26] (Abb. 5).

Darstellung von verschieden substituierten Cyclotrisiloxanen nach Syntheseroute 2

Eine zweite Methode zum Aufbau von Cyclosiloxanen ist die Umsetzung von Distannoxansilylverbindungen mit unterschiedlich substituierten 1,3-Disiloxanen [16]. Die Reaktion von Distannoxansilylverbindungen mit Halogeniden unter Abspaltung von Chlortrimethylstannan ist eine bekannte Synthesemethode zur Darstellung mehrdimensionaler Heteroatom-Siloxanverbindungen, wie z.B. im

Arbeitskreis um *H.W. Roesky* [26] gezeigt werden konnte. Hierbei lassen sich die Stannoxanausgangsverbindungen auf zwei präparativ unterschiedlichen Wegen erhalten. Die Umsetzung von Silanolen mit *Tris*(trimethylstannyl)amin unter Abspaltung von Ammoniak führt in einer quantitativen Reaktion zu den Sn–O–Si-Verbindungen [15, 26] (Gl. (5)).

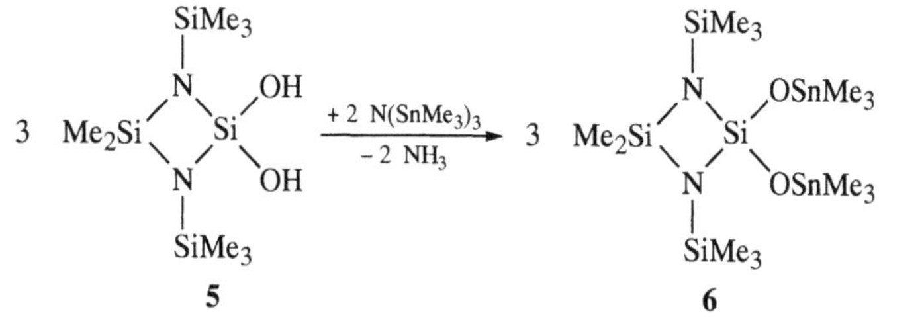

(5)

Die Ausgangsverbindung, das *Bis*(amino)silandiol **5** [15], läßt sich in Analogie zu den *Bis*(amino)chlorsilanolen **2a, b** durch Hydrolyse der Dichlorverbindung **1a** in Dioxan in Gegenwart von Triethylamin als Hydrochloridfänger darstellen. Trotz zweier Hydroxylgruppen und zwei Aminosubstituenten am gleichen Siliciumatom erweist sich **5** in Lösung als stabil und kann problemlos mit *Tris*(trimethylstannyl)-amin unter Ammoniakfreisetzung zu der Distannoxanverbindung **6** umgesetzt werden. **5** läßt sich auch als Substanz isolieren; die Ausbeuten liegen allerdings nur bei *ca.* 10%.

Der Nachteil bei der Darstellung von Si–O–Sn-Verbindungen nach Gl. (5) liegt darin, daß von Silanolen ausgegangen wird und somit nur hydrolysebeständige Reste am Siliciumatom eingesetzt werden können. *A. Tarassoli* und *D. A. Armitage* erhielten Hexamethyldisiloxan bzw. Hexamethyldigermoxan durch direktes Umsetzen von Chlorsilanen/Chlorgermanen mit Hexamethyldistannoxan, was eine fast beliebige Variation der am Siliciumatom bzw. Germaniumatom gebundenen Reste ermöglicht [27].

Unter Ausnutzung der bereitwilligen Bildung und der Flüchtigkeit von Chlortrialkylstannanen stellen wir eine weitere gezielte Synthesemethode zur Darstellung von funktionalisierten Cyclosiloxanen vor, die unter milden Bedingungen in homogener Phase mit Hilfe von Organodistannoxanen als Sauerstofflieferant entstehen. Die so erhaltenen cyclischen Siloxanverbindungen besitzen unterschiedliche Liganden an den Siliciumatomen und sind daher für weitere Modifizierungen der Substituenten wertvolle Ausgangsstoffe.

Die Synthese der Cyclen führt über zwei Stufen, wobei zunächst Dichlorsilane in die entsprechenden Distannoxane übergeführt werden (Gl. (6)), die anschließend mit einem Chlordisiloxan ($ClSiR_2$–O–SiR_2Cl, R = Me, Cl) unter Chlorstannaneliminierung zu den Siloxancyclen kondensieren (Gl. (7)).

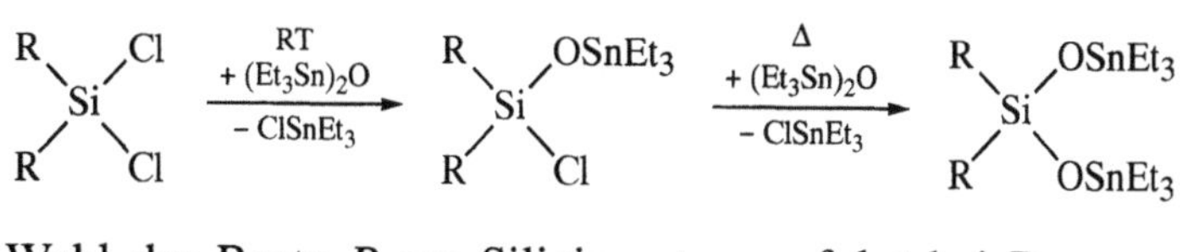

(6)

Je nach Wahl der Reste *R* am Siliciumatom erfolgt bei Raumtemperatur nur eine Monosubstitution eines Chloratoms durch eine $OSnEt_3$-Gruppe. Um eine vollständige Umsetzung der Dichlorsilane mit zwei Äquivalenten Hexaethyldistannoxan zu erhalten, muß der Reaktionsansatz, der ohne Lösungsmittel erfolgt, für vier Stunden auf 180°C erhitzt werden. Da auf Lösungsmittel verzichtet wird, ist das freigesezte Chlortriethylstannan entweder unter reduziertem Druck leicht destillativ entfernbar oder kann in eine Kühlfalle überkondensiert werden. Die Cyclotrisiloxane werden in einer anschließenden Umsetzung aufgebaut. Unter Austritt von zwei Molekülen Chlortriethylstannan erfolgt ein Ringschluß zu den cyclischen Siloxanen (Gl. (7)).

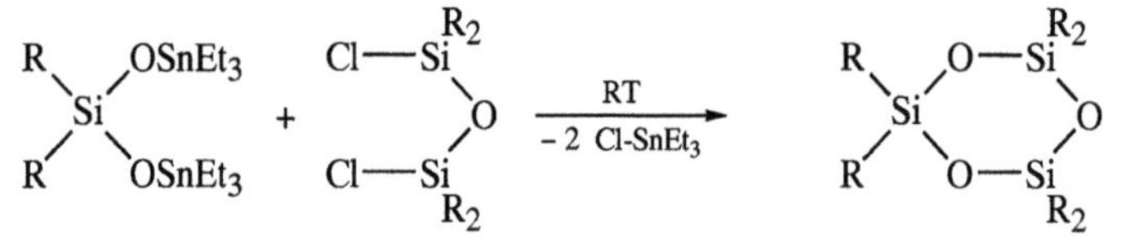

(7)

Durch Anwendung des *Ziegler*schen Verdünnungsprinzips wird einer möglichen Polymerisierung entgegengewirkt. Nach Entfernen des Lösungsmittels Diethylether und des Chlortriethylstannans unter reduziertem Druck lassen sich die cyclischen Siloxane als farblose Feststoffe durch Umkristallisieren erhalten. Die Ausbeuten sind mit *ca.* 50–70% noch befriedigend.

Nach dieser Darstellungsmethode wurden die in Abb. 6 aufgeführten funktionalisierten Cyclosiloxane hergestellt (Abb. 6).

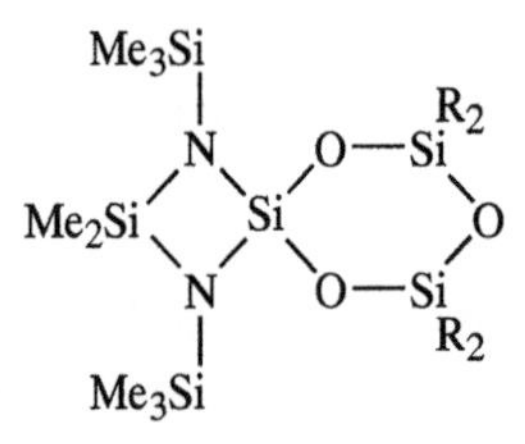

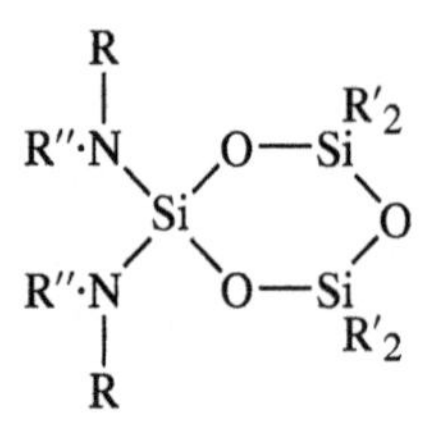

R = Me **7a** [16]
R = Cl **7b** [16]

R = ^{t}Bu, *R'* = Me, *R''* = H **8** [19]
R = ^{i}Pr, *R'* = Me **9a** [19]
R' = Cl **9b** [19]
R''–*R''* = CH_2–CH_2
R = Ph, *R'* = Me,
R''–*R''* = CH_2–CH_2 **10a** [19]

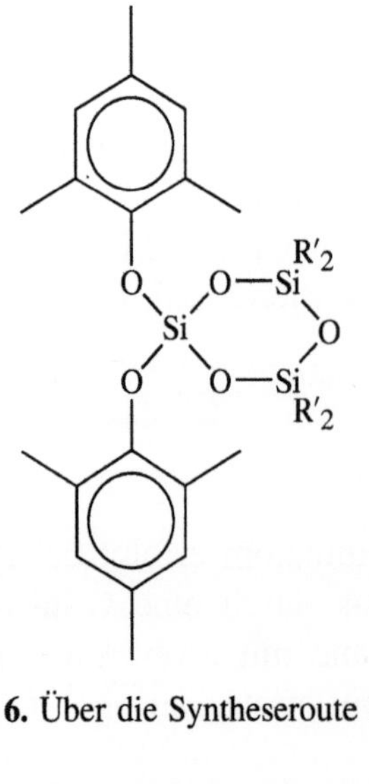

R' = Me **10b** [19]
R' = Cl **10c** [19]

Abb. 6. Über die Syntheseroute 2 dargestellte Cyclotrisiloxane

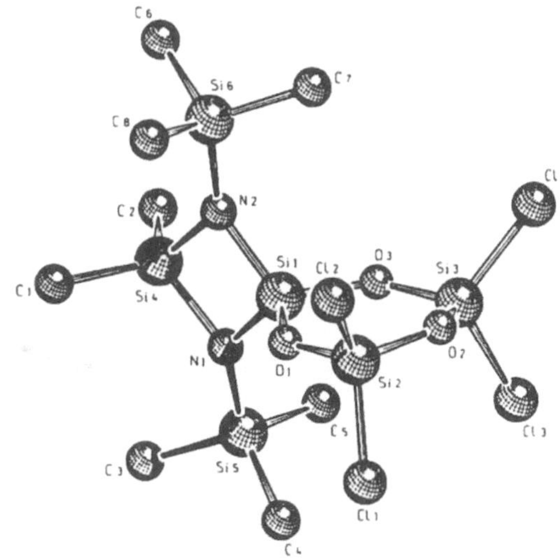

7b [15]

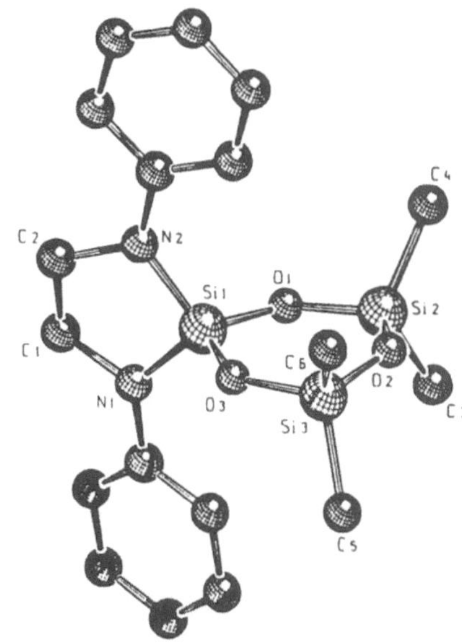

10a [19]

Abb. 7. Molekülstrukturen zweier amin-, chlor- und methylsubstituierter Cyclotrisiloxane; die Wasserstoffatome sind Übersichtlichkeit wegen weggelassen

Von zwei Vertretern sind in Abb. 7 die Ergebnisse der Röntgenstrukturanalysen dargestellt [15, 19]. In beiden Fällen wird das Molekülgerüst von Si_3O_3-Sechsringen aufgebaut, die im Fall von Verbindung **7b** spirocyclisch mit einem Si_2N_2-Vierring bzw. bei Verbindung **10a** mit einem SiN_2C_2-Fünfring verbunden sind.

Route 2 ist somit eine ausgezeichnete Methode zur Synthese einer Vielzahl unterschiedlich substituierter cyclischer Siloxane, wobei die Reste an den Siliciumatomen fast beliebig variiert werden können.

Additions- und Substitutionsreaktionen eines $Al_4(OH)_4$-Achtringes in einem molekularen Oligoalumosiloxan

In den letzten Jahren konnte insbesondere die Arbeitsgruppe von *H.W. Roesky* zeigen, daß in organischen Lösungsmitteln lösliche Alumopolysiloxane mit SiO_3-Baueinheiten überraschend einfach zugänglich sind [28a,b]. Wie wir vor kurzem berichten konnten, setzt sich Diphenylsilandiol mit *tert*-Butoxialan ((^{t}BuO-$AlH_2)_2$, **11** [29]) unter Wasserstoff- und *tert*-Butanoleliminierung in Diethylether in guter Ausbeute zu einem Alumopolysiloxan **12** um (Gl. (8)) [30].

Die Reaktion verläuft offenbar in zwei Etappen: Zunächst kondensiert $Ph_2Si(OH)_2$ zu $(HO)SiPh_2$–O–$Ph_2Si(OH)$ (wahrscheinlich unter dem Einfluß der *Lewis*-Säure ^{t}BuOAlH_2), welches dann mit dem *tert*-Butoxialan weiter reagiert.

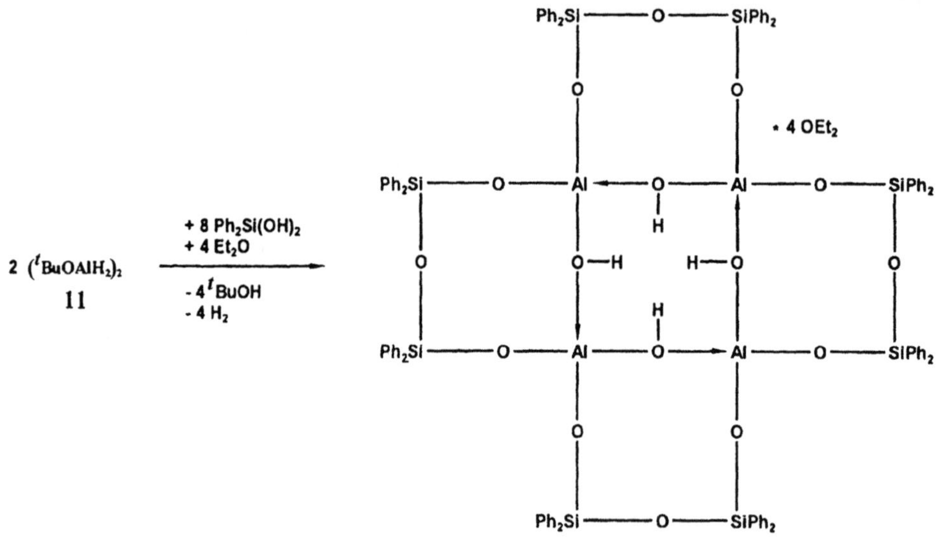

(8)

12

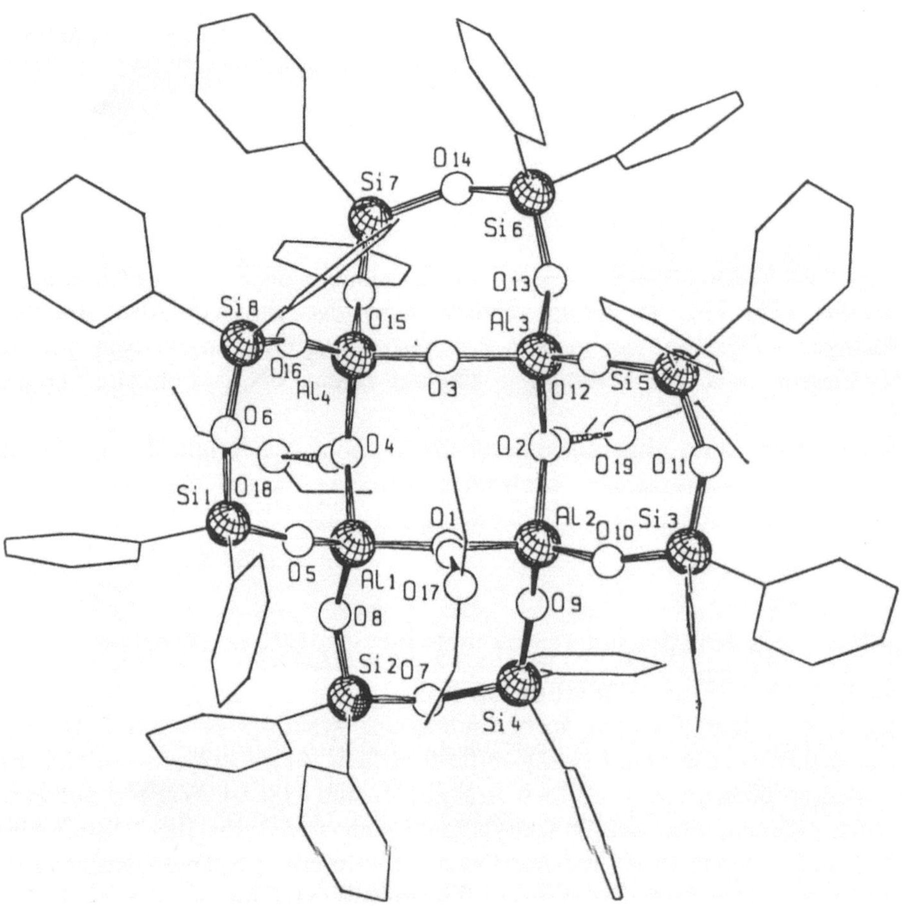

Abb. 8. Struktur des Oligoalumosiloxans **12**; die Phenylgruppen an den Siliciumatomen sowie die Kohlenstoffatome der Diethylethermoleküle sind der besseren Übersichtlichkeit wegen nur schematisch darstellt

Die molekulare Verbindung **12** besitzt eine polycyclische Struktur mit einer bemerkenswerten $Al_4(OH)_4$-Einheit im Zentrum des Moleküls. Nach der Röntgenstrukturanalyse kommt dem Polycyclus **12** annähernd die Punktsymmetrie -4(S_4) zu, wobei sich die vier Aluminiumatome in einer Ebene befinden und die anellierten Alumosiloxanringe alternierend nach oben und unten (fast senkrecht zur Al_4-Ebene) ausgerichtet sind. Die OH-Gruppen bilden Wasserstoffbrücken zu den Sauerstoffatomen der Diethylethermoleküle. Die Symmetrie wird allerdings dadurch gestört, daß aus Platzgründen nur drei Ethermolküle mit den OH-Gruppen koordinieren können; das vierte Ethermolekül wird in das *van der Waals*-Gitter des Kristalls eingebaut. Zusammen mit dem zentralen $Al_4(OH)_4$-Achtring entsteht ein Oligoalumosiloxangerüst aus fünf anellierten Achtringen (Abb. 8).

Die Hydroxylgruppen des zentralen $Al_4(OH)_4$-Achtrings bilden, wie bereits oben erwähnt, Wasserstoffbrückenbindungen zu den Ethermolekülen aus. Neben Additonsreaktionen von *Lewis*-Basen an die aciden Wasserstoffatome kann man auch Substitutionen der Protonen durch Lithiumkationen durchführen, wobei das Grundgerüst des Oligoalumosiloxans **12** erhalten bleibt [30, 31] (Gl. (9)).

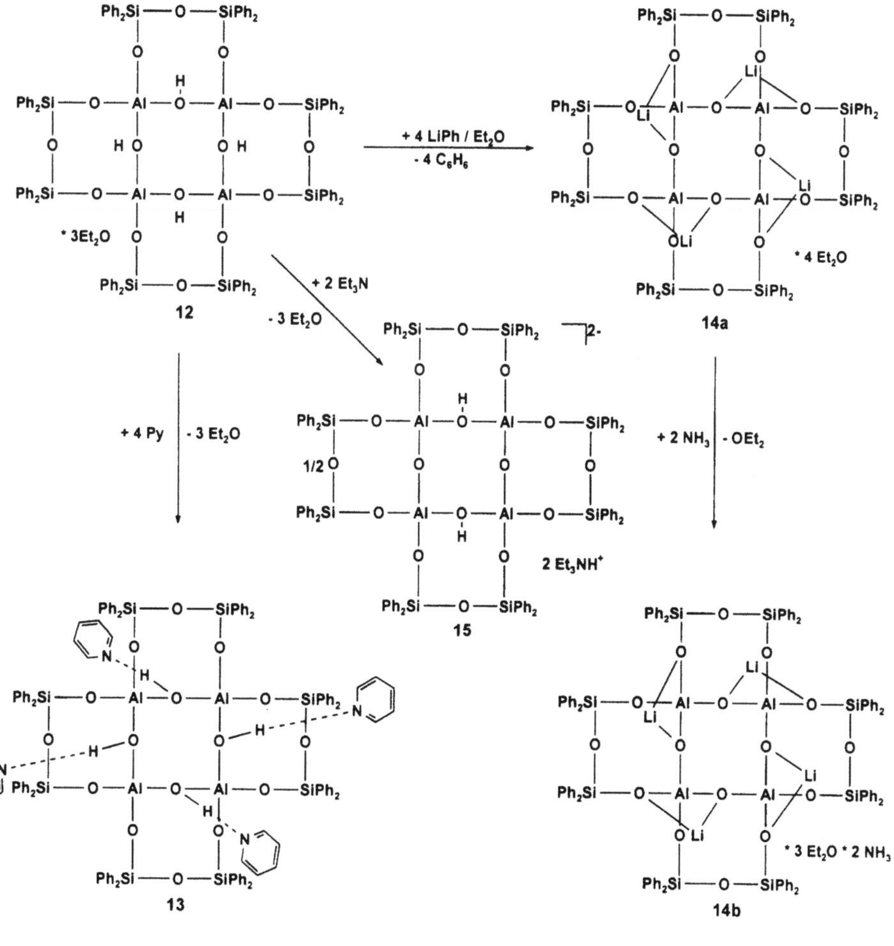

(9)

Mit Triethylamin reagiert **12** unabhängig vom Molverhältnis der zugesetzten Base immer nur mit zwei Moläquivalenten unter vollständiger Verdrängung der Diethylethermoleküle zum Aminaddukt **15** (Gl. (10), Abb. 9).

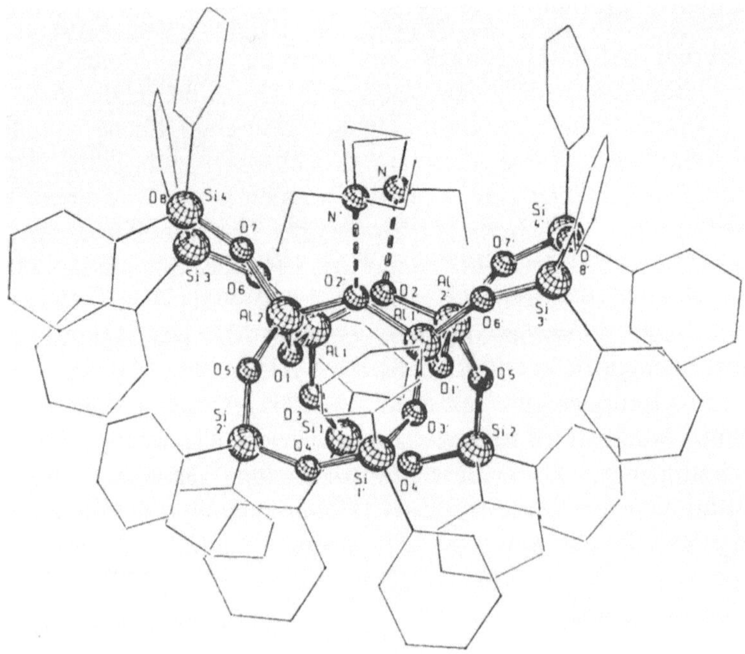

Abb. 9. Struktur der *Bis*(triethylamin)verbindung **15** [30]; die Phenylgruppen an den Siliciumatomen und die Ethylgruppen des Triethylamins sind der Übersichtlichkeit wegen nur schematisch dargestellt

$$\underset{\mathbf{12}}{[(Ph_2Si)_2O_3]_4Al_4(OH)_4 \cdot 4Et_2O} + 2Et_3N \rightarrow \underset{\mathbf{15}}{[(Ph_2Si)_2O_3]_4Al_4O_2(OH)_2]^{2-}} \cdot 2HNEt_3^+ + 4Et_2O \quad (10)$$

Die höhere Basizität des Stickstoffatoms im Triethylamin im Vergleich zum Sauerstoffatom des Diethylethers wirkt sich erwartungsgemäß auf die Bindungen des zentralen $Al_4(OH)_4$-Rings aus (nähere Einzelheiten hierzu siehe Lit. [30, 31]).

Mit Pyridin reagiert das Etheraddukt **12** ebenfalls unter völligem Basenaustausch, wobei die vier Pyridinmoleküle über Wasserstoffbrücken an den zentralen $Al_4(OH)_4$-Ring gebunden sind (Abb. 10).

Im Unterschied zum *Bis*(triethylamin)addukt **15** (nur zwei Moleküle pro vier OH-Gruppen) und zum Diethyletheraddukt **12** (drei Moleküle pro vier OH-Gruppen) sind in **13** alle Hydroxylgruppen koordinativ abgesättigt. Offensichtlich spielt der Platzbedarf der koordinierenden Basen eine entscheidende Rolle, wobei die sterisch anspruchsvollste Base das polycyclische Molekül derartig verformt, daß zwei OH-Gruppen koordinativ nicht mehr zugänglich sind; andererseits entstehen durch eine parallele Ausrichtung der Hälfte der peripheren Phenylgruppen zueinander Lücken, in die die Pyridinmoleküle exakt hineinpassen. Letztere werden von Phenylgruppen der Siliciumatome sandwichartig koordiniert.

Für die H···N-Wechselwirkung und die relative Säurestärke der OH-Gruppen sind besonders die Al-O(H)-Abstände aussagekräftig. Sie sind bei **13** innerhalb der Standardabweichungen gleich (Mittelwert 1.769 Å) und liegen zwischen den entsprechenden Abständen im Diethyletheraddukt **12** (Mittelwert 1.800 Å) und

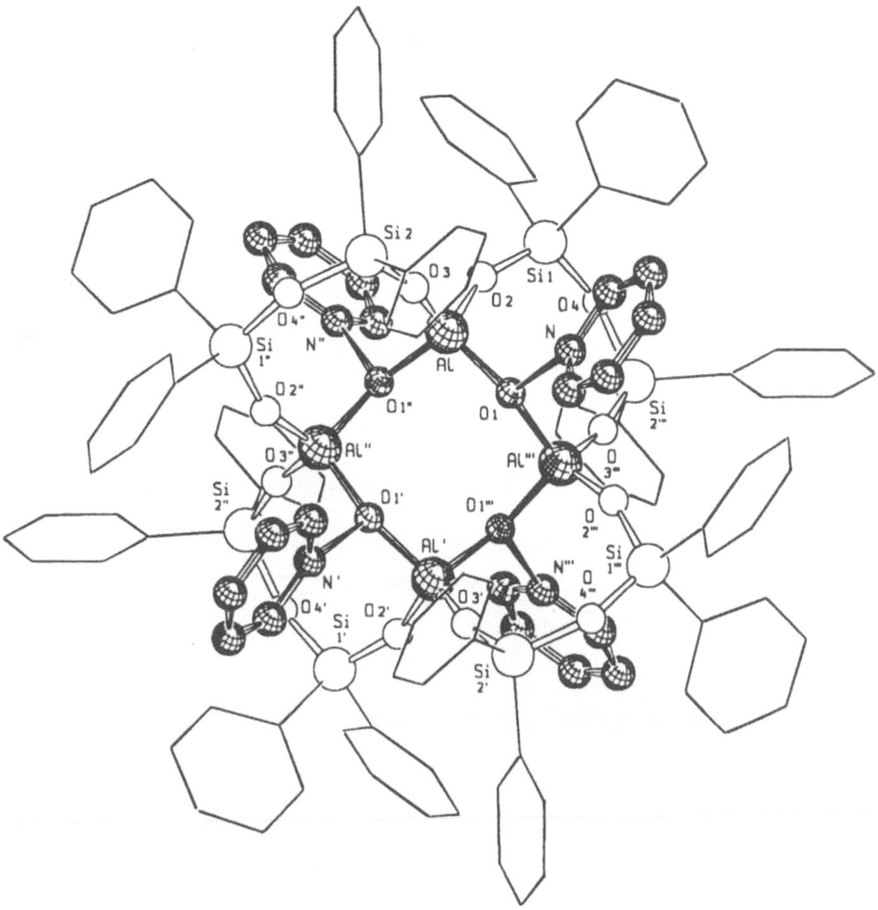

Abb. 10. Struktur von **13**; zur Verdeutlichung sind der $Al_4(OH)_4$-Ring und die Pyridinmoleküle hervorgehoben [31]; die Phenylgruppen an den Siliciumatomen sind der besseren Übersichtlichkeit wegen nur schematisch dargestellt

denen im Triethylaminaddukt **15** (Mittelwert 1.728 Å). In Einklang mit den pK_a-Werten von Diethylether (−3.0), Pyridin (5.2) und Triethylamin (10.8) [32] besteht eine Korrelation mit den Al–O-Abständen: erwartungsgemäß zieht die stärkste Base die Wasserstoffatome am stärksten an. Dies führt zu einer negativen Partialladung an den Sauerstoffatomen und damit zu einer Verkürzung der Al–O-Bindung. Dieser Effekt ist in der grafischen Auftragung (Abb. 11) gut zu erkennen.

Die Längen der Wasserstoffbrückenbindungen korrelieren nicht mit der Basenstärke; sie betragen im Etheraddukt **12** 2.65, im Triethylaminaddukt **15** 2.72 in **13** 2.55 Å. Anscheinend spiegelt sich in diesen Abständen die Raumerfüllung der Basen wieder, wobei Pyridin offenbar weniger Platz beansprucht als Diethylether oder Triethylamin, was nach dem Kalottenmodellen der Moleküle plausibel ist.

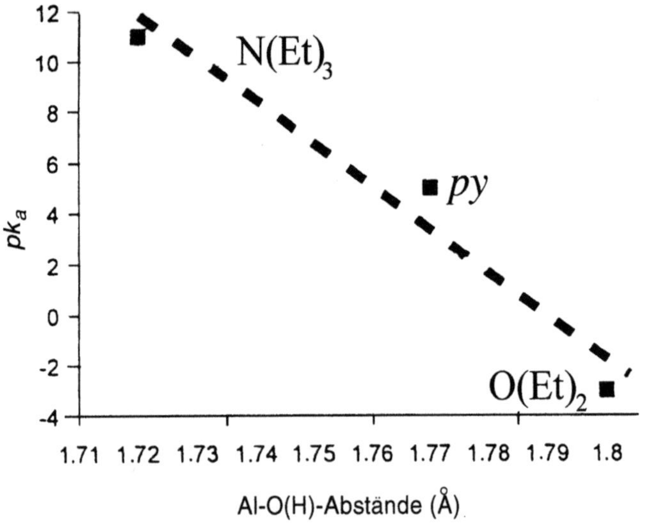

$[(Ph_2Si)_2O_3]Al_4(OH)_4$ 2 $N(Et)_2$	**15**
$[(Ph_2Si)_2O_3]Al_4(OH)_4$ 3 $O(Et)_2$	**12**
$[(Ph_2Si)_2O_3]Al_4(OH)_4$ 4 C_5H_5N	**13**

Abb. 11. Korrelation der pK_a-Werte der Basen $N(Et)_3$, $O(Et)_2$ und C_5H_5N mit den Al-O(H)-Abständen der Hydroxylgruppen, mit denen die Basen wechselwirken

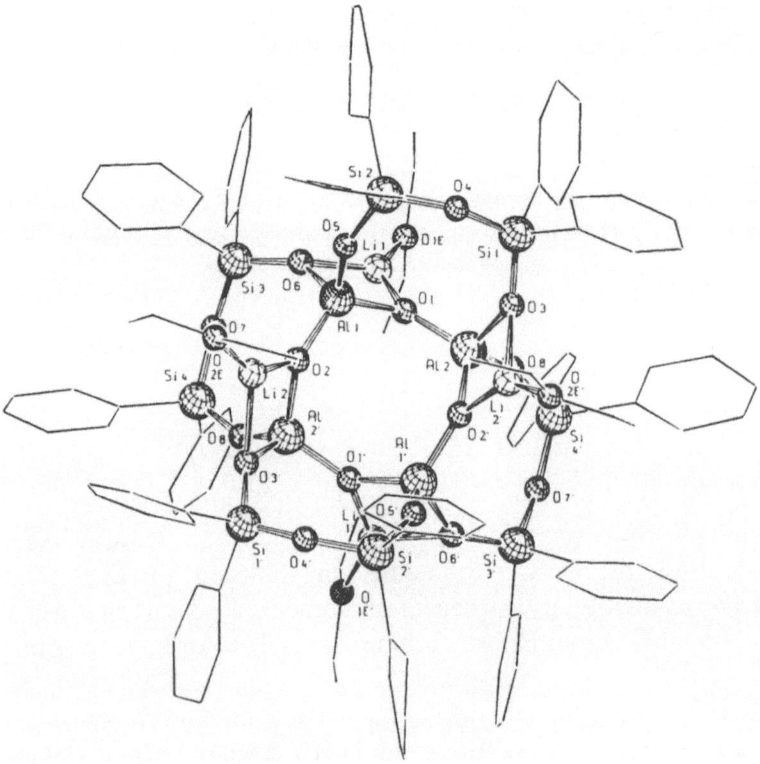

Abb. 12. Struktur des Lithiumalumosiloxans **14a** [31]; die Phenylgruppen an den Siliciumatomen sowie die Ethylgruppen der Diethylethermoleküle sind der besseren Übersichtlichkeit wegen nur schematisch dargestellt

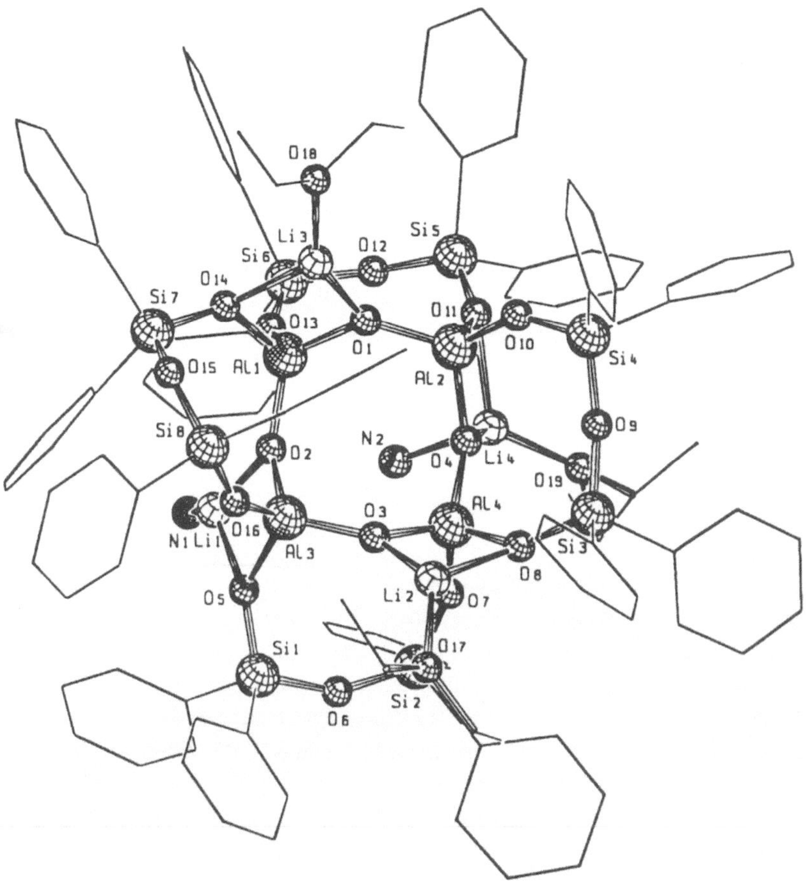

Abb. 13. Struktur von **14b** [31]; die Phenyl- und die Ethylgruppen an den Silicium- bzw. Sauerstoffatomen der Diethylethermoleküle sind der basseren Übersichtlichkeit wegen nur schematisch dargestellt

Durch Umsetzung von **12** mit Phenyllithium gelingt es, alle Wasserstoffatome der vier Hydroxylgruppen des $Al_4(OH)_4$-Achtringes durch Lithiumatome zu ersetzen [32] (Abb. 12).

Die gebildete Lithiumverbindung **14a**, bei der die Lithiumatome durch Diethylether koordiniert sind, reagiert mit Ammoniak in Toluol glatt zu der Verbindung **14b** weiter; zwei NH_3-Moleküle werden eingebaut und ein Diethylethermolekül wird freigesetzt (siehe Gl. (9)). Wie in **14a** weisen drei Lithiumatome durch eine zusätzliche Koordination mit dem Sauerstoffatom des Diethylethers oder dem Stickstoffatom eines Ammoniakmoleküls eine verzerrt trigonal-planare Koordinationsumgebung auf. Bei dem verbleibenden Lithiumatom bewirkt die doppelte Wechselwirkung mit einem Ammoniakmolekül und einem Ethermolekül eine Vierfachkoordination und damit eine tetraedrische Ligandensphäre (Abb. 13).

14a weist eine große Ähnlichkeit zu Lithiumalumosilikaten auf, wie sich aus der Formel $[(Ph_2SiO)_8]$ $[(LiAlO_2)_4] \cdot 4$ OEt_2 ergibt. Für **14a**, **b** scheint damit die

Loewenstein-Regel nicht zu gelten, nach der Alumosilikate nur existieren, wenn sich AlO_4- und SiO_4-Tetraeder im Gerüst abwechseln [33]; die Produktbildung ist offenbar eher kinetisch als thermodynamisch kontrolliert.

Dank

Dem *Fonds der Chemischen Industrie* und insbesondere der *Deutschen Forschungsgemeinschaft* (DFG) wird für die Unterstützung der vorliegenden Arbeit gedankt. Die Förderung fand im Rahmen des Schwerpunktprogramms der DFG *Spezifische Phänomene in der Silicium-Chemie: Neue experimentelle und theoretische Ansätze zum gezielten Aufbau and besseren Verständnis von mehrdimensionalen Systemen* statt.

Literatur

[1] Haiduc J (1970) The Chemistry of Inorganic Ring Systems, part 1. Wiley, London
[2] Seyferth D, Prud'Homme C, Wiseman GH (1983) Inorg Chem **22**: 2163
[3] Graalmann O, Klingebiel U, Clegg W, Haase M, Sheldrick GM (1984) Chem Ber **117**: 2988
[4] Schmidt-Bäse D, Klingebiel U (1989) Chem Ber **122**: 815
[5] Dippel K, Klingebiel U, Sheldrick GM, Stalke D (1987) Chem Ber **120**: 611
[6] Klingebiel U (1981) Angew Chem **93**: 696; (1981) Angew Chem Int Ed Engl **20**: 678
[7] Schmidt-Bäse D, Klingebiel U (1989) J Organomet Chem **364**: 313
[8] Schmidt-Bäse D, Klingebiel U (1989) Z Naturforsch Teil B **44**: 395
[9] Graalmann O, Klingebiel U, Clegg W, Haase M, Sheldrick GM (1984) Angew Chem **96**: 904; (1984) Angew Chem Int Ed Engl **23**: 891
[10] Graalmann O, Klingebiel U, Clegg W, Haase M, Sheldrick GM (1984) Z Anorg Chem **519**: 87
[11] Klingebiel U (1981) Chem Ber **114**: 3366
[12] Graalmann O, Klingebiel U (1984) J Organomet Chem **275**: C1–C4
[13] Wilkins CJ, Chambers DWS (1960) J Chem Soc 5088 sowie dort verzeichnete ältere Literatur
[14] Wannagat U, Bogedain G, Schervan A, Marsmann HC, Brauer DJ, Bürger H, Dörrenbach F, Pawelke G, Krüger C, Claus K-H (1991) Z Naturforsch **46b**: 931
[15] Veith M, Rammo A (1997) Phosphorus, Sulfur, and Silicon **123**: 75
[16] Veith M, Rammo A, Gießelmann M (1998) Z Anorg Allg Chem **624**: 49
[17] Bürger H, Geschwandtner W, Liewald GR (1983) J Organomet Chem **259**: 145
[18] Fink W (1964) Helv Chim Acta **47**: 498
[19] Veith M, Rammo A (unveröffentlichte Ergebnisse)
[20] Clegg W (1983) Acta Crystallogr **C39**: 901
[21] Clegg W (1982) Acta Crystallogr **B38**: 1648
[22] Lickiss PD, Redhouse AD, Thompson RJ, Stanczyk WA, Rozga K (1993) J Organomet Chem **435**: 13
[23] Seebach D, Armstutz R, Laube T, Scheizer WB, Dunitz JD (1985) J Am Chem Soc **107**: 5403
[24] Glidewell C, Robiette AG, Sheldrick GM (1970) J Chem Soc Chem Commun 931
[25] Oberhammer H, Zeil W, Fogarasie G (1973) J Mol Struct **18**: 309
[26] Winkelhofer N, Voigt A, Dorn H, Roesky HW, Steiner A, Stalke D, Reller A (1994) Angew Chem **106**: 1414; (1994) Angew Chem Int Ed Engl **33**: 1351
[27] Armitage DA, Tarassoli A (1973) Inorg Nucl Chem Letters **9**: 1225
[28a] Montero ML, Voigt A, Teichert M, Uson I (1995) Angew Chem **107**: 2761; (1995) Angew Chem Int Ed Engl **34**: 2504
[28b] Montero ML, Uson I, Roesky HW (1994) Angew Chem **106**: 2198; (1994) Angew Chem Int Ed Engl **33**: 2103
[29] Veith M, Faber S, Wolfanger H, Huch V (1996) Chem Ber **129**: 381

[30] Veith M, Jarczyk M und Huch V (1997) Angew Chem **109**: 140; (1997) Angew Chem Int Ed Engl **36**: 117
[31] Veith M, Jarczyk M, Huch V (1998) Angew Chem **110**: 109; (1998) Angew Chem Int Ed Engl **37**: 105
[32] Perrin DD, Dempsey B, Serjeant EP (1981) pKa Prediction for Organic Acids and Bases. Chapman & Hall, London
[33] Loewenstein W (1954) Am Mineral **39**: 92

Received July 15, 1998. Accepted (revised) October 2, 1998

Structure Determining Effect of Alcohols and Water on the Shape of the Hydrogen Bonded Network in Adducts with (9-Methyl-fluoren-9-yl)-silanetriol

Manuela Schneider, Beate Neumann, Hans-Georg Stammler, and **Peter Jutzi***

Fakultät für Chemie der Universität Bielefeld, D-33615 Bielefeld, Germany

Summary. (9-Methyl-fluoren-9-yl)-trichlorosilane (**1**) and the respective silanetriol **2** have been synthesized and characterized. In cocrystallization with **2**, ethanol, methanol, and water determine the morphology of the resulting hydrogen bonded network. Thus, incorporation of ethanol/water or methanol, respectively, leads to the tubular structures **2** · EtOH · H_2O (**2a**) and **2** · 2MeOH (**2b**), whereas the incorporation of water alone results in the double sheet structure **2** · H_2O (**2c**). The shapes of the different hydrogen bonded networks are discussed.

Keywords. Silanetriol; Cocrystallization; Hydrogen bonding.

Strukturbestimmender Einfluß von Alkoholen und Wasser auf die Gestalt des Wasserstoffbrückennetzwerks in Addukten mit (9-Methyl-fluoren-9-yl)-silantriol

Zusammenfassung. (9-Methyl-fluoren-9-yl)-trichlorsilan (**1**) und das entsprechende Silantriol **2** wurden synthetisiert und charakterisiert. In Kokristallisation mit **2** bestimmen Ethanol, Methanol und Wasser die Morphologie des resultierenden Wasserstoffbrückennetzwerks. Der Einbau von Ethanol/Wasser bzw. Methanol führt zu den Röhrenstrukturen **2** · EtOH · H_2O (**2a**) und **2** · 2MeOH (**2b**), während durch Einbau von Wasser allein die Doppelschichtstruktur **2** · H_2O (**2c**) erhalten wird. Die Gestalt der verschiedenen Wasserstoffbrückennetzwerke wird diskutiert.

Introduction

In recent years, silanetriols (RSi(OH)$_3$) have attracted a lot of attention because they represent versatile building blocks for the construction of polyhedral silsesquioxanes or metallasilsesquioxanes. Regarding the well known tendency of silanols towards selfcondensation, the isolation of stable silanetriols requires a proper balance of steric and electronic properties of the substituent R [1].

A general feature of silanols is the strong tendency to associate on the basis of intra- and intermolecular hydrogen bonding. In the case of silanetriols, this causes an astonishing diversity of network structures; the molecules adopt head-to-head and tail-to-tail arrangements forming hydrophobic (organic moieties) and hydrophilic areas (silanol groups) [2]. Crystal structures of silanetriols RSi(OH)$_3$ can be classified due to the shape of their hydrogen bonded network. In a double-

* Corresponding author

sheet structure, observed for $R = t$-Bu [2e], *c*-Hexyl [2a], and Cp^* [2g], hydrophobic and hydrophilic layers alternate. In the case of $R = 2,4,6\text{-}Me_3(C_6H_2)N\text{-}(SiMe_3)$ [2f] and $C_5H_4(SiMe_3)$ [2i], the molecules organize themselves in a tubular structure with a hydrophilic tube covered by a hydrophobic wrapping. A discrete-cage structure is typically found for the sterically more shielded silanetriols and can be characterized as a hydrophilic interior shielded by a hydrophobic outer layer. This structure type is observed for $R = [Co_3(CO)_9]C$ [2h], $(SiMe_3)_3C$, $(SiMe_3)_3Si$ [2b], and $(SiMe_2Ph)_3C)$ [2d]. Finally, the metal bound silanetriol bearing the bulky $OsCl(CO)(PPh_3)_2$ ligand shows no intermolecular hydrogen bonding at all [2c].

Molecules bearing Si–OH functionalities are capable to form adducts with a variety of compounds with potential hydrogen bonding sites. Thus, incorporation of ethers, alcohols, water, pyridine, pyridinium hydrochloride, pyridine-N-oxide, aromatics, and others into the hydrogen bonded network of silanols has been observed [2]. In the case of silanetriols we have previously reported the hemihydrate of $Cp^*Si(OH)_3$ [2g] as the only known adduct.

In this paper we report on the synthesis of (9-methyl-fluoren-9-yl)-trichlorosilane (**1**) and the corresponding silanetriol **2**. Furthermore, we show that **2** forms adducts in the cocrystallization process with alcohols and water. Depending on the cocrystallizing molecule, different structure types with respect to the hydrogen bonded network are realized. The tubular structures **2** · EtOH · H_2O (**2a**) and **2** · 2MeOH (**2b**) as well as the double sheet structure **2** · H_2O (**2c**) will be discussed on the basis of X-ray structural data.

Results and Discussion

Synthesis and characterization of **1** *and* **2**

The trichlorosilane **1** is obtained as colourless needles by the reaction of tetrachlorosilane with one equivalent of 9-methyl-fluorenyl lithium. Controlled

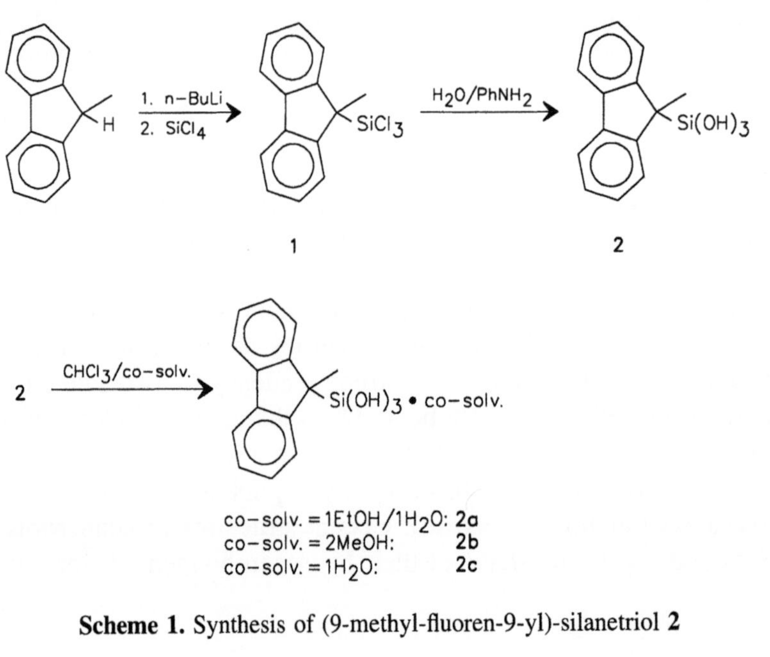

Scheme 1. Synthesis of (9-methyl-fluoren-9-yl)-silanetriol **2**

Table 1. Crystallographic data and parameters for X-ray structure determination of **2a–c**

	2a	**2b**	**2c**
Empirical formula	$C_{14}H_{14}O_3Si + C_2H_6O + H_2O$	$C_{14}H_{14}O_3Si + 2CH_4O$	$C_{14}H_{14}O_3Si + H_2O$
Formula weight	322.43	322.43	276.36
Crystal colour and habit	colourless needles	colourless needles	colourless plates
Crystal size	0.15 × 0.2 × 0.7 mm	0.2 × 0.4 × 0.8 mm	0.1 × 0.3 × 0.4 mm
Temperature	173(2) K	173(2) K	173(2) K
Wavelength	0.71073 Å (graphite monochromator)	0.71073 Å	0.71073 Å
Space group	Monoclinic C2/c	Orthorhombic Pnma	Monoclinic C2/c
Unit cell dimensions	$a = 17.860(10)$ Å, $\alpha = 90°$ $b = 6.603(4)$ Å, $\beta = 95.54(4)°$ $c = 28.476(13)$ Å, $\gamma = 90°$	$a = 7.021(3)$ Å, $\alpha = 90°$ $b = 17.027(7)$ Å, $\beta = 90°$ $c = 13.637(6)$ Å, $\gamma = 90°$	$a = 17.1847(11)$ Å, $\alpha = 90°$ $b = 6.9489(4)$ Å, $\beta = 103.8990(10)°$ $c = 24.1540(15)$ Å, $\gamma = 90°$
Volume	3342(3) $Å^3$	1630.3(12) $Å^3$	2799.9(3) $Å^3$
Z	8	4	8
Density (calculated)	1.281 Mg/m^3	1.314 Mg/m^3	1.311 Mg/m^3
Absorption coefficient	0.160 mm^{-1}	0.164 mm^{-1}	0.175 mm^{-1}
$F(000)$	1376	688	1168
Diffractometer used	Siemens R3mV	Siemens $P2_1$	Siemens Smart CCD
Completeness for $\theta = 27.00$	–	–	91.2%
θ range for data collection	2.29 to 27.07°	1.9 to 30.0°	2.44 to 27.11°
Index ranges	$0 \leq h \leq 22$, $0 \leq k \leq 8$, $-36 \leq l \leq 36$	$-9 \leq h \leq 0$, $0 \leq k \leq 23$, $-19 \leq l \leq 0$	$-21 \leq h \leq 22$, $-8 \leq k \leq 8$, $-30 \leq l \leq 30$
Reflections collected	3793	2454	12811
Independent reflections	3674 ($R_{int} = 0.0428$)	2454 ($R_{int} = 0.0000$)	3041 ($R_{int} = 0.0422$)
Absorption correction	Semi-empirical from Ψ-scans	Semi-empirical from Ψ-scans	Semi-empirical from equivalents
Max. and min. transmission	0.429 and 0.368	0.754 and 0.735	1.000000 and 0.888988
Refinement method	Full-matrix least-squares on F^2	Full-matrix least-squares on F^2	Full-matrix least-squares on F^2
Data/retraints/parameters	3674/0/210	2454/0/119	3041/0/192
Goodness-of-fit on F^2	1.041	1.102	1.030
Final R indices ($I > 2\sigma(I)$)	$R_1 = 0.1005$ for 1795 reflections	$R_1 = 0.0716$ for 1934 reflections	$R_1 = 0.0420$ for 2246 reflections
R indices (all data)	$R_1 = 0.2077$, $wR_2 = 0.2816$	$R_1 = 0.0888$, $wR_2 = 0.2394$	$R_1 = 0.0688$, $wR_2 = 0.1016$
Extinction coefficient	–	0.053(8)	–
Largest and mean Δ/σ	0.000 and 0.000	0.006 and 0.000	0.000 and 0.000
Largest diff. peak and hole	0.5 and $-0.5\,e \cdot Å^{-3}$	0.534 and $-0.530\,e \cdot Å^{-3}$	0.325 and $-0.331\,e \cdot Å^{-3}$
Remarks		Disorder of oxygen atom O(3) on two positions (50:50)	H(1*) to H(5*) were refined isotropically

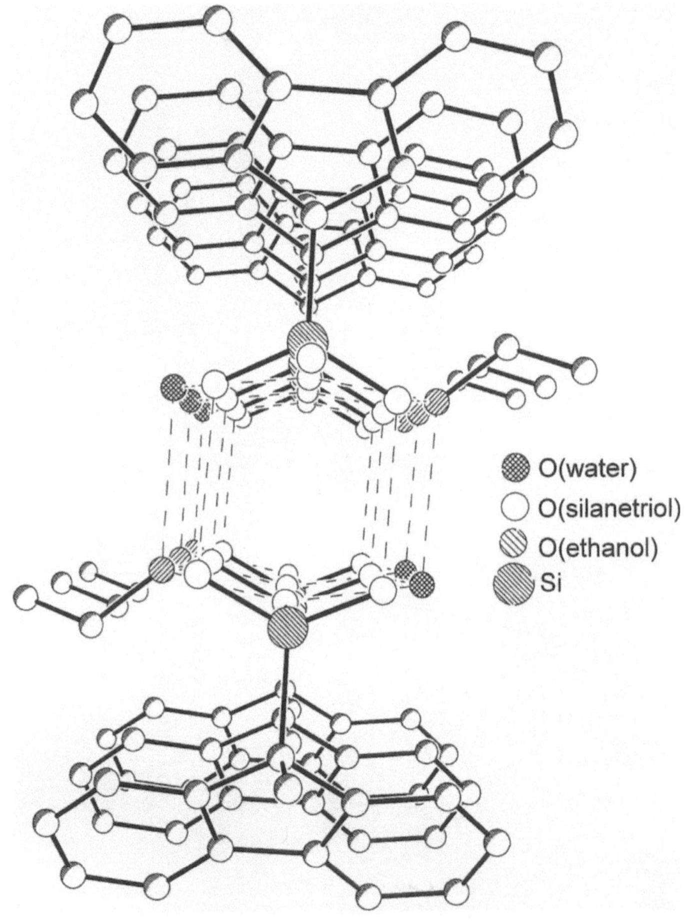

Fig. 1. Top view of the tubular structure of **2a**

hydrolysis of **1**, applying a method first described by *Takiguchi* [3], leads to the quantitative formation of the silanetriol **2** (Scheme 1). Compound **2** is readily soluble in polar solvents like chloroform, diethyl ether, *THF*, dimethyl sulfoxide, and nearly insoluble in aliphatic and aromatic solvents. When stored under an inert atmosphere at room temperature, **2** is stable for several months.

Characterization of the new compounds **1** and **2** was accomplished by means of elemental analysis as well as by mass spectrometry, IR, and multinuclear NMR spectroscopy.

Cocrystallization of **2** with alcohols and water leads to the structures **2a–c**. They can be described as polymeric arrangements of molecular subunits based on hydrogen bonding. The following discussion refers only to the different supramolecular assemblies. The bond distances and angles in the silanol molecules in **2a–c** are in the typical range for Si–C and Si–O units and are not discussed in this context. Crystallographic data are given in Table 1.

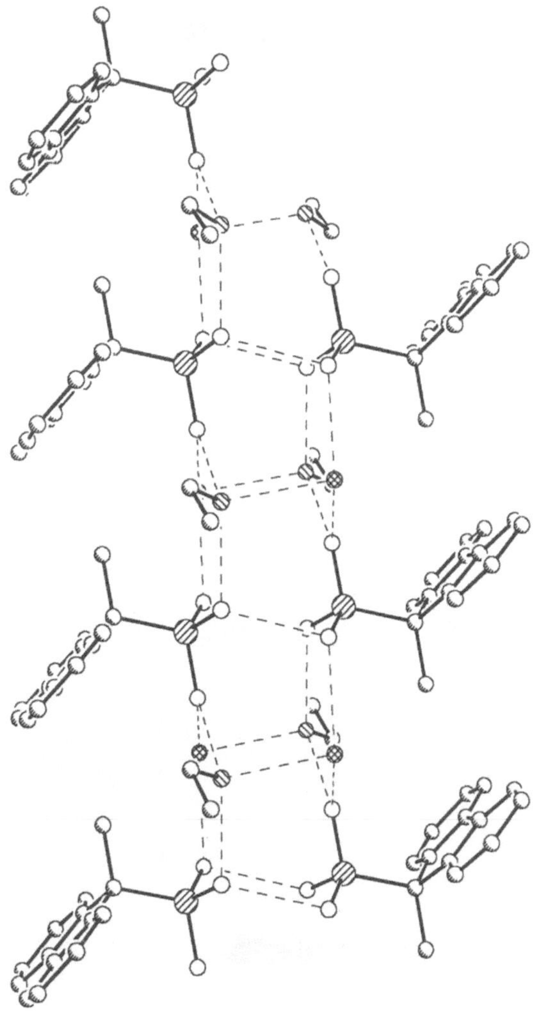

Fig. 2. Side view of the hydrogen bonded network of **2a**

Crystal structure of **2** · EtOH · H_2O (**2a**)

The structure of **2a** has to be described as a tubular hydrogen bonded network of silanetriol **2** incorporating one equivalent EtOH and one equivalent H_2O. As depicted in Fig. 1, each tube consists of two columns formed by the silanetriol molecules. The columns are created by stacking of the molecular units (one on top of the other). The network structure results from linking the silanetriol columns with two columns of water and ethanol molecules, respectively. A side view of the hydrogen bonded network, given in Fig. 2, clearly shows the opposite orientation of silanetriol molecules belonging to different columns.

Figure 3 gives also a side view of the tubular network, but with the tube rotated by 90 degrees. It can be seen that every oxygen and every hydrogen atom of the

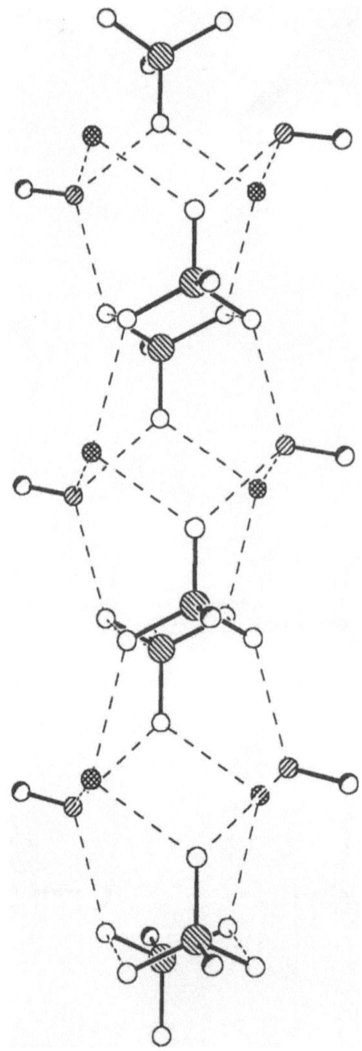

Fig. 3. Side view of the hydrogen bonded network of **2a** (for clarity, all organic moieties are indicated as single carbon atoms)

silanetriol units and of the water molecules is involved in one intermolecular hydrogen bond. In the case of the ethanol molecules, the oxygen atom forms two hydrogen bonds. The resulting hydrogen bonded network can also be understood as a regular arrangement of side-connected six-membered SiO_5 rings in boat conformation (each including one H_2O and two EtOH; hydrogen atoms are ignored). The oxygen-oxygen distances of the O–H–O units range from 2.71 to 2.88 Å.

Crystal structure of **2** · 2MeOH (**2b**)

Like **2a**, adduct **2b** also reveals a tubular structure. In this case, two equivalents of methanol are involved in the hydrogen bonded network. A tube of **2b** consists of

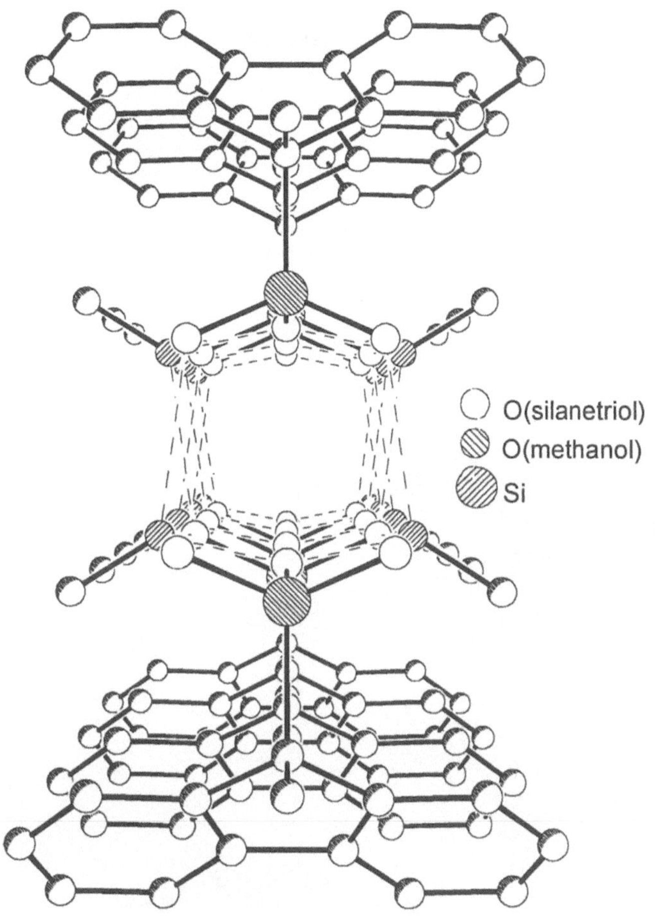

Fig. 4. Top view of the tubular structure of **2b**

two silanetriol columns opposing each other which are linked by four methanol columns (Fig. 4). In contrast to **2a**, the silanetriol molecules of opposite columns point into the same direction, but are shifted with respect to each other as can be seen from a side view (Fig. 5).

An interesting aspect of structure **2b** is the disorder of the methanol oxygen atoms on two positions with equal probability. As demonstrated in Fig. 6, a different hydrogen bonding situation results depending on the position of the methanol oxygen. In a), all methanol oxygen atoms are fixed at positions where bonding to three hydrogen atoms can be realized, whereas in b) they are fixed at the other position with bonding to only two hydrogen atoms. The corresponding bond lengths, given in Fig. 6, indicate that the formation of an additional hydrogen bond leads to a prolongation of the bonds. From an energetical point of view, the situation in a) and b) is equivalent. In Figs. 4 and 5, the oxygen atom of methanol is fixed on the position depicted in a) for reasons of clarity.

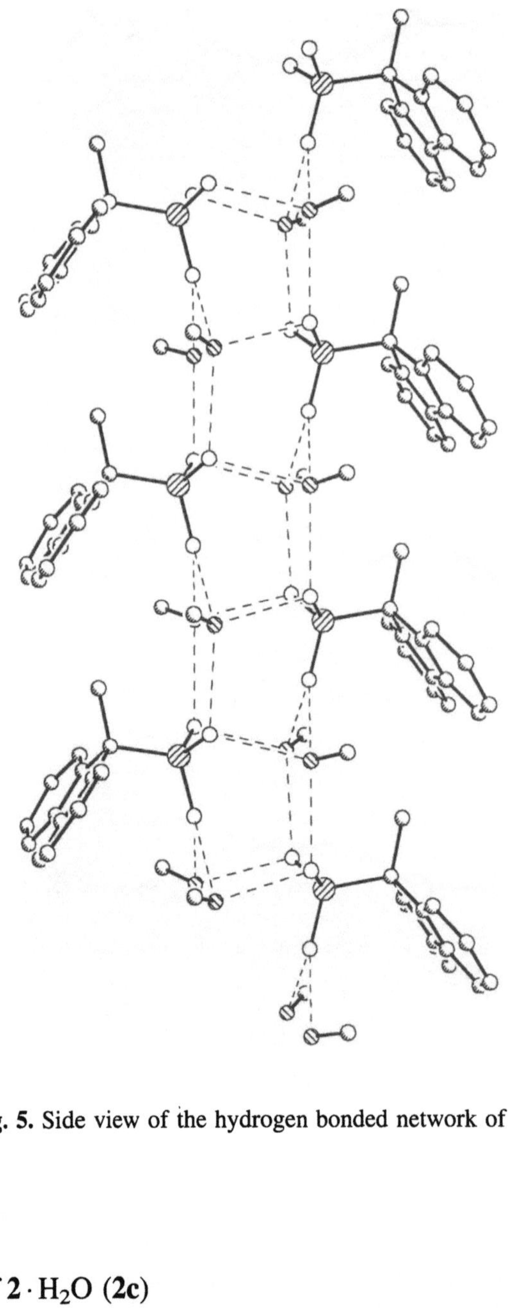

Fig. 5. Side view of the hydrogen bonded network of **2b**

Crystal structure of **2** · H_2O (**2c**)

It could be shown that with incorporation of one equivalent H_2O the silanetriol **2** crystallizes in the shape of a double sheet structure (**2c**). As depicted in Fig. 7, the Si–OH units and the water molecules form a hydrophilic double sheet surrounded by a hydrophobic layer formed by the organic moieties. The hydrogen bonded network is given in Fig. 8. Hydrogen bonding between the silanol groups leads to the formation of chains which are connected by hydrogen bonds to the water molecules. The oxygen-oxygen distances of the hydrogen bonds range from 2.66 to 2.85 Å.

An interesting comparison can be made with the double-sheet structure of the *Cp**Si(OH)$_3$ hemihydrate [2i]. In this case, all silanetriol molecules form

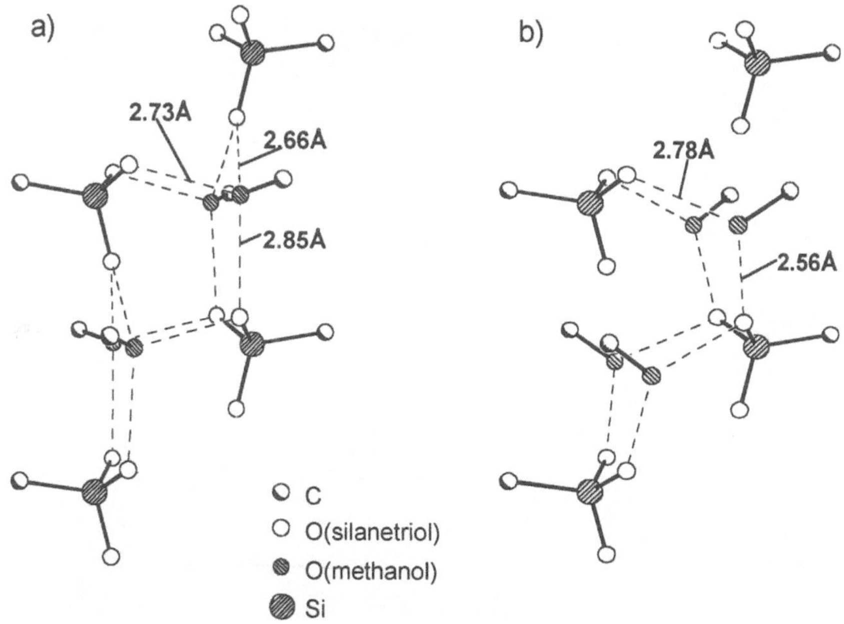

Fig. 6. Disorder of methanol oxgyen atoms demonstrated in a section of **2b**; all methanol oxygen atoms are binding to three (a) or two (b) hydrogen atoms (for clarity, the methyl-fluorenyl moietes are indicated as single carbon atoms)

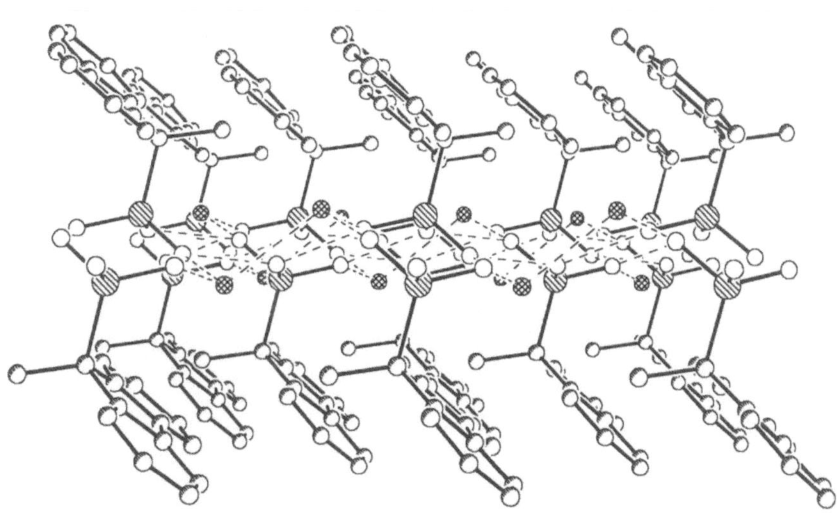

Fig. 7. Side view of the double-sheet structure of **2c**

hydrogen bonds to other silanetriol molecules, and only one half equivalent of water is incorporated in the network. Because of the sterically more demanding (methyl)fluorenyl group, one equivalent of water is required to connect the hydrogen bonded chains of the silanetriol molecules yielding the network structure **2c**.

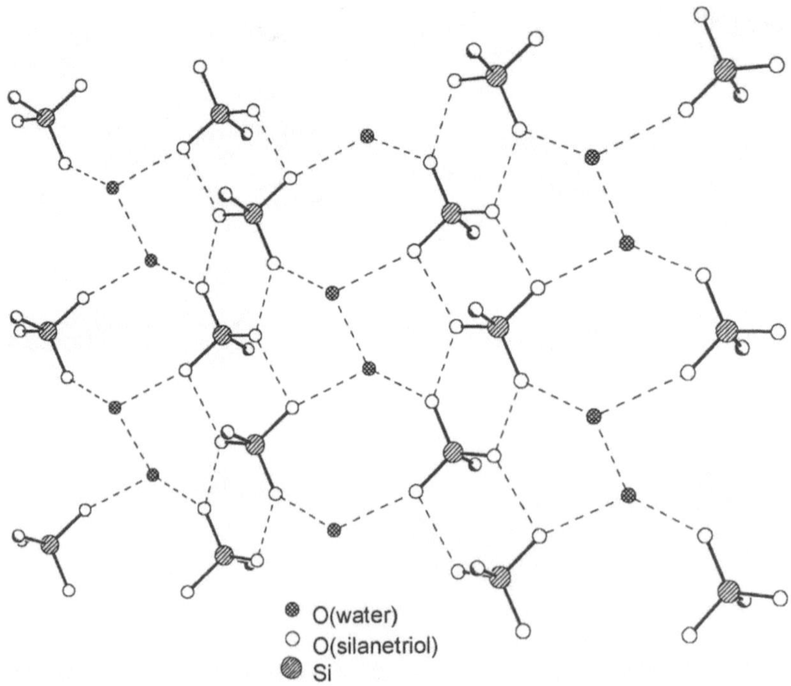

Fig. 8. Top view of the hydrogen bonded network of **2c** (for clarity, the methyl-fluorenyl moietes are indicated as single carbon atoms)

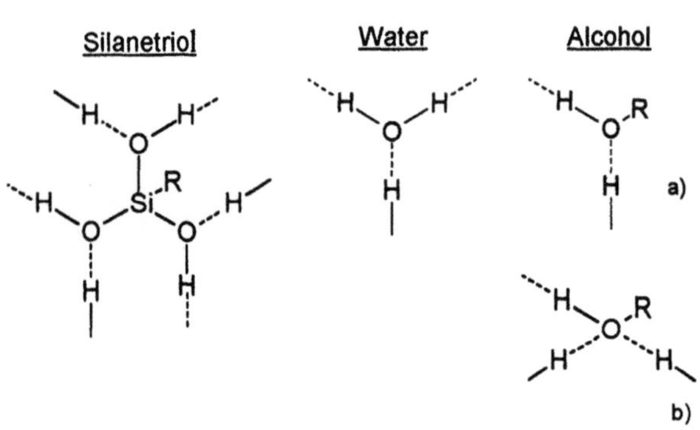

	Silanetriol	Water	Alcohol
Potential hydrogen bonding sites	6	3	$2^{a)}/3^{b)}$
Proton acceptor sites	3	2	$1^{a)}/2^{b)}$
Proton donor sites	3	1	1

Scheme 2. Potential hydrogen bonding sites of silanetriols, alcohols, and water

Conclusion

It could be demonstrated that even small structural changes of the cocrystallizing molecule lead to different structure types in terms of the hydrogen bonded network. The tubular structures **2** · EtOH · H_2O (**2a**) and **2** · 2MeOH (**2b**) as well as the double sheet structure **2** · H_2O (**2c**) have been characterized. The discussed hydrogen bonded networks of **2a–c** differ from each other concerning the specific connectivity pattern of the building blocks based on their potential hydrogen bonding sites (Scheme 2).

Further research work will be directed towards a systematic structural design of such hydrogen bonded assemblies. The structure determining influence of other molecules with potential hydrogen bonding sites in cocrystallization with silanetriols will also be investigated.

Experimental

General comments

All experiments were carried out under an atmosphere of purified and dry argon using standard *Schlenk* techniques. Solvents were dried and distilled prior to use by standard methods. Melting point determination was performed using a Büchi 510 melting point apparatus. The elemental analyses were carried out by the Microanalytical Laboratory of the faculty of chemistry at the university of Bielefeld. NMR spectra were recorded on a Bruker Avance DRX 500 spectrometer (^{1}H, 500.1 MHz; $^{13}C\{^{1}H\}$, 125.8 MHz; $^{29}Si\{^{1}H\}$, 99.4 MHz). Chemical shifts are reported relative to internal *TMS* using the respective solvent as secondary standard (^{1}H, ^{13}C) or to external *TMS* (^{29}Si). IR spectra were obtained on a Bruker Vector 22 spectrometer (only strong absorption values are given). Mass spectrometry (EI, 70 eV) was performed using a VG Autospec spectrometer; only characteristic fragments and isotopes of highest abundance are mentioned.

$C_{14}H_{12}$ was prepared following a reported procedure [4]. *n*-Butyllithium was used as a 1.6 *M* solution in *n*-hexane (Aldrich). Tetrachlorosilane (Aldrich) was purified by distillation and stored over magnesium filings. Aniline (Aldrich) was purified by distillation.

(9-Methyl-fluoren-9-yl)-trichlorosilane (**1**)

A stirred solution of 5.91 g (32.8 mmol) 9-methyl-fluorene in 45 ml of *THF* was treated dropwise with 21.0 ml (32.8 mmol) of *n*-BuLi at 0°C. The mixture was allowed to warm to room temperature while stirring was continued for further 2 h. Then the mixture was slowly added to a solution of 11.1 g (65.6 mmol) $SiCl_4$ in 30 ml of *THF* at −78°C. After stirring for 12 h and warming to room temperature the solvent was removed *in vacuo*, and the residue was extracted with 50 ml of *n*-pentane. The volume of the extract was reduced *in vacuo*. Crystallisation at −30°C yielded 7.20 g of $C_{14}H_{11}Si_1Cl_3$ (70%) as colourless needles.

M.p.: 131°C; ^{1}H NMR ($CDCl_3$): $\delta = 1.86$ (s, 3H, Me), 7.38 (t, $J_{HH} = 7.4$ Hz, 2H, aromat. H), 7.45 (t, $J_{HH} = 7.5$ Hz, 2H, aromat. H), 7.70 (d, $J_{HH} = 7.6$ Hz, 2H, aromat. H), 7.84 (d, $J_{HH} = 7.6$ Hz, 2H, aromat. H) ppm; ^{13}C NMR ($CDCl_3$): $\delta = 18.4$ (Me), 48.2 (allyl-C), 120.3, 124.9, 127.2, 127.9, 141.1, 144.2 (aromat. C) ppm; ^{29}Si NMR ($CDCl_3$): $\delta = 6.78$ ppm; IR (KBr): $\nu = 531$(vs), 549(vs), 576(vs), 601(vs), 728(vs), 746(vs), 763(s), 1442(s), 2864(m), 2926(m), 2969(m), 3066(m) cm^{-1}; MS: *m/z* (rel.int. in %, [fragment]) = 312 (7, [M^+]), 179 (100, [M^+-$SiCl_3$]); $C_{14}H_{11}Cl_3Si_1$ (313.71); calcd.: C 53.60, H 3.53; found: C 53.43, H 3.49.

(9-Methyl-fluoren-9-yl)-silanetriol (**2**)

A solution of 6.58 g of **1** in 100 ml of Et_2O was slowly added to a stirred solution of 5.86 g (62.9 mmol) aniline and 1.13 g (62.9 mmol) H_2O in 300 ml of Et_2O at 0°C. Stirring was continued for 18 h, allowing the reaction mixture to warm to room temperature. The anilinium hydrochloride was filtered off, and the solvent was removed *in vacuo*. The residue was washed with pentane yielding 4.86 g (90%) of $C_{14}H_{11}Si_1(OH)_3$ as a colourless solid.

M.p.: 148°C (dec.); ^{1}H NMR (*DMSO*-d_6): $\delta = 1.54$ (s, 9H, Me), 6.12 (s, 3H, OH), 7.25 (m, 4H, aromat. H), 7.62 (d, $J_{HH} = 7.3$ Hz, 2H, aromat. H), 7.82 (d, $J_{HH} = 7.2$ Hz, 2H, aromat. H) ppm; ^{13}C NMR (*DMSO*-d_6): $\delta = 20.9$ (Me), 43.5 (allyl-C), 119, 124, 125, 126, 140, 151 (aromat. C) ppm; ^{29}Si NMR (*DMSO*-d_6): $\delta = -50.7$ ppm; IR (KBr): $\nu = 495$(s), 729(s), 744(vs), 767(vs), 801(s), 858(vs), 874(s), 928(vs), 1026(s), 1110(s), 1262(m), 1438(s), 2870(m), 2928(m), 2962(m), 3032(m), 3058(s), 3378 (vs, br), 3559(vs) cm^{-1}; MS: *m/z* (rel.int. in %, [fragment]) = 258 (44, [M^+]), 243 (4, [M^+-CH_3]), 240 (9, [M^+-H_2O]), 179 (100, [M^+-$Si(OH)_3$]); $C_{14}H_{14}Si_1O_1 \cdot H_2O$ (276.37); calcd.: C 60.84, H 5.84; found: C 61.33, H 6.05.

X-ray structure determination

Suitable crystals of **2a–c** were grown from a mixture of **2** with the respective alcohol and/or water in $CHCl_3$ at -30°C. The single crystals were coated with a layer of hydrocarbon oil, attached to a glass fibre, and cooled to 173(2) K for data collection on a Siemens P2(1) diffractometer. Crystallographic programs used for structure solution are Siemens SHELXTL plus and SHELXL-93. The structure was solved by direct methods and was refined using full-matrix least squares with anisotropic thermal parameters for all non-hydrogen atoms. All hydrogen atoms of **2a** and **2b** were fixed at the calculated positions using a riding model.

Additional material to the structure determination may be ordered from the Cambridge Crystallographic Data Centre as supplementary publication no. CCDC-102594 (**2a**), 102595 (**2b**), and 102596 (**2c**). Copies of the data can be obtained free of charge on application to The Director, CCDC, 12 Union Road, Cambridge CB2 1EZ, UK (Fax: internat. (+)441223 336–033; e-mail: deposit@chemcrys.cam.ac.uk).

Acknowledgements

Financial support by the *Deutsche Forschungsgemeinschaft* is gratefully acknowledged. *Manuela Schneider* wishes to thank the federal state *Nordrhein-Westfalen* for a research fellowship.

References

[1] a) Murugavel R, Chandrasekhar V, Roesky HW (1996) Acc Chem Res **29**: 183; b) Murugavel R, Voigt A, Walawalkar MG, Roesky HW (1996) Chem Rev **96**: 2205

[2] For a review on this subject see: Lickiss PD (1995) Adv Inorg Chem **42**: 147; a) Ishida H, Koenig JL, Gardner KC (1982) J Chem Phys **77**: 5748; b) Al-Juad SS, Buttrus NH, Damja RI, Derouiche Y, Eaborn C, Hitchcock PB Lickiss PD (1989) J Organomet Chem **371**: 287; c) Rickard CEI, Roper WR, Salter DM, Wright LJ (1992) J Am Chem Soc **114**: 9682; d) Whittaker SM PhD (1993) Thesis, University of Salford, United Kingdom; e) Winkhofer N, Roesky HW, Noltemeyer M, Robinson WT (1992) Angew Chem **104**: 670; f) Murugavel R, Chandrasekhar V, Voigt A, Roesky HW, Schmidt H-G, Noltemeyer M (1995) Organometallics **14**: 5298; g) Jutzi P, Strassburger G, Schneider M, Stammler H-G, Neumann B (1996) Organometallics **15**: 2842; h) Ritter U, Winkhofer N, Schmidt H-G, Roesky HW (1996) Angew Chem **108**: 59; i) Jutzi P, Schneider M, Stammler H-G, Neumann B (1997) Organometallics **16**: 5377

[3] Takiguchi TJ (1959) Am Chem Soc **81**: 2359

[4] Murdoch JR, Bryson JA (1982) J Am Chem Soc **104**: 600

Received June 30, 1998. Accepted (revised) August 13, 1998

Silsesquioxane Chemistry II. Tin(IV) and Hafnium(IV) Compounds of Silsesquioxanes

Yurii K. Gun'ko[1], **László Nagy**[2], **Wolfgang Brüser**[1], **Volker Lorenz**[1], **Axel Fischer**[1] **Stephan Gießmann**[1], **Frank T. Edelmann**[1,*], **Klaus Jacob**[3], and **Attila Vértes**[4]

[1] Chemisches Institut, Otto-von-Guericke-Universität Magdeburg, D-39106 Magdeburg, Germany

[2] Department of Inorganic and Analytical Chemistry, Attila József University, H-6701 Szeged, Hungary

[3] Martin-Luther-Universität Halle-Wittenberg, Fachbereich Chemie, Institut für Anorganische Chemie, D-06217 Merseburg, Germany

[4] Department of Nuclear Chemistry, Eötvös University, H-1518 Budapest, Hungary

Summary. Four complexes of silsesquioxanes with di- and triorganotin(IV) cations have been prepared; their composition was determined by standard analytical and spectroscopic methods. The results show that complexes containing an organotin(IV) moiety and the ligand in a 1:1 ratio are formed. FTIR spectra are consistent with the presence of Sn–O vibrations in the compounds. The structure of the complexes is discussed on the basis of *Mößbauer* and multinuclear (^1H, ^{13}C, ^{119}Sn, and ^{29}Si) NMR spectroscopic measurements. Comparison of the experimental quadrupole splitting (*QS*) with that calculated on the basis of the partial quadrupole splitting (*PQS*) concept revealed that the complexes formed with dialkyltin(IV) cations as well as the trialkyl derivatives have a regular tetrahedral structure. A new organohafnium silsesquioxane derivative ($[Cy_7Si_7O_{12}Hf(C_5Me_5)]$, *Cy* = cyclohexyl, **5**), was prepared by treatment of $[Cy_7Si_7O_9(OH)_3]$ with $(C_5Me_5)HfMe_3$ under elimination of methane. The molecular structure of **5** was determined by X-ray crystallography.

Keywords. Silsesquioxanes; Organotin(IV) compounds; *Mößbauer* spectroscopy; Organohafnium(IV) compounds; X-Ray crystallography.

Silsesquioxanchemie II. Zinn(IV)- und Hafnium(IV)-Verbindungen von Silsesquioxanen

Zusammenfassung. Vier Komplexe von Silsesquioxanen mit Di- und Triorganozinn(IV)-Kationen wurden hergestellt; ihre Zusammensetzung wurde mit analytischen und spektroskopischen Standardmethoden bestimmt. Die Ergebnisse zeigen, daß in den gebildeten Komplexen Organozinneinheit und Ligand im Verhältnis von 1:1 vorhanden sind. FTIR-Spektren stehen im Einklang mit dem Auftreten von Sn–O-Schwingungen in den Verbindungen. Die Struktur der Komplexe wird auf der Grundlage von *Mößbauer*-Spektren und multinuklearen NMR-Experimenten (^1H, ^{13}C, ^{119}Sn, ^{29}Si) diskutiert. Ein Vergleich der experimentell beobachteten Quadrupolaufspaltung (*QS*) mit nach dem Konzept der partiellen Quadrupolaufspaltung (*PQS*) berechneten Werten ergab sowohl für die mit dem Dialkyzinn(IV)-Kation gebildeten Komplexe als auch für die Trialkylderivate eine regelmäßige tetraedrische Struktur. Ein neues Organohafniumsilsesquioxan-

* Corresponding author

derivat ([$Cy_7Si_7O_{12}Hf(C_5Me_5)$], *Cy* = Cyclohexyl, **5**) wurde durch Umsetzung von [$Cy_7Si_7O_9(OH)_3$] mit $(C_5Me_5)HfMe_3$ unter Abspaltung von Methan hergestellt; seine Molekülstruktur wurde mittels Röntgenstrukturanalyse bestimmt.

Introduction

In a recently published review article by *Fehér et al.* [1] it was pointed out that incompletely condensed polyhedral silsesquioxanes (POSS) are an interesting class of ligands for both main group and transition metal elements. They possess a sufficient degree of oligomerization as to be relevant models for highly siliceous materials; furthermore, they retain reactive Si–OH functionalities which allow for their exploitation as ligands in a wide variety of main group and transition metal complexes. The electrochemical behavior of oligometallic sandwich complexes of cyclosiloxanolate ligands has also recently been studied [2].

The most common starting material in this field is the cyclohexyl substituted trisilanol [$Cy_7Si_7O_9(OH)_3$], which is generally prepared by the kinetically controlled hydrolysis of cyclohexyltrichlorosilane.

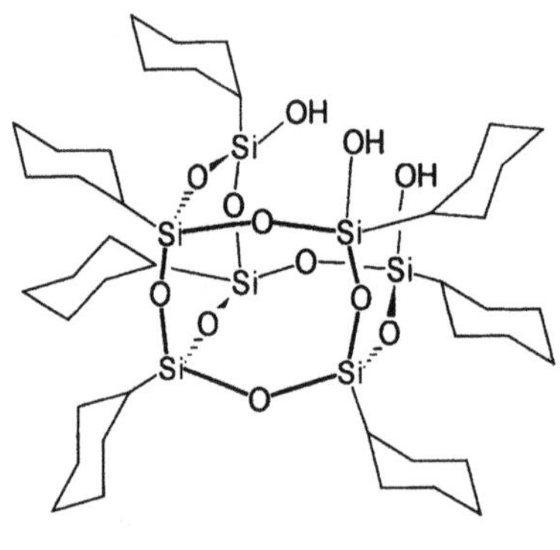

One or two OH groups in [$Cy_7Si_7O_9(OH)_3$] can be selectively blocked by trimethylsilyl substitution to afford the corresponding di- and monosilanols [$Cy_7Si_7O_9(OSiMe_3)(OH)_2$] and [$Cy_7Si_7O_9(OSiMe_3)_2(OH)$], respectively. The amine assisted metathesis reaction of Si–O–H with $L_xM{-}X$ bonds is quite convenient and has proven to be the most effective method for the synthesis of POSS complexes [1]. Utilizing this method, the mono- and disilanol have been capped with tri- and dialkyltin(IV) cations. The complexes thus formed were characterized by FTIR, multinuclear (^1H, ^{13}C, ^{119}Sn, and ^{29}Si) NMR, and *Mößbauer* spectroscopic methods. The latter was applied for the first time to investigate this type of compounds. Further, a new synthetic route has been developed to cap the incompletely condensed silsesquioxane [$Cy_7Si_7O_9(OH)_3$] with a $(C_5Me_5)Hf$ fragment.

Results and Discussion

Preparative results

Reactions of [$Cy_7Si_7O_9(OSiMe_3)(OH)_2$] and [$Cy_7Si_7O_9(OSiMe_3)_2(OH)$] with organotin(IV) chlorides were carried out in benzene solution according to Eq.

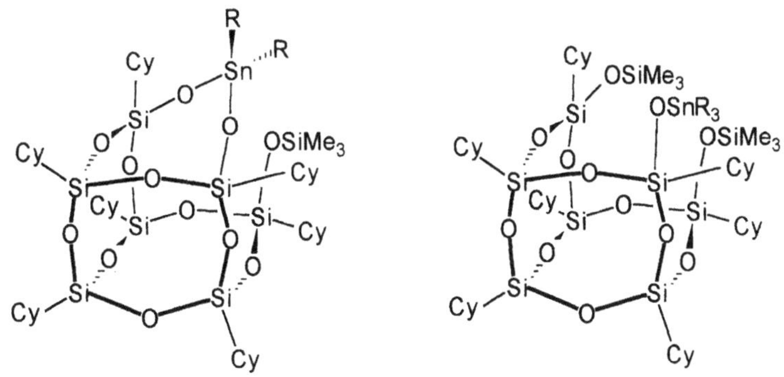

1: $R = n$-Bu
2: $R = $ Me

3: $R = $ Me
4: $R = CH_2Ph$

Fig. 1. Proposed structures of complexes **1–4**

(1) and (2). Appropriate amounts of triethylamine were added in order to trap the eliminated HCl.

$$[Cy_7Si_7O_9(OSiMe_3)(OH)_2] + R_2SnCl_2 \xrightarrow[-2\,[Et_3NH]Cl]{2\,Et_3N} [Cy_7Si_7O_9(OSiMe_3)O_2SnR_2] \quad (1)$$

1 : $R = n$-Bu
2 : $R = $ Me

$$[Cy_7Si_7O_9(OSiMe_3)_2(OH)] + R_3SnCl \xrightarrow[-[Et_3NH]Cl]{Et_3N} [Cy_7Si_7O_9(OSiMe_3)_2OSnR_3] \quad (2)$$

3 : $R = $ Me
4 : $R = CH_2Ph$

The analytical data of **1–4** clearly demonstrate that complexes with the required stoichiometry were formed *via* substitution of the chloride atoms in the alkyltin(IV) chlorides by hydrogen atoms of the free hydroxyl groups of the ligands, HCl molecules being eliminated (Fig. 1).

Related group 4 compounds derived from $[Cy_7Si_7O_9(OH)_3]$ have been reported earlier by *Feher et al.* [1]. The standard synthetic route is similar to that described here for the Sn derivatives and involves treatment of $[Cy_7Si_7O_9(OH)_3]$ with cyclopentadienyl metal trichlorides of Ti, Zr, and Hf in the presence of a base. However, this method has the drawback that it requires the separation of the product from the organoammonium chloride by-product. We report here an alternative route involving treatment of $[Cy_7Si_7O_9(OH)_3]$ with the corresponding trimethyl complexes as exemplified by the reaction with $(C_5Me_5)HfMe_3$ in diethylether solution. The reaction proceeded smoothly at room temperature under elimination of methane and afforded the hitherto unknown (pentamethylcyclopentadienyl)hafnium(IV) complex **5** in 78% yield (Scheme 1).

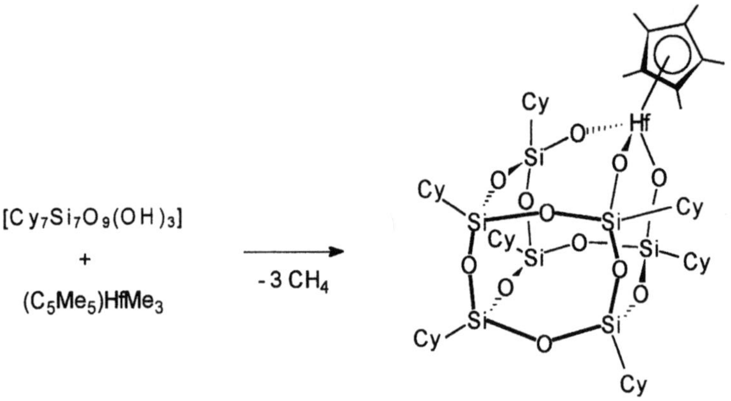

Scheme 1. Synthesis of hafnium complex **5**

Mößbauer measurements

In the case of **1–4** *Mößbauer* parameters (*IS* and *QS*) give information about the coordination polyhedra surrounding the tin atom. Certain ranges of *QS* associated with a particular stereochemistry are as following: tetrahedral $\mathrm{Sn}R_2X_2$: 1–2.4 mm/s, trigonal bipyramidal *cis*-$\mathrm{Sn}R_2X_3$: 3.0–3.5 mm/s, octahedral *cis*-$\mathrm{Sn}R_2X_4$: *ca.* 2 mm/s, and octahedral *trans*-$\mathrm{Sn}R_2X_4$: *ca.* 4 mm/s. However, it should be emphasized that most compounds have a structure which deviates more or less from these regular polyhedra, and the observed *QS* depends on the angle of R–Sn–R (Θ) according to Eq. (3)

$$QS = -4[R](1 - (3/4)\sin^2\Theta)^{1/2} \tag{3}$$

A similar treatment may be applied to tetrahedral $R_3\mathrm{Sn}X$ compounds, many of which show distorted structures. In such a case, *QS* becomes

$$QS = 2[X]_{\mathrm{tetr}} - 3[R]_{\mathrm{tetr}}(1 - 3\cos^2\Theta) \tag{4}$$

so that a reduction in Θ increases the contribution of the R groups [4].

For exemplification, a *Mößbauer* spectrum is shown in Fig. 2. All spectra exhibited *IS* and *QS* values which clearly indicated the presence of tin(IV) species within the complexes **1–4**. All spectra of the complexes studied comprised only one well developed doublet, suggesting the presence of equivalent environments of tin(IV) centres in these compounds. The *Mößbauer* parameters determined by computer evaluation of the spectra measured at liquid nitrogen temperature are collected in Table 1 together with the suggested configurations. To determine the stereochemical arrangement of the coordination sphere in these organotin(IV) compounds, the experimental *QS* values were calculated on the basis of a simple but general molecular orbital model, according to the *PQS* concept [4, 5] for the symmetries of four-coordinated tin(IV) atoms binding two or three alkyl groups and one or two negatively charged (deprotonated) oxygen atoms from the ligands. The *PQS* values of the different functional groups used in calculations and the calculated *QS* values for tin(IV) in tetrahedral stereochemical arrangements are

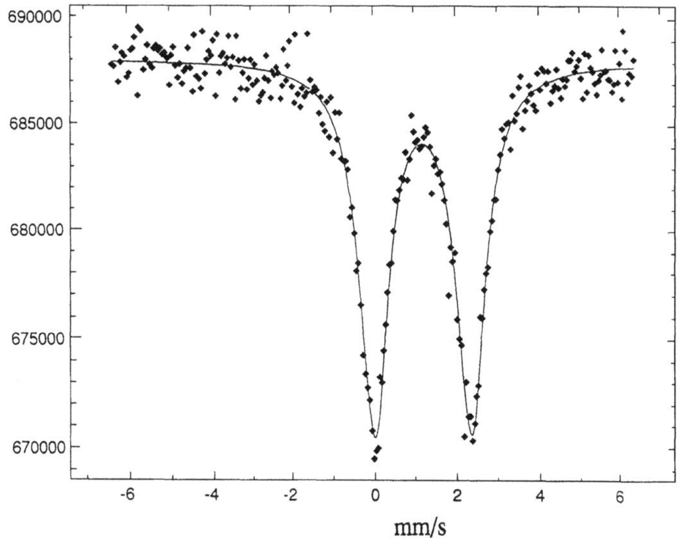

Fig. 2. Mößbauer spectrum of **1** measured at 77 K

Table 1. Experimental and calculated *Mößbauer* parameters for compounds **1–4**

	1	**2**	**3**	**4**
IS (mm/s)	1.18(4)	1.07(2)	1.20(4)	1.35(4)
QS_{exp} (mm/s)	2.35(8)	2.29(4)	2.13(9)	1.91(8)
QS_{calc} (mm/s)	2.30	2.30	2.00	1.78
Line width (mm/s)	0.86	0.86	0.83	0.88
Asymmetry (left/right)	1.008	1.021	1.155	1.116
Θ (C–Sn–C)	111.4°	109.2°	108.5°	108.4°

given as *PQS* (mms^{-1}) $\{Alk\}^{tetr} = -1.37$, $\{Ph, Bz) = -1.26$, $\{-O^-\}^{tetr} = -0.37$, and $\{-OH\}^{tetr} = -0.25$.

$$QS(T_d) = 1.15(2\{R\}^{tetr} - 2\{O^-\}^{tetr}) \quad (5)$$

Some *PQS* values were taken from the literature [5, 6]. The value of $\{O^-\}^{tetr}$ was calculated *via* equation (6) in Ref. [7].

The evaluation of the *Mößbauer* measurements on the basis of the *PQS* concept and chemical evidence showed that the tin(IV) in the studied alkytin(IV) complexes **1–4** is four-coordinated and that all four complexes have a regular tetrahedral structure with a small distortion (Table 1). These findings are consistent with the result obtained by single crystal X-ray diffraction studies on the *bis*(triphenyltin) derivative [$Cy_8Si_8O_{11}(OSnPh_3)_2$] [3].

NMR investigations

The ^1H NMR spectra of **1** and **2** show several broad signals in the range from 2.19 to 0.9 which correspond to the protons of the *c*-C_6H_{11} groups. In the ^1H NMR spectrum of **2** there are two separated signals for the methyl groups bonded to the central tin atom (at 0.53 and 0.47 ppm) with a coupling constant of 19.62 Hz. It is impossible to see an analogous picture in the spectrum of **1**, because the proton signals of the *n*-butyl groups are very close to those of the c-C_6H_{11} protons. Due to coupling with the tin, the ^{13}C NMR spectra for both **1** and **2** show also two signals with coupling constants of 42.9 and 42.4 Hz, respectively. This may be associated with some steric influence of the bulky $SiMe_3$ group on the alkyl substituents at the Sn atoms in **1** and **2**. The ^{29}Si NMR spectra of the two complexes are very similar and contain six (two signals at −67.4 and −67.5 ppm are very close) singlets for **1** and five singlets for **2** (two signals are too close to be resolved in this case). The shifts observed in the ^{29}Si NMR spectra of **1** and **2** are slightly different from those of the starting silsesquioxane [3]. A ^{119}Sn NMR spectrum of **1** shows just one singlet at −46.14 ppm. In contrast, the ^{119}Sn NMR spectrum of **2** displays a multiplet (signals are very close to each other) at −8.07 ppm. This difference suggests that in contrast to monomeric **1** (the *n*-butyl groups are probably too bulky) complex **2** adopts a dimeric structure in solution. The ^1H, ^{13}C, and ^{29}Si NMR spectra of **3** and **4** show two separated signals for the $SiMe_3$ groups, which are however quite close to each other. The ^{29}Si NMR spectra of **3** and **4** also contain a number of signals in a very narrow range corresponding to the siloxane skeleton. The ^{119}Sn NMR spectra of **3** and **4** show singlets at 113.43 and −20.41 ppm, respectively. As expected, the chemical shift in for **3** (113 ppm) is essentially different from those of **1** and **2**. This can be explained by different electronic effects of monosilanol and disilanol ligands on the tin atom. Complexes **3** and **4** have obviously monomeric structures in solution. The NMR data of **5** are also consistent with a monomeric structure in which the incompletely condensed silsesquioxane cube is completed by a $(C_5Me_5)Hf$ fragment.

FTIR investigations

The IR data are also consistent with the formation of well-defined compounds. The IR spectra of the complexes formed with alkyltin(IV) cations (**1–3**) show clearly the presence of alkyl groups in the molecules. The characteristic frequencies of the di-*n*-butyltin(IV) oxide **1** in the region of 2300–700 cm^{-1} are weak, and the absorptions related to these in the complexes are generally masked by the strong absorptions of the ligands. In the IR spectra of the complexes, the most characteristic changes occur in the spectral regions of 600–400 cm^{-1}, 1200–1000 cm^{-1}, and 3600–2700 cm^{-1}. Vibration bands characteristic of $(Sn{-}C)_{as}$ at 544–562 cm^{-1}, $(Sn{-}C)_s$ at 453–502 cm^{-1} [8], and $(Sn{-}O)_{as}$ at 640–643 cm^{-1} as well as $(Sn{-}O)_s$ at 576–619 [9] appeared (Table 2). The changeover from the -C–O vibration of the -COH groups in the ligands to those of the -COSn groups in the complexes caused alterations in the region 1200–1000 cm^{-1}. The interaction of the

Table 2. FTIR data for compounds **1–4** (cm^{-1})

	1	**2**	**3**	**4**
$\nu(Sn{-}C)_{as}$	548	559	544	549
$\nu(Sn{-}C)_s$	501	502	501	501
$\nu(Sn{-}O)_{as}$	642	643	641	640
$\nu(Sn{-}O)_s$	619	618	576	614

alkyltin(IV) cation with the ligands clearly involved the OH groups of the ligands. Accordingly, the free OH stretching absorptions disappeared in the vibration spectra of the complexes.

X-ray structure determination of **5**

The measurement was performed using a Siemens SMART CCD diffractometer with MoK_α X-radiation ($\lambda = 0.71073$ Å). A selected crystal of **5** was coated with mineral oil, mounted on a glass fibre, and transferred to a cold nitrogen stream (Siemens LT-2 attachment). Full hemispheres of the reciprocal space were scanned by ω in three sets of frames of 0.3°. Crystal data of **5**: $C_{52}H_{92}HfO_{12}Si_7$, $M = 1284.38$, orthorhombic, space group Pca2_1, $a = 2522.92(1)$, $b = 1184.84(1)$ $c = 4216.6(3)$ pm, $V = 12.5927(2)$ nm^3, $Z = 8$, $D_x = 1.355$ Mg · m^{-3}, $F(000) = 5360$, $\lambda = 71.073$ pm, $T = 153(2)$ K, μ(MoK_α) $= 1.843$ mm^{-1}. Absorption correction by SADABS; $T_{min}/T_{max} = 0.64/1.00$. The structure was solved by direct methods and refined on F^2 to give R(conventional) $= 0.055$ (wR^2, all data $= 0.103$), and GoF $= 1.034$ for 82486 (observed) and 27852 (independent) reflections; $R_{int} = 0.077$ [10]. All non-hydrogen atoms were refined anisotropically. The hydrogen atoms were included using a riding model, except for rigid methyls (*Elack* parameter $x = -0.003(4)$ [11]).

The X-ray structure determination of **5** confirms the presence of a closed silsesquioxane cage in which the open corner of the precursor [$Cy_7Si_7O_9(OH)_3$] has been capped by a (pentamethylcyclopentadienyl)hafnium fragment. The average Hf–O distance is 195.8 pm, whereas the average Hf–C bond lengths to the C_5Me_5 ligand have been determined as 250.5 pm (Fig. 3).

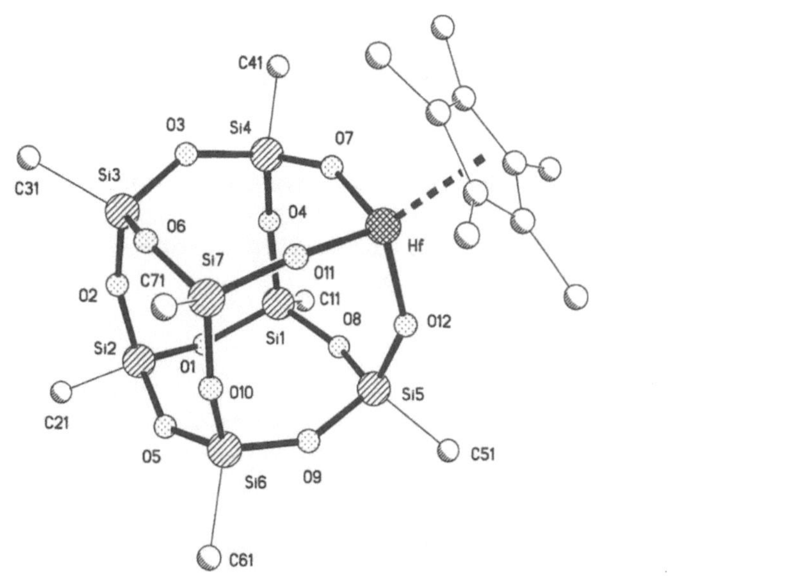

Fig. 3. Molecular structure of one of the two independent molecules of **5**; all carbon atoms except for the *ipso*-carbons of the cyclohexyl groups and the *Cp** ring carbons as well as all hydrogen atoms are omitted for clarity; selected bond lengths and angles; Hf-O7 194.9(4), Hf-O11 196.7(4), Hf-O12 195.9(4), Hf-C81 250.3(7), Si1-O1 163.3(5), Si1-C11 185.9(7), average Hf-C distance 250.5(7) pm; O7-Hf-O11 102.7(2), O1-Si2-O5 108.5(2), Si4-O7-Hf 153.2(3), Si2-O5-Si6 144.9(3)°.

Experimental

Preparation of the ligands and the complexes

All manipulations were performed using *Schlenk* and dry box techniques. Solvents were appropriately dried and distilled under nitrogen prior to use. The literature method was used for preparation and purification of the ligand [3]. Di- and trimethyltin(IV)-chlorides as well as $(C_5Me_5)HfCl_3$ were obtained from Aldrich and used as received.

[$Cy_7Si_7O_9(OSiMe_3)O_2SnBu_2$] (**1**, *Cy* = *c*-$C_6H_{11}$)

0.82 g (0.78 mmol) of [$Cy_7Si_7O_9(OSiMe_3)(OH)_2$] were added to a solution of Bu_2SnCl_2 (0.24, 0.78 mmol) and 0.5 ml of triethylamine in 25 ml of benzene. The mixture was stirred for 3 h and then evaporated in vacuum. The residue was extracted with 30 ml of toluene, the extract was filtered and the filtrate was evaporated in vacuum to give white crystals (0.83 g, 82%).

$C_{53}H_{104}O_{12}Si_8Sn$; calcd.: C 49.81, H 7.92; found: C 49.15, H 8.03; ^1H NMR (300 MHz, C_6D_6, 20°C): $\delta = 2.19$–1.24 (m, 82H, CH_2(Bu)+CH_2(*c*-C_6H_{11})), 1.10–0.89 (m, 13H, CH_2(Bu)+CH (*c*-C_6H_{11})), 0.37 (s, 9H, $SiMe_3$) ppm; $^{13}C\{^1H\}$ NMR (75 MHz, C_6D_6, 20°C): $\delta = 28.26$, 28.16, 27.93, 27.84, 27.69, 27.60, 27.53, 27.42, 27.24, 27.13, 27.07 (CH_2, *c*-C_6H_{11}), 26.04, 25.75, 25.25, 24.28, 23.86 (CH, *c*-C_6H_{11}), 21.57, 21.01 (CH_2, Bu), 13.78 (CH_2, Bu), 2.09 ($SiMe_3$), 1.37 (CH_3, Bu) ppm; $^{29}Si\{^1H\}$ NMR (60 MHz, C_6D_6, 22°C): $\delta = 9.04$ (s, $SiMe_3$), −64.21, −66.66, −67.50, −69.91 (s, silox) ppm; $^{119}Sn\{^1H\}$ NMR (112 MHz, C_6D_6, 22°C): $\delta = -46.14$(s) ppm.

[$Cy_7Si_7O_9(OSiMe_3)O_2SnMe_2$] (**2**)

1.33 g (1.27 mmol) of [$Cy_7Si_7O_9(OSiMe_3)(OH)_2$] were added to a solution of 0.28 g (1.27 mmol) of Me_2SnCl_2 and 0.7 ml of triethylamine in 30 ml of benzene. The mixture was stirred for three hours; then the mixture was evaporated in vacuum. Work-up as described above yielded a white solid (1.13 g, 75%).

$C_{47}H_{92}O_{12}Si_8Sn$; calcd.: C 47.35, H 7.72; found: C 47.27, H 7.64; ^1H NMR (300 MHz, C_6D_6, 20°C): $\delta = 2.14$–1.26 (m, 70H, CH_2(*c*-C_6H_{11})), 1.09–0.96 (m, 7H, CH(*c*-C_6H_{11})), 0.53 and 0.47 (s, 6H, CH_3), 0.34 (s, 9H, $SiMe_3$) ppm; $^{13}C\{^1H\}$ NMR (75 MHz, C_6D_6, 20°C): $\delta = 28.51$, 28.43, 28.36, 28.21, 28.13, 27.97, 27.93, 27.79, 27.71, 27.67, 27.56 (CH_2, *c*-C_6H_{11}), 26.11, 25.99, 25.48, 24.56, 24.15 (CH, *c*-C_6H_{11}), 2.94 ($SiMe_3$), 0.79, 0.22 (CH_3) ppm; $^{29}Si\{^1H\}$ NMR (60 MHz, C_6D_6, 22°C): $\delta = 9.31$ (s, $SiMe_3$), −63.75, −66.80, −67.61, −69.91 (s, silox) ppm; $^{119}Sn\{^1H\}$ NMR (112 MHz, C_6D_6, 22°C): $\delta = -8.07$(s) ppm.

[$Cy_7Si_7O_9(OSiMe_3)_2OSnMe_3$] (**3**)

0.84 g (0.75 mmol) of [$Cy_7Si_7O_9(OSiMe_3)_2(OH)$] were added to a mixture of 0.15 g (0.75 mmol) of Me_3SnCl and 0.25 ml of triethylamine in 30 ml of benzene. The mixture was stirred for 3 h, evaporated to dryness, and the residue was extracted with 40 ml of hexane. The extract was filtered, and the filtrate was evaporated in vacuum to give a white solid (0.74 g, 77%).

$C_{51}H_{104}O_{12}Si_9Sn$; calcd.: C 47.85, H 8.13; found: C 47.32, H 7.96; ^1H NMR (300 MHz, C_6D_6, 24°C): $\delta = 2.17$–1.31 (m, 70H, CH_2(*c*-C_6H_{11})), 1.07–0.93 (m, 7H, CH(*c*-C_6H_{11})), 0.52–0.23 (m, 27H, CH_3+$SiMe_3$) ppm; $^{13}C\{^1H\}$ NMR (75 MHz, C_6D_6, 24°C): $\delta = 27.78$, 27.70, 27.61, 27.44, 27.40, 27.34, 27.30, 27.19, 27.16, 27.07, 27.02 (CH_2, *c*-C_6H_{11}), 26.89, 26.74, 26.69, 25.21, 23.80 (CH, *c*-C_6H_{11}), 1.86 and 0.80 ($SiMe_3$), −2.31 and −2.85 (CH_3) ppm; $^{29}Si\{^1H\}$ NMR (60 MHz, C_6D_6, 22°C): $\delta = 9.88$, 8.75 (s, $SiMe_3$), −58.97, −65.09, −67.07, −67.92, −68.06, −68.24, −69.06, −69.78, −70.09 (s, silox) ppm; $^{119}Sn\{^1H\}$ NMR (112 MHz, C_6D_6, 25°C): $\delta = 113.43$(s) ppm.

[$Cy_7Si_7O_9(OSiMe_3)_2OSn(CH_2Ph)_3$] (**4**)

0.96 g (0.86 mmol) of [$Cy_7Si_7O_9(OSiMe_3)_2(OH)$] were added to 0.37 g (0.86 mmol) of $(PhCh_2)_3SnCl$ and 0.5 ml of triethylamine in 30 ml of benzene and stirred for 3 h. The mixture was evaporated in vacuum and the residue was extracted with 50 ml of hexane. The extract was filtered and the solution was evaporated in vacuum to give a slightly yellow wax-like solid (0.96 g, 74%).

$C_{69}H_{116}O_{12}Si_9Sn$; calcd.: C 54.94, H 7.7; found: C 54.65, H 7.87; ^1H NMR (300 MHz, C_6D_6, 24°C): $\delta = 7.11–6.69$ (m, 15H, C_6H_5), 2.73 (s, 6H, CH_2), 2.14–1.22 (m, 70H, $CH_2(c\text{-}C_6H_{11})$), 1.08–0.93 (m, 7H, $CH(c\text{-}C_6H_{11})$), 0.36 (s, 18H, $SiMe_3$) ppm; $^{13}C\{^1H\}$ NMR (75 MHz, C_6D_6, 24°C): $\delta = 139.49$, 129.02, 124.64 (C_6H_5), 46.75 (CH_2), 28.37, 28.23, 28.15, 27.93, 27.89, 27.86, 27.74, 27.70, 27.63, 27.59, 27.49 (CH_2, $c\text{-}C_6H_{11}$), 27.32, 27.25, 25.61, 24.77, 24.37 (CH, $c\text{-}C_6H_{11}$), 2.15 ($SiMe_3$) ppm; $^{29}Si\{^1H\}$ NMR (60 MHz, C_6D_6, 22°C): $\delta = 9.89$, 9.15 (s, $SiMe_3$), −58.93, −65.03, −67.06, −68.04, i 68.22, −68.8, −69.13, −69.70 (s, silox) ppm; $^{119}Sn\{^1H\}$ NMR (112 MHz, C_6D_6, 25°C): $\delta = -20.41$(s) ppm.

[$Cy_7Si_7O_{12}Hf(C_5Me_5)$] (**5**)

56 mmol MeLi (35 ml 1.6 *M* ether solution) were added to a suspension of 7.7 g (18.4 mmol) *Cp**$HfCl_3$ in 200 ml ether at −78°C, stirred for 5 h, and warmed up slowly to 20°C. The mixture was cooled down again to −78°C, and 17.9 g (18.4 mmol) [$Cy_7Si_7O_9(OH)_3$] were added during 2 h. The solution was stirred overnight; then the solvent was evaporated in vacuum, and the residue was extracted with 150 ml of pentane. The extract was filtered and the solvent was removed partially in vacuum to give colorless crystals (17 g, 78%).

$C_{52}H_{92}HfO_{12}Si_7$; calcd.: C 48.63, H 7.17; found: C 48.05, H 6.94; ^1H NMR (400 MHz, C_6D_6, 24°C): $\delta = 2.09$ (s, 15H, C_5Me_5), 1.73 (m, 35H, $CH_2(c\text{-}C_6H_{11})$), 1.24 (m, 35H, $CH_2(c\text{-}C_6H_{11})$), 0.71 (m, 7H, $CH(c\text{-}C_6H_{11})$) ppm; $^{13}C\{^1H\}$ NMR (75 MHz, C_6D_6, 24°C): $\delta = 120.50$ (C_5Me_5), 28.50, 28.36, 28.30, 27.93, 27.84, 27.75, 27.65 (CH_2, $c\text{-}C_6H_{11}$), 25.04, 24.36, 24.33 (3:3:1, CH, $c\text{-}C_6H_{11}$), 10.64 (C_5Me_5) ppm; IR (KBr): $\nu = 2922$ vs, 2849 vs, 2666 vw, 1448 m, 1354 w, 1347 w, 1275 m, 1268 m, 1196 s, 1107 vs, 1027 vs, 1001 m, 930 s, 894 m, 848 m, 825 w, 759 w, 742 w, 689 w, 670 w, 648 w, 618 w, 556 m, 546 m, 515 s, 491 m, 467 s cm^{-1}.

Measurements

Microanalyses were carried out at the Department of Chemistry, Otto-von-Guericke-University, Magdeburg. The analytical data showed that complexes with a stoichiometry of ligand: organotin(IV) = 1:1 were formed. IR spectra in the region of 4000–200 cm^{-1} were recorded in KBr pellets using a Perkin Elmer 2000 FTIR spectrometer. The ^{119}Sn *Mößbauer* spectra were recorded at 77 K on a RANGER spectrometer in constant accelerating mode with a source activity of 0.4 GBq. Computer evaluation was used to determine isomer shift (*IS*) and quadrupole splitting (*QS*) values. The spectra were analyzed as *Lorentzian* lines by least-squares fitting. The reproducibility of the *Mößbauer* parameters was found to be ±0.02 mm/s (*IS*) and 0.04 mm/s (*QS*), respectively, in each measurement. The *IS* values are referred to that of $CaSnO_3$.

Acknowledgements

This work was supported financially by the Hungarian Research Foundation (T 022909 OTKA). *L Nagy* and *Y. Gun'ko* are grateful to DAAD (*Deutscher Akademischer Austauschdienst*) and *Alexander von Humboldt–Stiftung*, respectively, for providing fellowships in 1995–1996 to Germany. *V. Lorenz* and F. T. Edelmann thank the *Deutsche Forschungsgemeinschaft* (*Schwerpunktprogramm Siliciumchemie*) and the Otto-von-Guericke-Universität Magdeburg for financial support.

References

[1] Fehér FJ, Budzichowski TA (1995) Polyhedron **14**: 3239 and references therein

[2] Borsari M, Gavioli G, Zucci C, Pályi G, Pesaro R, Ugo R, Shchegolikhina OI, Zhdanov AA (1997) Inorg Chim Acta **258**: 139

[3] Fehér FJ, Newman DA, Walzer JF (1989) J Am Chem Soc **111**: 1741

[4] Parish RV (1984) Structure and Bonding in Tin Compounds. In: Long GJ (ed) Mößbauer Spectroscopy Applied to Inorganic Chemistry. Plenum Press, New York London

[5] Bancroft GM, Kumar Das VG, Sham TK, Clark MG (1976) J Chem Soc Dalton Trans 643

[6] Korecz L, Saghier AA, Burger K, Tzschach A, Jurkschat A (1982) Inorg Chim Acta **58**: 243

[7] Bancroft GM, Sham TK (1976) J Chem Soc Dalton Trans 467

[8] Pellerito L, Ruisi G, Barbieri R, Lo Giudice MT (1977) Inorg Chim Acta **21**: L33

[9] Bucher FK, Gerrard W, Money EF, Rees RG, Williams HA (1964) Spectrochim Acta **20**: 51

[10] Sheldrick GM (1995) SHELXTL, Structure Determination Software Programs, Version 5.03 (PC), Siemens Analytical X-Ray Instruments. Madison, WI. Additional material to the structure determination may be ordered from Fachinformationszentrum Karlsruhe, Gesellschaft für wissenschaftlich-technische Information mbH, D-76344 Eggenstein-Leopoldshafen, Federal Republic of Germany, referring to the deposition number CSD-410524

[11] Flack HD (1983) Acta Cryst **A39**: 876

Received June 12, 1998. Accepted (revised) September 30, 1998

Photolumineszenzeigenschaften von substituierten Silsesquioxanen der Zusammensetzung $R_n(SiO_{1.5})_n$

Christoph Ossadnik[1], **Stan Veprek**[1,*], **Heinrich C. Marsmann**[2] und **Eckhard Rikowski**[2]

[1] Institut für Chemie Anorganischer Materialien der TU-München, Lichtenbergstr. 4, D-85747 Garching, Deutschland

[2] Institut für Anorganische Chemie, Universität-Gesamthochschule Paderborn, D-33095 Paderborn, Deutschland

Zusammenfassung. Um ein tieferes Verständnis der Photolumineszenz von nanokristallinem Silizium zu gewinnen, haben wir als Modellverbindungen substituierte Silsesquioxane ($R_n(SiO_{1,5})_n$, $n=6$, 8, 10, 12) mit verschiedenen Liganden R untersucht. Alle Silsesquioxane zeigen eine Photolumineszenz im blauen Spektralbereich, deren Intensität und Spektralverteilung von der Käfiggröße n sowie von der Größe und Elektronegativität der Liganden abhängig sind. Als Mechanismus der Photolumineszenz wird ein *charge transfer* zwischen dem Käfig und den Liganden vorgeschlagen.

Photoluminescent Properties of Substituted Silsesquioxanes of the Composition $R_n(SiO_{1.5})_n$

Summary. To achieve a better insight into photoluminescence phenomena of nanocrystalline silicon, silsesquioxanes ($R_n(SiO_{1.5})_n$, $n=6$, 8, 10, 12) with various Ligands R have been investigated. All silsesquioxanes show photoluminescence in the blue spectral region; its intensity depends on the cage size n and on size and electronegativity of the ligands. A charge transfer process between cage and ligands is proposed as the mechanism responsible for photoluminescence.

Keywords. Photoluminescence; Silsesquioxane; Nanocrystalline Silicon.

Einleitung

Der Wunsch, opto-elektronische Baugruppen in die Siliziummikroelektronik zu integrieren, hat eine intensive Suche nach leuchtenden kompatiblen Siliziumstrukturen ausgelöst. Dabei wurden insbesondere Anstrengungen bezüglich der Dimensionsreduzierung von Halbleiterstrukturen unternommen. Das *quantum confinement model* [1], das das Phänomen des leuchtenden Siliziums zu erklären versucht, basiert auf der theoretischen Voraussage, daß die Bandlücke von

* Corresponding author

Siliziumnanopartikeln mit kleiner werdender Größe aufgeweitet wird und in eine direkte Bandlücke übergeht. Die Passivierung der Oberfläche der Nanokristalle ist dabei notwendig, um die Wahrscheinlichkeit nichtstrahlender Rekombinationen an Defekten zu verringern. Viele Organosilane und abgesättigte molekulare Siliziumstrukturen sind daher sehr stark photolumineszierende Materialien. Das *surface state model* [2, 3] geht von dem Ansatz aus, daß das im Kristallit photogenerierte Elektron-Loch-Paar in Oberflächenzuständen, die energetisch in der Bandlücke liegen, eingefangen wird und von dort rekombiniert. Noch immer ungeklärt sind dabei der Transportmechanismus der beteiligten Anregungen und die für die Lumineszenz verantwortlichen Oberflächenzustände, die stark an ihre chemische Umgebung gekoppelt sind.

Die Untersuchung der optischen Eigenschaften synthetisierter Siliziumstrukturen ist hilfreich, um die elektronischen Eigenschaften und den mikroskopischen Photolumineszenzmechanismus in Nanokristallen zu verstehen. Lokale Siliziumstrukturen mit relevanter chemischer Umgebung können durch eine kontrollierte Synthese modelliert und untersucht werden. Die optischen Eigenschaften von linearen Organosilanen beruhen auf einem erlaubten HOMO-LUMO-Übergang ($\sigma \rightarrow \sigma^*$) [4, 5]. Die scharfen Absorptionsbanden langer Ketten im Bereich von 3 bis 4 eV sind für die Zustandsdichteverteilung in eindimensionalen elektronischen Systemen typisch. Die Bindung in linearen Polysilanen wird mit zueinander ausgerichteten sp^3-hybridisierten Molekülorbitalen zweier Nachbaratome und entsprechenden Orbitalen der angelagerten Substituenten beschrieben. Die starke Überlagerung der sp^3-Bindungsorbitale entlang der Siliziumkette führt zu einer Delokalisierung der Bindungselektronen. Der Einfluß der Liganden auf dieses System kann je nach deren Elektronegativität zu einer stärkeren oder schwächeren Delokalisierung führen. Im chemischen Kontext bedeutet eine stärkere Delokalisierung die Destabilisierung des HOMO-Zustands der σ-Kettenelektronen, der dadurch energetisch steigt, und die Stabilisierung des LUMO-Zustands, der somit energetisch absinkt. Die Veränderung der HOMO-LUMO-Lücke führt zur Verschiebung der Absorptionsbanden und der Lumineszenz. Sinkt die Kettenlänge unter 20 Siliziumatome, werden Absorption und Photolumineszenz blauverschoben, und die Quantenausbeute sinkt. Ursache hierfür ist ein *quantum confinement*-Effekt. Die chemisch delokalisierten Bindungselektronen werden physikalisch-räumlich lokalisiert, die HOMO-LUMO-Lücke wird aufgeweitet, und die Anzahl der Zustände sinkt. Ein analoges Verhalten erhält man auch bei dreidimensionalen Teilchen, wenn deren Größe unter *ca.* 5 nm bzw. 20 Atome sinkt. Der Einfluß von Struktur und Form sowie von Spannungen im Gitter auf die Lumineszenz ist groß. In zweidimensionalen Strukturen ist die Situation noch komplizierter. Die Bandlücke wird *quasi*-direkt. Die Exzitonen sind stark an das Netzwerk gekoppelt; dies führt zu einer großen *Stokes*-Verschiebung der Photolumineszenz.

Experimente an gezielt synthetisierten Organosiliziumverbindungen mit definiertem Sauerstoffeinbau ermöglichen Rückschlüsse auf Grenz- und Oberflächeneffekte an sauerstoff-bedeckten Siliziumkristalliten und den von uns dargestellten *nc*-Si/*a*-SiO_2-Systemen [3]. Käfigartige Siloxane mit polyedrischer Gerüststruktur, sogenannte Silsesquioxane ($R_n(SiO_{1.5})_n$), sind akzeptierte Modellverbindungen für oberflächenmodifizierte Kieselgele. Für den Vergleich zu unseren Systemen sind der Aufbau und die variable Belegung mit organischen

Tabelle 1. Untersuchten Silsesquioxane, Abkürzungen und Literaturverweise

	R	Abkürzung	Lit.
$R_6(SiO_{1,5})_6$	$(CH_3)_3SiO$; Trimethylsiloxy		[6]
$R_8(SiO_{1,5})_8$	H		[7]
	CH_3; Methyl	Me	[7]
	C_2H_5; Etyl	Et	[7]
	C_3H_7; Propyl	Prop	[7]
	C_4H_9; Butyl	Bu	[7]
	C_4H_9; iso-Butyl	iso-Bu	[7]
	C_5H_{11}; Pentyl	Pent	[7]
	C_6H_{13}; Hexyl	Hex	[7]
	C_7H_{15}; Heptyl	Hept	
	C_8H_{17}; Octyl	Oct	
	C_9H_{19}; Nonyl	Non	
	$C_{10}H_{21}$; Decyl	Dec	[8]
	ClC_3H_6; 3-Chloro-*n*-Propyl	(Cl-Pr)	[10]
	IC_3H_6; 3-Iodo-*n*-Propyl	(I-Pr)	[10]
	BrC_3H_6; 3-Bromo-*n*-Propyl	(Br-Pr)	[10]
	NSC_4H_6; isothiocyanato-*n*-Propyl	(NCS-Pr)	[10]
	$(CH_3)_3SiO$; Trimethylsiloxy		[6, 7]
$R_{10}(SiO_{1,5})_{10}$	H		[9]
	ClC_3H_6; 3-chloro-*n*-Propyl	(Cl-Pr)	[11]
	$(CH_3)_3SiO$; Trimethylsiloxy		
$R_{12}(SiO_{1,5})_{12}$	ClC_3H_6; 3-Chloro-*n*-Propyl	(Cl-Pr)	[11]
	$(CH_3)_3SiO$; Trimethylsiloxy		

Substituenten genauso von Vorteil wie die hohe thermische und chemische Stabilität und die gute Löslichkeit in organischen Lösungsmitteln, die den experimentellen Zugang erleichtern. Bis auf wenige Ausnahmen beschränken sich die Synthesekonzepte für Silsesquioxane auf das Octamer, wobei bevorzugt homofunktionalisierte Octasilsesquioxane mit acht gleichen Resten R am Käfiggerüst dargestellt werden. Prinzipielle Synthesestrategien für Silsesquioxane sind:

a) die hydrolytische Polykondensation trifunktioneller Monomere des Typs $XSiY_3$ mit $X = H$, Halogen, Alkyl- oder funktionalisierter Alkylrest, $Y =$ Halogen oder Alkoxyrest,
b) Reaktionen mit Modifikation der Substituenten bei existierenden Silsesquioxanen,
c) die Thermolyse von Polyorganylsilsesquioxanen und
d) die Silylierung von Tetraalkylammoniumsilicaten.

Einen Überblick über die von uns untersuchten Silsesquioxane gibt Tabelle 1. Darin wird auch die entsprechende Literatur bezüglich deren Beschreibung und Darstellung zitiert. Alle Silsesquioxane wurden an der Universität-Gesamthochschule Paderborn synthetisiert. Durch die unterschiedlichen Ringgrößen der Käfige werden unterschiedliche Oberflächenstrukturen erschlossen. In Abhängigkeit von der Kettenlänge der Substituenten R sind die meisten Silsesquioxane farblose kristalline Festkörper. Ab $R = C_6H_{13}$ sind sie oft zähflüssige und transparente Substanzen. Bis auf $Me_n(SiO_{1,5})_n$, das unlöslich ist, sind sie in den meisten organischen Lösungsmitteln gut löslich. Mit zunehmender Länge der Alkylkette am Gerüst nehmen Dichte und Schmelzpunkt tendenziell ab und die Löslichkeit zu.

Ergebnisse und Diskussion

Die grundlegende Überlegung bei der Auswahl der Versuchsreihen war, die Ergebnisse nach Effekten der Käfiggröße, der Molekülgröße und der Belegung mit unterschiedlichen Substituenten zu trennen. Die Silsesquioxane wurden daher in verschiedene experimentelle Reihen aufgeteilt.

In der Alkylreihe bleibt das octamere, homofunktionalisierte Grundgerüst erhalten; die Liganden R gehen von Methyl bis zu Decyl. Zum Vergleich wird das octamere Hydridosilsesquioxan hinzugezogen. Von $H_8(SiO_{1,5})_8$ aufwärts wird also der Substituent um je ein CH_2-Kettenelement länger. Der bestimmende Parameter ist, neben eventuell auftretenden elektronischen Effekten, die Kettenlänge der Substituenten und damit die Größe des Silsesquioxans.

Auch in der Halo-Propyl-Reihe ändern sich nur die Substituenten, wobei zusätzlich zu der Veränderung der Molekülgröße die elektronische Struktur verändernde Effekte auftreten. Die Silsesquioxane $Prop_8(SiO_{1,5})_8$ und (NCS-Pr)$_8(SiO_{1,5})_8$ sind zum weiteren Vergleich in dieser Reihe enthalten.

Der Einfluß der Käfiggröße wird an den Versuchsreihen der Trimethylsiloxysilsesquioxane $((CH_3)_3SiO)_n(SiO_{1,5})_n$ der Käfiggrößen $n = 6$, 8, 10 und 12 sowie an der Chloro-Propyl-Reihe und den beiden Hydridosilsesquioxanen studiert. Hierbei bleiben die Substituenten gleich, und nur die Käfiggröße ändert sich. Die Käfig- und Molekülgrößen wurden dabei aufgrund ihrer geometrischen Anordnung mit Hilfe von Bindungswinkeln und -abständen aus der Literatur abgeschätzt.

Abbildung 1 zeigt die Photolumineszenzspektren der Alkylreihe als Kristallite auf einem Siliziumsubstrat. Charakteristisch sind die breiten Banden mit geringer Intensität relativ zur Photolumineszenz in Lösung (s. Abb. 2). Im allgemeinen steigt die Intensität der Photolumineszenz mit zunehmender Molekülgröße. Einige Abweichungen von diesem Trend werden später diskutiert. Eine sehr starke Bande bei 2.8 eV, die in einigen Fällen beobachtet wird, resultiert mit großer Wahrscheinlichkeit aus angelagerten OH-Gruppen durch die Lagerung der Proben an feuchter Luft. Die Proben sind dünne Schichten mit unregelmäßiger Struktur und Oberfläche, die entstehen, wenn die gelösten Silsesquioxane auf ein Siliziumwaferstück aufgebracht werden und das Lösungsmittel verdampft. An der Oberfläche werden OH-Gruppen physisorbiert, die als eigenständige Anregungs- und Rekombinationszentren fungieren. Diese Bande ist bei den Spektren aus gelösten Silsesquioxanen nicht mehr zu sehen (s. Abb. 2). Sie zeigt eine gute Übereinstimmung mit der Photolumineszenz von angelagerten OH-Gruppen an *nc*-Si/SiO_2-Schichten [3, 16, 17]. Das gleiche Lumineszenzverhalten erhält man z.B. auch, wenn man Metalloxide oder Gläser in hochreinem Wasser kocht. Die OH-

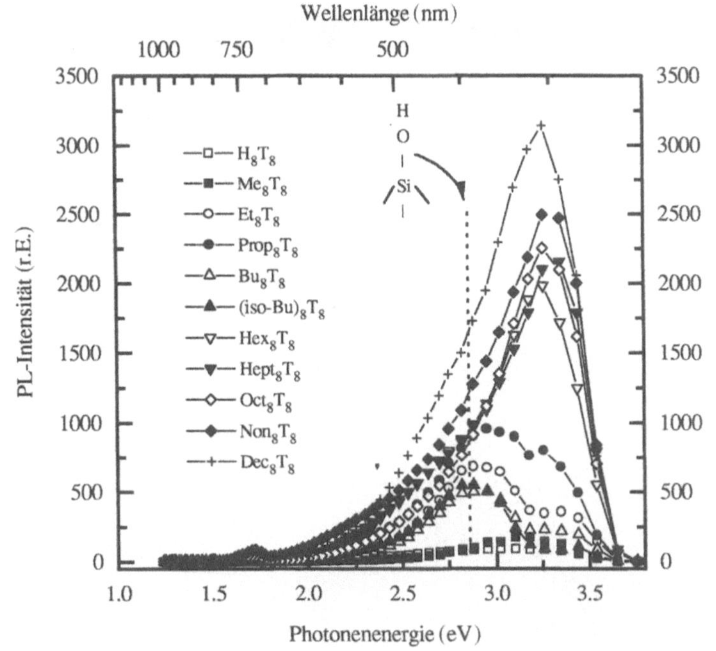

Abb. 1. Photolumineszenzspektren homofunktionalisierter Ottasilsesquioxane; Kristallite auf Si-Substrat, Variation der Liganden, Anregung: He–Cd-Laser, 325.4 nm, 3.8 eV

Lumineszenz verschwindet jedoch beim Tempern der Schichten, wobei die OH-Gruppen desorbieren [16, 17]. Die Peakposition der Lumineszenz ist durch den spektralen *cut-off* der Apparatur bei hohen Energien und durch die geringe Anregungsenergie von 3.8 eV verfälscht. Tendenziell geht die Peaklage jedoch mit zunehmender Kristallitgröße zu kleineren Energien.

Abbildung 2 zeigt die Photolumineszenzspektren und die molaren Absorptionskoeffizienten der Alkylreihe in Pentan. Die Photolumineszenzspektren haben wesentlich schmälere Banden. Die den Si–OH-Gruppen zugeordnete Bande bei 2.8 eV wird nicht mehr beobachtet. Die integrale Intensität steigt einheitlich mit der Molekülgröße und folgt dem Absorptionsverhalten. Eine Ausnahme hiervon sind $Bu_8(SiO_{1,5})_8$, dessen integrale Intensität zu niedrig ist, und $Hept_8(SiO_{1,5})_8$ dessen integrale Intensität zu hoch ist. $H_8(SiO_{1,5})_8$ als kleinstes und kein C-Atom enthaltendes Silsesquioxan paßt in das beschriebene Intensitätsverhalten. Eine Verschiebung des Peakmaximums ist wiederum wegen der geringen Anregungsenergie nicht feststellbar. Die Absorptionsspektren zeigen eine einheitliche Struktur. Nach einem leichten Anstieg der Absorption im Bereich von 2 eV bis 4.2 eV folgt eine stärkere Zunahme bis *ca.* 5.6 eV mit auflösbaren Absorptionsbanden bei *ca.* 4.4 eV, 4.6 eV und 5.4 eV. Bei höheren Energien (um 6 eV bis 6.5 eV) befindet sich jeweils die stärkste Absorptionsbande. Wie vorausgehend erwähnt, müssen bei diesen hohen Energien der spektrale *cut-off* und der Meßartefakt bei 6.7 eV beachtet werden.

Der starke Einfluß der Länge der Substituenten auf das Absorptions- und Lumineszenzverhalten äußert sich bei der Alkylreihe aus einer Zunahme der Intensitäten von Lumineszenz und Absorption mit zunehmender Molekülgröße.

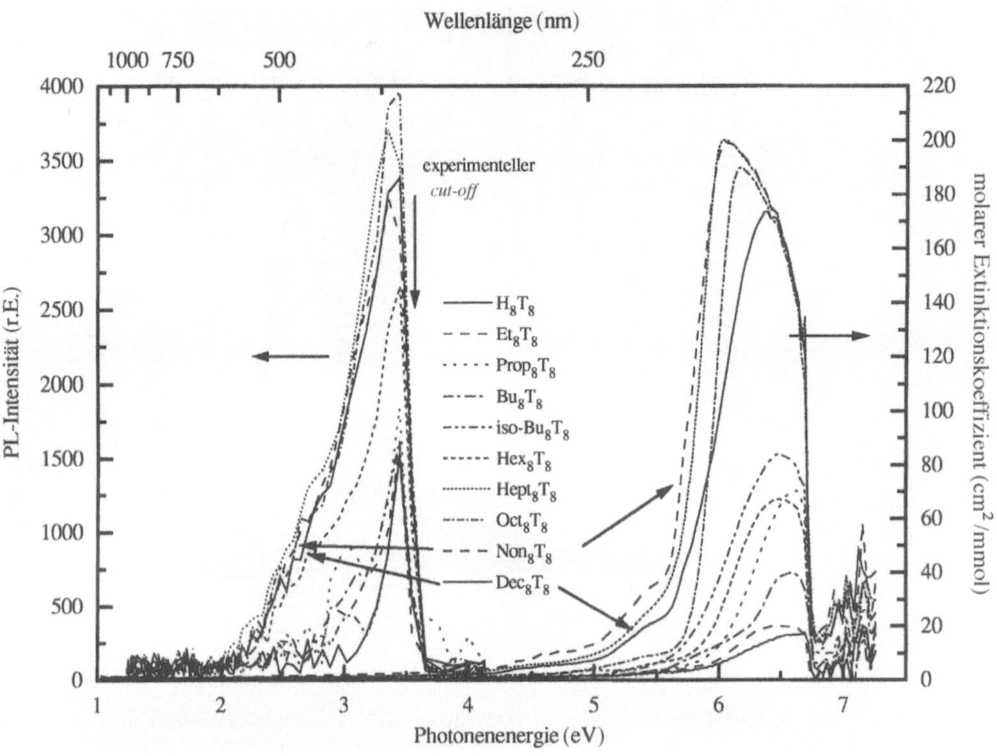

Abb. 2. Photolumineszenz und Absorptionsspektren homofunktionalisierter Silsesquioxane in einer 0.01 molaren Pentanlösung; Variation der Liganden, Anregung: He–Cd-Laser, 325.4 nm, 3.8 eV

Ein solches Verhalten spricht, ungeachtet der genauen Natur des strahlenden Übergangs, für einen *Charge-transfer*-Effekt zwischen den Zuständen des $(SiO_{1,5})_n$ Käfigs und den unbesetzten Orbitalen der Liganden. Dabei werden die Elektronen über eine relativ große Entfernung bewegt, und das Übergangsdipolmoment und somit auch die Absorption nehmen mit zunehmender Ligandengröße zu. Dies erlaubt die Interpretation der Intensitätszunahme durch den elektronenschiebenden Effekt der Alkylsubstituenten. Die Charakterisierung des strahlenden Zustandes ist jedoch wesentlich schwieriger. Nach Lit, [18] ist der energetisch tiefste erlaubte strahlende Übergang durch das *lone-pair* des Sauerstoffs im Si–O–Si-Käfigelement gegeben. Dessen Übergangsenergie liegt nach numerischer Simulation (*Hartree-Fock*) jedoch bei *ca.* 8 eV und sollte von den angelagerten Substituenten oder durch verschiedene Käfiggrößen weder in der Intensität noch in der Energie verändert werden. Demgegenüber zeigt sich in den Messungen sehr wohl eine Lumineszenz, die auch bei Energien von 3.8 eV angeregt werden kann. Als möglicher Mechanismus kann für die Lumineszenz der Übergang vom nichtbindenden (*lone-pair*) Zustand am Sauerstoff in den σ^*-Zustand einer Si–O-Bindung des Käfigs verantwortlich gemacht werden. Als der wahscheinlichere Mechanismus der Photolumineszenz erscheint jedoch ein Übergang zwischen dem *lone-pair* und einem unbesetzten Zustand der Liganden (z.B. σ^*). Berücksichtigt man die Tatsache, daß die *Hartree-Fock*-Rechnungen bekanntlich die HOMO-LUMO-Energiedifferenz stark übeschätzen, so dürfte die Energie eines solchen *Charge-*

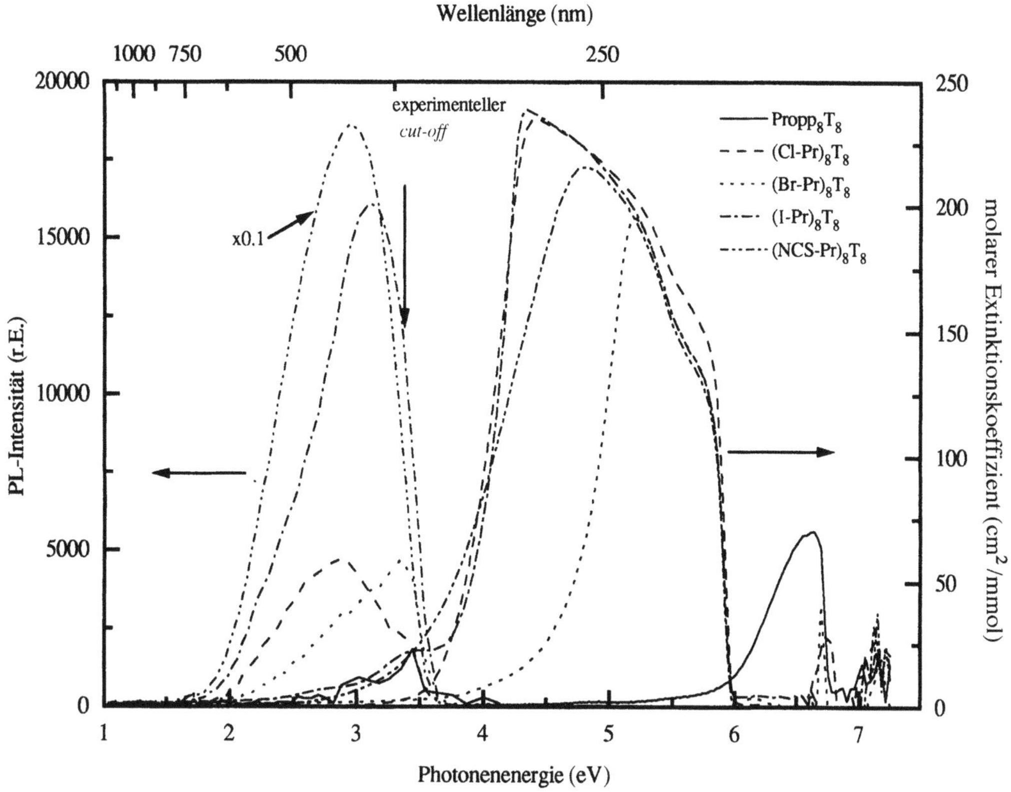

Abb. 3. Photolumineszenz und Absorptionsspektren homofunktionalisierter Silsesquioxane in einer 0.01 molaren *THF*-Lösung; Variation der Liganden, Anregung: He–Cd-Laser, 325.4 nm, 3.8 eV

transfer-Übergangs durchaus in dem beobachteten Bereich der Photolumineszenz von 3.8–4 eV liegen. Ein stark elektronenschiebender Ligand führt zu einer hohen Elektronendichte im O-Bindungsorbital; dies wiederum hebt den, *Lone-pair*-Zustand energetisch an und senkt den σ^*-Zustand ab. Dieser Effekt dürfte zur gemessenen rotverschiebenden Wirkung führen.

Abbildung 3 zeigt die Absorptions- und Photolumineszenzspektren der in *THF* gelösten Silsesquioxane der *R*-Propyl-Reihe. Sehr deutlich sind die Auswirkungen der Substituenten zu sehen. $Prop_8(SiO_{1,5})_8$ hat sein Absorptionsmaximum bei 6.5 eV und eine nur geringe Extinktion bei der Anregungsenergie von 3.8 eV. Die Photolumineszenz ist demzufolge auch nur sehr gering. Ersetzt man das terminale H-Atom durch Cl, Br, J oder NCS, so werden die Absorptionsbanden zu kleineren Energien verschoben. Bis auf $(Br\text{-}Pr)_8(SiO_{1,5})_8$ (von 5 bis 6 eV) sind diese sehr breit (von 4 bis 6 eV) und ihre Anstiegsflanken liegen schon im Bereich der Anregungswellenlänge der Photolumineszenz. Die Intensität der Photolumineszenz ist direkt an die Extinktion bei 3.8 eV gekoppelt. Aus den Spektren der Photolumineszenz ist eine eindeutige Verschiebung der Peaklage zu erkennen. Bei $Prop_8(SiO_{1,5})_8$ und $(Br\text{-}Pr)_8(SiO_{1,5})_8$ liegen die Peakpositionen noch nahe der Nachweisgrenze bei 3.6 eV, bei $(J\text{-}Pr)_8(SiO_{1,5})_8$ liegt der Peak bei 3.1 eV, bei $(NCS\text{-}Pr)_8(SiO_{1,5})_8$ bei 2.9 eV und bei $(Cl\text{-}Pr)_8(SiO_{1,5})_8$ bei 2.8 eV. Die starke Beeinflussung der Intensität und der Lage der Photolumineszenzbande kann hier nicht mehr allein durch die Veränderung der Molekülgröße erklärt werden. Eine Überlagerung

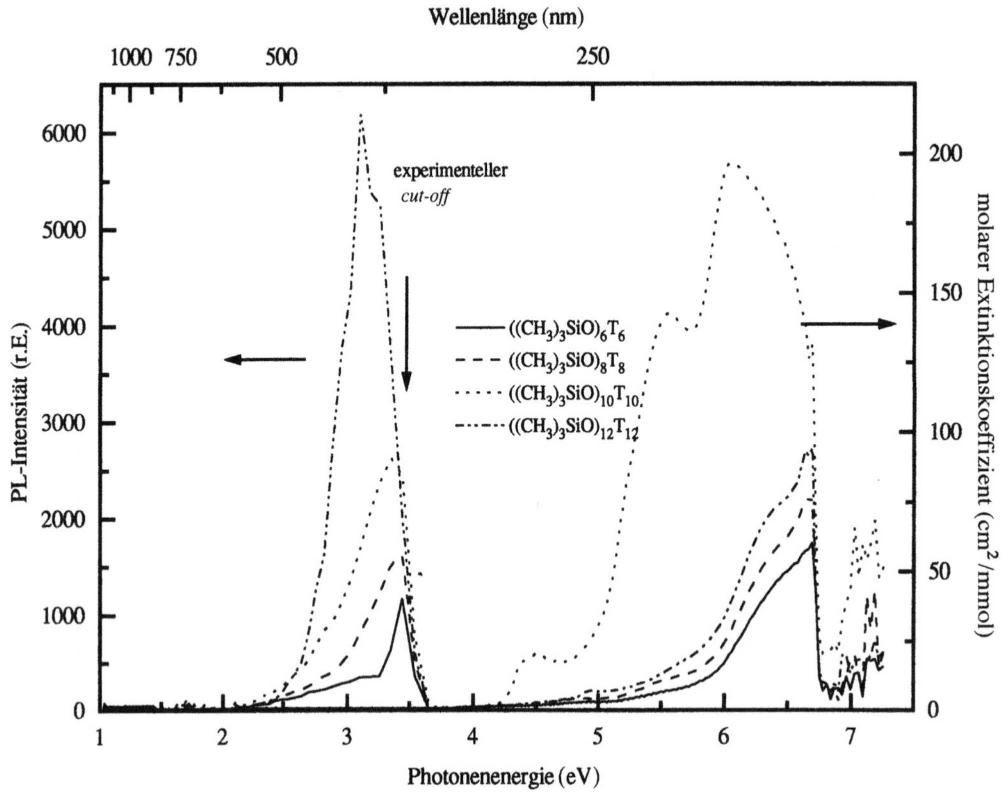

Abb. 4. Photolumineszenz und Absorptionsspektren homofunktionalisierter Silsesquioxane in einer 0.01 molaren *THF*- Lösung; Variation der Käfiggröße, Anregung: He–Cd-Laser 325.4 nm, 3.8 eV

von weiteren chemischen Effekten neben dem elektronenschiebenden Effekt erscheint zwingend.

Eine weitere charakteristische Auffälligkeit ist der Zusammenhang der Peaklage der Photolumineszenz der einzelnen Silsesquioxane mit der Lage der jeweils zugehörigen Absorptionsbande. Die Verschiebung von Absorptionskante und Lumineszenzmaximum bleibt für alle Silsesquioxane fast konstant. Dies legt den Schluß nahe, daß jeweils die gleichen Transportmechanismen zu den Übergängen führen. Die relativ starke Verschiebung von *ca.* 1 eV deutet auf eine starke Phononen- bzw. Schwingungsbeteiligung hin.

Abbildung 4 zeigt die Spektren der Trimethylsiloxysilsesquioxane (*THF*-Lösung). Die integrale Intensität der Photolumineszenz pro Si–O-Bindung steigt von $n = 6$ zu $n = 8$ um den Faktor 1,4, von $n = 8$ zu $n = 10$ um 1.5 und von $n = 10$ zu $n = 12$ um den Faktor 1.6. Dabei verschiebt sich die Peaklage von *ca.* 3.4 eV beim 6er-Käfig auf *ca.* 3.1 eV beim 12er-Käfig. Das Absorptionsverhalten von $((CH_3)_3SiO)_{10}(SiO_{1,5})_{10}$ unterscheidet sich völlig vom dem der übrigen Silsesquioxane. Mit zunehmender Energie nimmt die Absorption stetig zu und zeigt nach einem langsamen Anstieg bis 4 eV eine deutliche Bandenstruktur mit sehr hoher Absorption, während die restlichen Silsesquioxane eine geringer ausgeprägte Bandenstruktur mit kleiner Absorption zeigen, die mit der Käfiggröße steigt. Die Ursache für dieses Verhalten kann nur in einer starken Veränderung der elektronischen Struktur aufgrund chemischer Effekte am 10er-Käfig liegen.

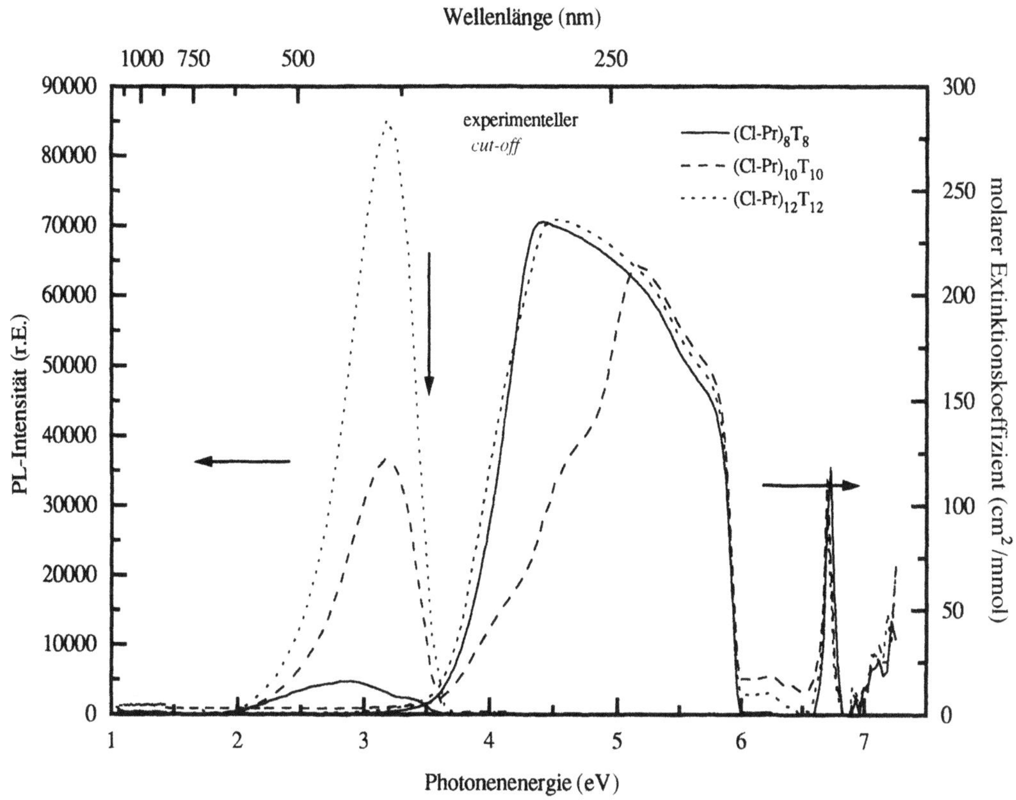

Abb. 5. Photolumineszenz und Absorptionsspektren homofunktionalisierter Silsesquioxane in einer 0.01 molaren *THF*-Lösung; Variation der Käfiggröße, Anregung: He–Cd-Laser, 325.4 nm, 3.8 eV

Auch bei der Chloro-Propyl-Reihe, deren Absorptions- und Photolumineszenzspektren Abbildung 5 zeigt, ist das Absorptionsverhalten des 10er-Käfigs uncharakteristisch. Die Absorption liegt bis *ca.* 3.5 eV deutlich über denen des 8er- und 12er-Käfigs, ab 3.5 eV jedoch weit darunter. Die Absorptionsbanden der ganzen Reihe sind gegenüber der Hydrido- und der Trimethylsiloxysilsesquioxane um *ca.* 1 eV rotverschoben. Die *Stokes*-Verschiebung zwischen Photolumineszenz- und Absorptionsbanden ist kleiner, und die Peakmaxima der Photolumineszenz zeigen mit steigender Käfigröße eine eindeutige Blauverschiebung. Die Verschiebung zu höheren Energien ist im vorgeschlagenen Übergang vom *Lone-pair*-Zustand des Sauerstoffatoms zum σ^*-Zustand der Liganden (oder der Si–O-Bindung?) mit einem energetischen Absinken des *Lone-pair*-Zustands oder einer Anhebung des σ^*-Zustands gleichzusetzen. Die Intensität der Photolumineszenz pro Si–O-Bindung steigt um den Faktor 4.4 bzw. 1.83 für die Käfiggrößen 8/10 bzw. 10/12. Insgesamt zeigt die Chloro-Propyl-Reihe sehr starke Lumineszenz.

Um eine weiterführende Interpretation der gemessenen Spektren zu ermöglichen, wurde die Photolumineszenz der Silsesquioxane auf das Vorhandensein eines Polarisationsgedächtnisses untersucht. Ein Polarisationsgedächtnis bedeutet, daß ein großer Anteil der emittierten Photonen die gleiche Polarisation hat wie die anregenden Photonen. Dies ist ein Hinweis auf eine sehr starke Kopplung bzw. Identität des Anregungszentrums mit dem Rekombinationszentrum. Alle vermes-

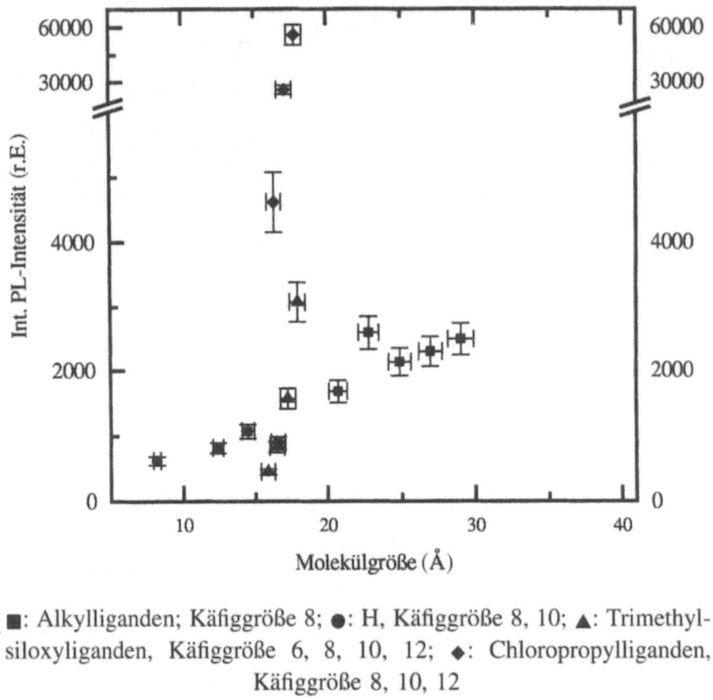

Abb. 6. Integrale Intensität der Photolumineszenzspektren der homofunktionalisierten Silsesquioxane *vs.* Molekülgröße (Durchmesser); Anregung: He–Cd-Laser, 325.4 nm, 3.8 eV

senen Silsesquioxane zeigen jedoch kein Polarisationsgedächtnis. Die angeregten Zustände sind also über einen Transportmechanismus an die Rekombinationszentren gebunden, bei dem die Information über die Polarisierung verlorengeht. Der Transportmechanismus und anschließende radiative Prozeß ist sehr schnell. Die vermessene Abklingzeit der Photolumineszenz liegt bei allen Silsesquioxanen unterhalb der Zeitauflösung unserer Apparatur (3 ns). Auch dies spricht für einen *Charge-transfer*-Mechanismus der Photolumineszenz.

Um einen Überblick über den Einfluß der Molekülgröße auf die integrale Intensität der Photolumineszenz zu erhalten, wurde beides gegeneinander aufgetragen. Abbildung 6 zeigt den linearen Anstieg der Intensität mit der Molekülgröße bei der Alkylreihe. $Bu_8(SiO_{1,5})_8$ liegt leicht unterhalb dieser Relation, $Hept_8(SiO_{1,5})_8$ liegt darüber. Bei dem Herausfallen dieser Silsesquioxane aus der Intensität-Größen-Relation könnte es sich um Effekte kritischr Kettenlängen der Alkylsubstituenten handeln, bei denen stärkere Veränderungen der elektronischen Struktur auftreten (siehe auch Diskussion der Chloro-Propyl-Reihe).

Die Zunahme der Photolumineszenzintensität der Trimethylsiloxysilsesquioxanreihe ist ebenfalls deutlich. Beide experimentellen Reihen zeigen neben der Intensitätssteigerung von Lumineszenz und Absorption mit steigender Größe bzw. Länge der Substituenten eine tendenzielle Rotverschiebung der Photolumineszenz und der Anstiegsflanke der Absorption. In der Absorption sind zusätzlich Bandenstrukturen zu erkennen.

Alle diese experimentellen Fakten deuten auf einen „*Charge-transfer*"-Effekt hin. Mit steigender Länge der Substituenten wird Elektronendichte vom Liganden in den Silsesquioxankäfig geschoben und damit dort die Elektronendichte erhöht. Die Polarisierbarkeit der Si–O–Si-Bindungsstruktur sowie das gesamte Dipolmoment des Moleküls werden größer und sind die Ursache der steigenden Intensität von Absorption und Lumineszenz. Typisch für einen *Charge-transfer*-Effekt ist auch die Rotverschiebung der beeinflußten Zustände. Durch die höhere Elektronendichte in der C–Si–O-Struktur wird die Orbitalüberlappung der C–Si und Si–O-Zustände stärker. Sie nehmen in der Folge eine intermediäre Stellung ein. Die Zustandslücke der Si–O-Bindung wird kleiner, die der C–Si-Bindung größer. Eine weiterführende Interpretationsmöglichkeit bei der Trimethylsiloxyreihe ist eine einsetzende Delokalisierung der Bindungsorbitale über die gesamte Käfigstruktur. Die größere Orbitalüberlappung führt zur Aufhebung der Zustandsentartung der Bindungszustände. Die entstehenden Zustandsbänder haben eine kleinere HOMO-LUMO-Lücke. Die tendenzielle Blauverschiebung und Intensitätseinbuße bei kleinen Käfigen verhält sich dann nach einem analogen Modell, wie es im Abschnitt über Lumineszenz aus Organosilanen erklärt wurde. Die delokalisierten Elektronen werden physikalisch-räumlich lokalisiert.

Im Absorptionsspektrum kann die starke Bande bei hohen Energien dem optisch erlaubten $\sigma \rightarrow \sigma^*$-Übergang zugeordnet werden. Das Intensitätsverhalten und die Rotverschiebung werden durch den *Charge-transfer*-Effekt erklärt. Die Photolumineszenzanregung bei 3.8 eV liegt jedoch energetisch weit unterhalb dieser Bande bei nur geringer Absorption. Dies legt die Vermutung nahe, daß es sich um einen $n \rightarrow \sigma^*$-Übergang handelt, bei dem Elektronen aus den *Lone-pair*-Zuständen am Sauerstoff angeregt werden. Die Übergangswahrscheinlichkeiten sollten relativ gering sein. Der Einfluß des *charge transfers* auf die energetische Lage der *lone pairs* sollte jedoch (analog zu oben Gesagtem) zu einer höheren Übergangswahrscheinlichkeit und Rotverschiebung der Photolumineszenz führen.

Abbildung 7 zeigt die Photolumineszenzintensität der *R*-Propylreihe gegen die Molekülgröße. Der Elektronegativität der terminalen Atome folgend, sollte der elektronenschiebende Effekt zwischen den Liganden und dem $(SiO_{1.5})_n$ Käfig gegenüber dem reinen Propylsubstituenten sinken. In der Reihenfolge H, I, Br, Cl (mit steigender Elektronegativität) liegt der NCS-Rest wahrscheinlich zwischen H und I. Der *Charge-transfer*-Effekt auf den Käfig wird mit zunehmender Elektronegativität kleiner. Dies unterstützt den vorgeschlagenen *Charge-transfer*-Mechanismus der Photolumineszenz, bei dem Elektronen aus dem *lone pair* am Sauerstoff in unbesetzte Orbitale der Liganden angeregt werden.

Eine weitere Stützung erfährt diese Interpretation bei der Analyse der Intensität und der Peaklage der Photolumineszenz der Chloropropylsilsesquioxane bei Variation der Käfiggöße. Abbildung 6 zeigt die Photolumineszenzintensität bei Variation der Käfiggrröße. Der Intensitätszuwachs pro Si–O-Bindung beim Übergang vom 8er- auf den 10er- Käfig beträgt 4.4 und vom 10er- auf den 12er-Käfig 2.4. Gleichzeitig stellt man eine sehr starke Blauverschiebung der Photolumineszenz beim Übergang vom 8er- auf den 10er-Käfig und eine wesentlich geringere Blauverschiebung beim Übergang vom 10er- auf den 12er-Käfig fest. Diese Verschiebung könnte mit dem oben vorgeschlagenen Modell erklärt werden. Sowohl die Peakverschiebung als auch der Intensitätszuwachs mit

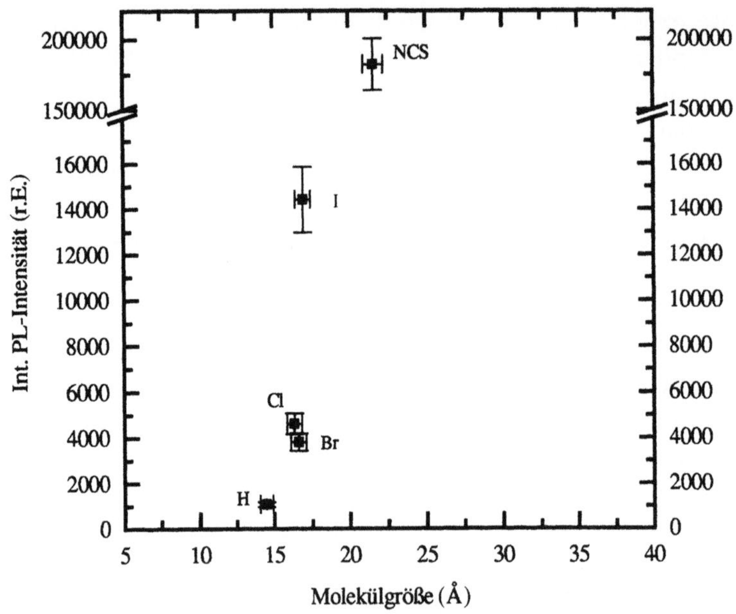

Abb. 7. Integrale Intensität der Photolumineszenzspektren der homofunktionalisierten Silsesquioxane *vs.* Moleküldurchmesser; Anregung: He–Cd-Laser, 325.4 nm, 3.8 eV; Käfiggröße:8, Variation der *R*-Propyl Liganden ($R =$ H, Cl, Br, J, NCS)

der Käfiggröße kann durch die veränderte Elektronendichte in der C–Si–O-Bindung erklärt werden.

Schlußfolgerungen

Zusammenfassend kann man sagen, daß Silsesquioxane ausgeprägte und interpretierbare opto-elektronische Eigenschaften zeigen. Einflußnehmende Größen sind die Käfiggröße und die angelagerten Substituenten. Bei der verwendeten Anregungsenergie von 3.8 eV kann aber kein erlaubter $\sigma \rightarrow \sigma^*$-Übergang angeregt werden, sondern nur ein verbotener $n \rightarrow \sigma^*$-Übergang. Das *Charge-transfer*-Modell läßt eine Interpretation der experimentellen Daten zu. Eine höhere Elektronendichte in der C–Si–O-Bindung führt zur Delokalisierung der Bindungsorbitale und somit zur Absenkung der Energielücke. Dabei liegt noch keine vollständige Delokalisierung über den Käfig vor, so daß noch molekülartige Spektren mit bereits breiten optischen Banden entstehen. Eine starke Phononenbeteiligung, die auch im Absorptionsspektrum sichtbar wird, sichert eine meßbare Übergangswahrscheinlichkeit des vorgeschlagenen $n \rightarrow \sigma^*$-Übergangs. Eine Rotverschiebung bis in den Bereich von 2.8–2.9 eV durch die Wahl entsprechender Substituenten ist möglich. In Bezug auf Lumineszenzphänomene in reinen Si-Systemen und *nc*-Si/*a*-SiO_2-Grenzflächen ist somit ein wichtiges Bindeglied verfügbar, das zur Klärung der roten Lumineszenz aus den Grenzschichten beiträgt. Der hier vorgeschlagene Mechanismus der Photolumineszenz muß trotz seiner guten Übereinstimmung mit den vorhandenen experimentellen Daten als vorläufig betrachtet werden.

Es stellt sich die Frage, ob die beobachtete Photolumineszenz nicht durch Verunreinigungen oder Defekte an einem Teil der Silsesquioxane (z.B. Si–Si Bindungen im $(SiO_{1.5})_n$-Käfig) erklärt werden könnte. Dies kann aus folgenden Gründen ausgeschlossen werden: Alle Substanzen wurden mittels NMR-Spektroskopie sorgfaltig auf mögliche Verunreinigungen überprüft. Berücksichtigt man die Nachweisgrenze der NMR-Spektroskopie von *ca.* 1 mol% und die Tatsache, daß die relative Effizienz der Photolumineszenz im Vergleich zum nanokristallinem Silizium (*nc*-Si) bis um einen Faktor von fast 1000 (R = NCS-Pr) höher ist und die Quantenausbeute der Photolumineszenz von *nc*-Si *ca.* 3–5% beträgt, so müßte die Quantenausbeute solcher Verunreinigungen bzw. Defekte den Wert von mehreren tausend erreichen, was eindeutig auszuschließen ist.

Um das Verständnis des Mechanismus der Photolumineszenz weiter zu verbessern, sollen denmächst die Anregungsspektren gemessen sowie der Effekt einer Variation des Lösungsmittels untersucht werden. Im Hinblick auf eine mögliche Anwendung der Silsesquioxane für opto-elektronische Devices kann jedoch bereits jetzt die Tatsache, daß die Photolumineszenz keinen $\sigma \rightarrow \sigma^*$-Übergang einschließt, als sehr günstig hervorgehoben weden. Die beobachtete Photolumineszenz erfolgt eindeutizg zwischen Energieniveaus, deren Energieunterschied niedriger als der $\sigma \rightarrow \sigma^*$-Übergang ist. Demzufolge sollte beim Langzeiteinsatz keine Photodegradation der Moleküle auftreten.

Experimentelles

Dem Interesse an den optischen Eigenschaften der Silsesquioxane als Modellsubstanz unserer *nc*-Si/*a*-SiO_2-Schichten folgend, wurden deren UV Vis-Absorption und Photolumineszenz vermessen. Die Messungen der Photolumineszenzspektralverteilungen, der Abklingzeit und des Polarisationsgedächtnisses wurden an Apparaturen durchgeführt, die in Lit. [12, 13] beschrieben werden. Als Anregungsquellen wurden ein He-Cd-Laser (Omnichrom, Modell 3056-M-AO1, $\lambda = 325.4$ nm, 22 mW) für die Lumineszene und ein N_2-Laser (Laser Science In-corp., Modell Vsl 337ND, $\lambda = 337.1$ nm) für die Bestimmung der Abklingzeit verwendet. Der Laserstrahl wird auf die Probe justiert und die Lumineszenz senkrecht zur Einstrahlrichtung mit einem Monochromator (Spex, Modell 1681) spektralzerlegt gemessen. Die Detektion erfolgte mit einer Si-Diode (Hamamatsu, K3413-05) zur Aufnahme der Spektren und mit einem Photomultiplier (Burle, Modell C31034, Anstiegszeit 2.5 ns) zur Messung der Abklingzeit. Dabei wurden kristalline Silsesquioxanschichten und eine 0.01 molare Lösung in Tetrahydrofuran (*THF*) oder Pentan vermessen. Zur Bestimmung des Polarisationsgedächtnisses wurde der He–Cd-Laser verwendet. Dessen Polarisationsebene wird difiniert gedreht und die Intensität der Lumineszenz mit einem Analysator parallel und senkrecht zur ursprünglichen Polarisationsebene gemessen. Ein Depolarisator vor dem Monochromator macht die Korrektur auf dessen polarisationsabhängige Transmission unnötig. Das Verhältnis $\rho(\theta) = (I_{\parallel} - I_{\perp})/(I_{\parallel} + I_{\perp})$ der Intensitäten bei senkrechter ($I_{\perp}$) und paralleler ($I_{\parallel}$) Analysatorstellung geben in Anhängigkeit von der Polarisatorstellung (θ) an, ob sich ein detektiertes Photon an das anregende Photon erinnert.

Die Absorptionsspektren wurden an einem Lambda 9 (Perkin Elmer) Spektrometer im Bereich von 170–1200 nm repetitiv vermessen und zur Verbesserung des Signal-zu-Rausch-Verhältnisses gemittelt. Man erhält den molaren Extinktionskoeffizienten ε aus dem *Lambert-Beer*schen Gesetz ($A = \varepsilon[J]d = -\log(I_E/I_A)$) mit $A =$ Extenction, $[J]$ = Konzentration der untersuchten Substanz, d = durchstrahlte Dicke der Probe und I_E bzw. I_A = transmittierte bzw. Anregungsintensität.

Hochreines und UV-spektroskopietaugliches Pentan und *THF* werden als Standardlösungsmittel analog zu ähnlichen Substanzen verwendet [14, 15]. Die Lösungen wurden durch Wägung der

Silsesquioxane und Pipettierung des Lösungsmittels hergestellt. Der Fehler der Molarität bei der Herstellung und durch Verdampfen des Lösungsmittels während der Messung ist kleiner 1%. Zur Bestimmung der Extinktion wurde der Absorptionsanteil des Lösungsmittels und der verwendeten UV-tauglichen Küvette aus Spektrosil abgezogen.

Da der Moleküldurchmesser der Silsesquioxane im Bereich von 0.8 nm bis zu 4 nm liegt, wurde auch ein eventueller Einfluß der *Rayleigh*-Streuung abgeschätzt. Selbst für das größte untersuchte Silsesquioxan ($Q_8M_8^{32Allyl}$, Durchmesser 4 nm) liegt der Effekt unterhalb der Meßfehler.

Bei den Absorptionsspektren der Silsesquioxane fiel bei hohen Energien (ab *ca.* 6.3 eV) ein systematisches Absinken der Extinktion und bei *ca.* 6.7 eV ein Peak auf Dabei handelt es sich um einen gerätebedingten spektralen *cut-off* bzw um ein Meßartefakt. Mit einer Eichmessung wurde dieser Effekt bestätigt und muß bei der Diskussion beachtet werden.

Dank

Diese Arbeit wurde von der *Deutschen Forschungsgemeinschaft* (DFG) unterstützt.

Literatur

[1] Canham LT (1990) Appl Phys Lett 57: 1046
[2] Koch F, Petrova-Koch V, Muschnik T, Nikolov A, Gavrilenko V (1993) Mater Res Soc Symp Proc **283**: 197
[3] Veprek S, Wirschem T (1997) In: Hummel E (ed) Handbook of Optical Properties, Vol II, Optics of Small Particles, Interfaces and Surfaces. CRC Press, Boca Raton
[4] Miller RD, Michl J (1989) Chem Rev **89**: 1359
[5] Kanemitzu Y (1995) Physics Reports **263**: 1
[6] Dittmar V (1993) Dissertation, Universität-GH Paderborn
[7] Voronkov MG, Lavrentyev V (1982) Top Curr Chem **102**: 199
[8] Bassindale AR, Gentle TA (1993) J Mater Chem **3**: 1319
[9] Frye CL, Collins WT (1970) J Am Chem Soc **92**: 5586
[10] Dittmar U, Hendan BJ, Flörke U, Marsmann HC (1995) J Organomet Chem **489**: 185
[11] Rikowski E, Marsmann HC (1997) Polyhedren **16**: 3357
[12] Verprek S, Wirschem T, Rückschloß M, Perna S, Ossadnik C, Dian J, Gregora I (1996) Mater Res Soc Symp Proc 405
[13] Veprek S, Rückschloß M, Landkammer B, Ambacher O (1993) Mater Res Soc Proc **298**: 117
[14] Kanemitzu Y (1995) Physics Reports **263**: 1
[15] Kanemitzu Y, Suzuki K, Kyushin S, Matsumoto H (1995) Phys Rev B **51**: 13103
[16] Tamura H, Rückschloß M, Wirschem T, Veprek S (1994) Appl Phys Lett **65**(12)
[17] Rückschloß M, Wirschem T, Tamura H, Ruhl G, Oswald J, Veprek S (1995) Lumin J **63**: 279
[18] Schneider R (1997) Universität Bielefeld, Diplomarbeit und persönliche Mitteilung

Received June 19, 1998. Accepted August 5, 1998

Gezielte Synthese siloxenähnlicher Polymerer

Alois Kleewein und **Harald Stüger***

Institut für Anorganische Chemie, Technische Universität Graz, A-8010 Graz, Österreich

Synthesis of Siloxene-Like Polymers

Summary. Polysiloxane structures closely related to *Kautsky*'s siloxene have been prepared starting from well defined cyclosilane units. The products exhibit moderate to strong fluorescence ranging from 400 to 550 nm, depending on the ring size and the number and nature of substituents. No fluorescence was observed for products derived from linear silane units. The fluorescence is influenced by the functionality of the siloxane networks as shown in a comparison of cyclosilane compounds with one functionality per ring up to species with two functionalities per each silicon atom of the ring.

Keywords. Cyclosilanes; Fluorescence spectra; Siloxene; Polymer; Polysiloxanes.

Einleitung

Mit der Entdeckung der außerordentlichen Lumineszenzeigenschaften von porösem Silicium, die unter anderem auch auf das Vorhandensein von Siloxenschichten zurückgeführt wurden [1], standen Chemie und Physik des Siloxens wieder im Zentrum zahlreicher Untersuchungen [2]. Siloxene werden durch Einwirkung von verdünnter Salzsäure auf Calciumdisilicid erhalten, wobei die Methoden nach *Kautsky* [3] und *Wöhler* [4] sich in den Reaktionsbedingungen unterscheiden und Produkte mit verschiedenartigen Eigenschaften liefern, was möglicherweise auf vorhandene Strukturunterschiede zurückzuführen ist.

Moderne Strukturuntersuchungen an frisch hergestelltem *Wöhler*-Siloxen mittels Pulverröntgendiffraktometrie und Einkristallröntgenstrukturanalyse [5] zeigen eindeutig gewellte Si_6-Ringe, die analog zum $CaSi_2$ über Si–Si-Bindungen zu Schichten verknüpft sind. Die Absättigung der restlichen Valenzen der Si-Atome erfolgt alternierend über H-bzw. OH-Substituenten (Abb. 1a). Im Gegensatz dazu ist die Struktur des *Kautsky*-Siloxens noch immer nicht eindeutig geklärt. Im *Kautsky*-Modell des Siloxens [6] sind Siliciumsechsringe über Sauerstoffbrücken zu einem Flächenpolymer verknüpft, die 4. Valenz des Siliciums ist in diesem Falle durch Wasserstoff abgesättigt (Abb. 1b). Nach einem weiteren Strukturmodell besteht Siloxen aus linearen Polysilanketten, die über Sauerstoffbrücken vernetzt sind (Abb. 1c). Für beide Strukturmodelle gibt es jedoch bis jetzt keinerlei experimentelle Beweise.

* Corresponding author

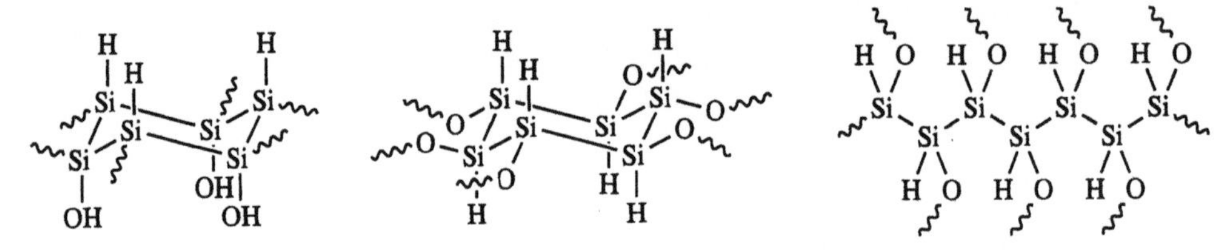

a b c

Abb. 1. Strukturmodelle für Siloxen

Betrachtet man das Strukturmodell nach *Kautsky* (Abb. 1b), so handelt es sich um ein Poly (1,2,3,4,5,6-Hexaoxocyclohexasilan), das formal durch Selbstkondensation aus dem entsprechenden Hexahydroxocyclohexasilan unter Ausbildung von Si–O–Si-Brücken entstanden ist. Eine gezielte Synthese analoger Strukturelemente sollte also ausgehend von Hexahalogencyclohexasilanen möglich sein, da Si-Halogen-Bindungen in der Regel extrem leicht zu Si–OH Gruppen hydrolysiert werden können, aus denen unter Wasserabspaltung meist spontan die entsprechenden Siloxane entstehen (vgl. Schema 1).

Dieser Ausgangspunkt für unsere Untersuchungen wird vor allem dadurch unterstützt, daß von Pentaiodpentaphenylcyclopentasilan bereits lange bekannt ist, daß es zu einem gelben, fluoreszierenden, polymeren Hydrolyseprodukt reagiert [7], das analog dem Modell des *Kautsky*-Siloxens als Poly (1,2,3,4,5-pentaoxo-

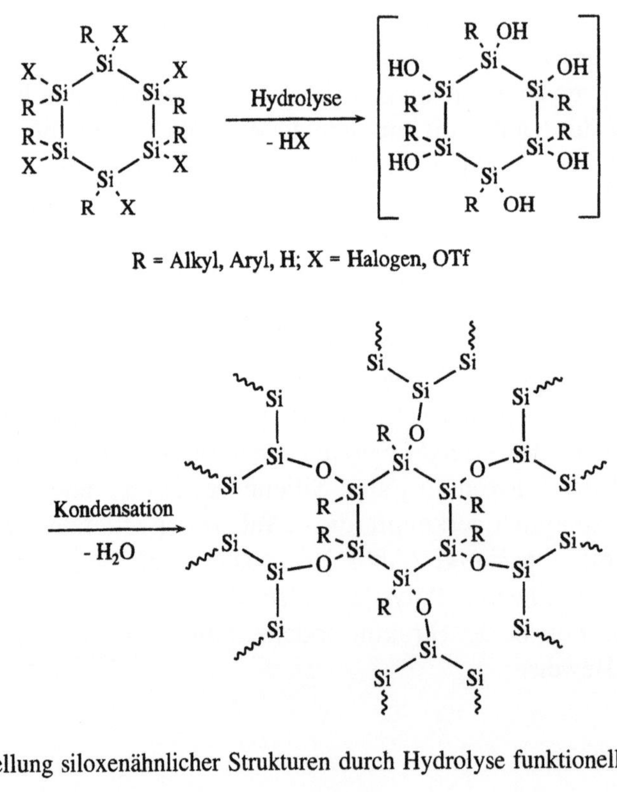

Schema 1. Darstellung siloxenähnlicher Strukturen durch Hydrolyse funktioneller Cyclohexasilane

pentaphenylcyclopentasilan) beschrieben werden kann. Mit der Zielsetzung, einen indirekten Strukturbeweis für *Kautsky*-Siloxen zu liefern, lag es daher nahe, die Eigenschaften der Hydrolyseprodukte geeignet funktionalisierter Oligosilane, besonders ihr Photolumineszenzverhalten, genauer zu untersuchen, wobei durch das Ergebnis beim Pentaiodpentaphenylcyclopentasilan zu erwarten war, daß sich die Fluoreszenz nicht nur auf Siliciumsechsringe wie im Siloxen beschränken wird. Zusätzlich wird durch die im Pentaiodpentaphenylcyclopentasilan vorhandenen Phenylgruppen (anstelle des Wasserstoffes im Siloxen) ersichtlich, daß das Vorhandensein von Si–H-Funktionen nicht als unbedingte Voraussetzung für die Fluoreszenz der Materialien notwendig ist.

Ergebnisse und Diskussion

Die für die Hydrolyseexperimente verwendeten Ausgangsmaterialien sowie ausgewählte Eigenschaften der erhaltenen Reaktionsprodukte sind in Tabelle 1 zusammengefaßt.

Um Rückschlüsse über den Einfluß der Struktur des Si–Si-Gerüstes auf die Eigenschaften der Hydrolyseprodukte zu erlauben, wurden sowohl offenkettige als auch cyclische Oligosilane verschiedener Kettenlänge hydrolysiert. Als hydrolyselabile Abgangsgruppen fanden dabei je nach Verfügbarkeit der entsprechenden Substrate Cl, Br oder CF_3SO_3 (=OTf) Verwendung. Um möglichst hohe Kondensationsgrade zu erreichen, wurden die primär anfallenden Hydrolysate mehrere Stunden im Hochvakuum bei 150°C getrocknet. Danach konnten nach Extraktion mit *THF* oder anderen organischen Lösungsmitteln massenspektroskopisch keine löslichen Bestandteile mehr nachgewiesen werden. Wird auf diesen Trocknungsschritt verzichtet, bleiben die Hydrolyseprodukte zumindest teilweise in *THF* löslich. Untersuchungen der nach Hydrolyse von **11** erhaltenen löslichen Fraktionen mittels GPC ergaben Werte von 1700 bis 2000 g/mol als Zahlenmittel (relativ zu Polystyrol).

Die nach vollständiger Kondensation erhaltenen Produkte fallen in der Regel in Form polymerer, pulverförmiger Feststoffe an, deren IR-Spektren für Silanole bzw. Siloxane charakteristische Banden um 3400 cm^{-1} (ν(O–H)) bzw. um 1030 cm^{-1} (ν_{as}(Si–O–Si)) zeigen. Entsprechende Banden der in den Ausgangsstoffen vorhandenen Si-X-Gruppierungen wie die in Polysilanen im Bereich von 500–550 cm^{-1} auftretenden Si–Cl-Valenzschwingungen sind verschwunden. Nachweise auf vorhandenes Halogen nach einem *Parr*-Bombenaufschluß mit Natriumperoxid mittels Silbernitrat verliefen ebenfalls negativ. Untersuchungen der erhaltenen Polymeren in Hinblick auf eine bessere physikalische Charakterisierung einschließlich Festkörper-^{29}Si-NMR Spektroskopie werden derzeit durchgeführt; über die Ergebnisse wird in einer nachfolgenden Arbeit berichtet werden.

Die Hydrolyseprodukte offenkettiger Ausgangsmaterialien sind farblos und nicht photolumineszenzaktiv, während die Hydrolyseprodukte cyclischer Ausgangsmaterialien meist farbig sind und bemerkenswerte, teils intensive Photolumineszenz im sichtbaren Spektralbereich zeigen (vergleiche Tabelle 1 sowie Abb. 2).

Wie aus Abb. 2 ersichtlich ist, sind dabei, analog zum Siloxen, ausgeprägte Substituenteneinflüsse auf die Lage der Fluoreszenzmaxima sowie Abhängigkeiten des Photolumineszenzverhaltens von der Größe der Siliciumringe im Ausgangs-

Tabelle 1. Ausgangsmaterialien für die Hydrolyseexperimente sowie ausgewählte Eigenschaften der erhaltenen Reaktionsprodukte

	Ausgangsmaterial	Eigenschaften der Hydrolyseprodukte	
		Produktbeschreibung	Fluoreszenzmaximum (nm)
1	$ClMe_2SiSiMe_2Cl$	farblos, ölig	keines
2	$ClMe_2SiSiMeClSiMe_2Cl$	farblos, ölig	keines
3	$ClMe_2Si(SiMeCl)_2SiMe_2Cl$	weißer Feststoff	keines
4	$Cl_2MeSi(SiMeCl)_2SiMeCl_2$	weißer Feststoff	keines
5	$SiMeCl_2$ | $Cl_2MeSi\text{-}SiMe\text{-}SiMeCl_2$	weißer Feststoff	keines
6	*cyclo*-Si_4Br_8	weißer Feststoff	438
7	*cyclo*-Si_5Cl_{10}	weißer Feststoff	400
8	*cyclo*-Si_6Cl_{12}	weißer Feststoff	432
9	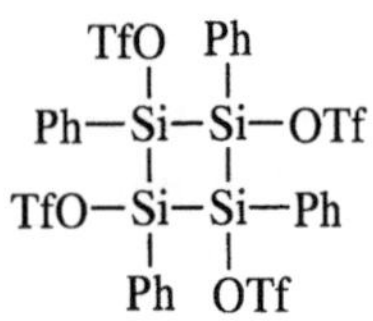	gelber Feststoff	553
10	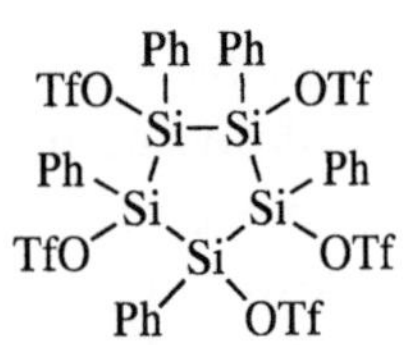	gelber Feststoff	522
11	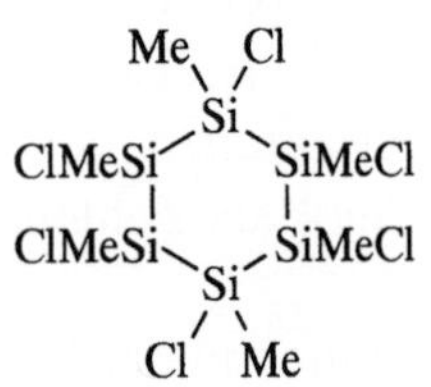	gelb-grüner Feststoff	436
12	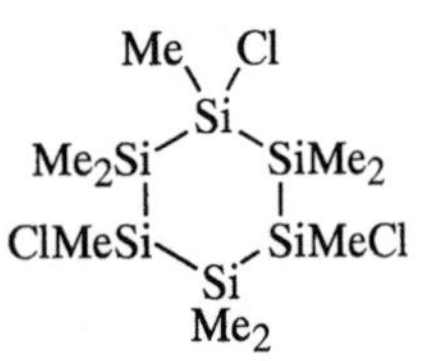	weißer Feststoff	420, schwach
13	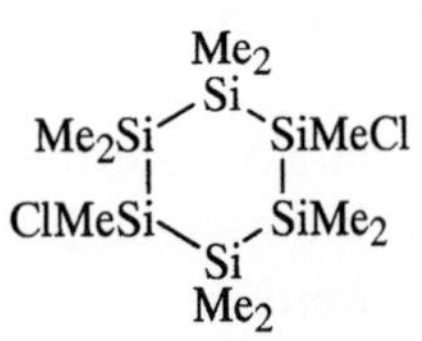	weißer Feststoff	keines
14	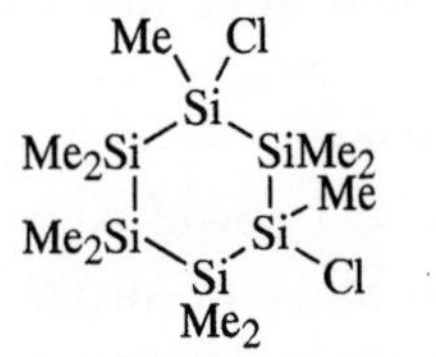	weißer Feststoff	keines

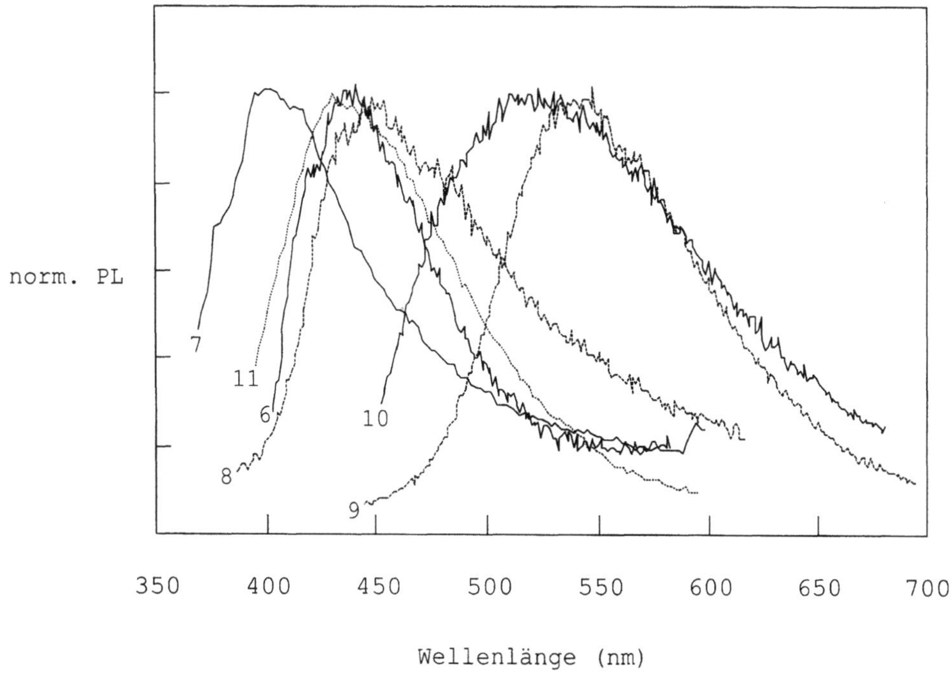

Abb. 2. Photolumineszenzspektren der Hydrolyseprodukte der Cyclosilane **6–11**

material zu beobachten. Eine genauere Diskussion dieser Phänomene zu diesem Zeitpunkt erscheint jedoch verfrüht.

Interessante Hinweise möglicher Struktureinflüsse auf das Photolumineszenzverhalten der untersuchten Verbindungen ergeben sich aus den Resultaten der Hydrolyse der Cyclohexasilanderivate **11–14** (vergleiche Tabelle 1 und Schema 2).

So zeigen die Hydrolyseprodukte der difunktionellen Verbindungen **13** und **14**, für die ein linearer Aufbau aus sauerstoffverknüpften Cyclohexasilaneinheiten wahrscheinlich ist, keine Photolumineszenz, während im Falle des Hydrolyseproduktes von 1,3,5-Trichlornonamethylcyclohexasilan (**12**), das bereits eine vernetzte Struktur besitzen sollte, das Auftreten von schwacher Photolumineszenz beobachtet wird. Das Hydrolyseergebnis der hexafunktionellen Verbindung **11** zeigt weiters, daß die Intensität dieser Photolumineszenz durch eine Erhöhung des Vernetzungsgrades noch beträchtlich gesteigert werden kann. Keine Photolumineszenzaktivität hingegen beobachtet man für das dimere Strukturelement *Bis*(undecamethylcyclohexasilanyl)ether.

Kriterien für das Auftreten intensiver Photolumineszenz in den untersuchten Verbindungen sind demnach sowohl Subeinheiten aus cyclischen Oligosilanen als auch polymere, mehrdimensionale Strukturen mit hohem Vernetzungsgrad, wodurch ein Aufbau aus siloxenähnlichen Struktureinheiten zumindest nicht auszuschließen ist. An dieser Stelle muß jedoch mit Nachdruck darauf hingewiesen werden, daß die erhaltenen Hydrolyseprodukte zwangsläufig komplexe Isomerengemische darstellen, da von Stereoisomerengemischen der Halogensilane ausgegangen wird. Inwieweit die Stereochemie der Ausgangsmaterialien die Eigenschaften der gebildeten Polymeren beeinflussen kann, wird in nachfolgenden

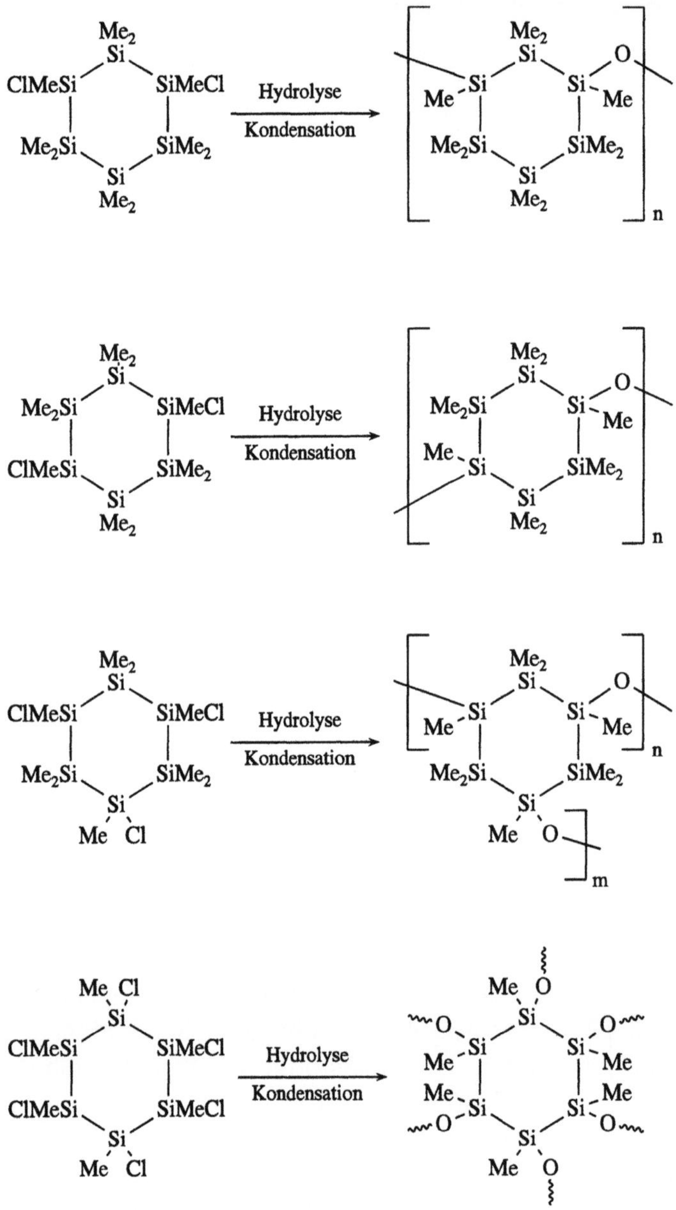

Schema 2. Wahrscheinliche Strukturen der Hydrolyseprodukte der Cyclohexasilanderivate **11–14**

Untersuchungen z.B. durch Hydrolyse isomerenreiner polyfunktioneller Halogensilane zu klären sein.

Experimentelles

Wegen der extremen Hydrolyseempfindlichkeit der meisten Ausgangsprodukte wurden Synthesen und spektroskopische Charakterisierung fast ausschließlich unter Verwendung von Standard-*Schlenk*-Techniken in einer Atmosphäre von hochreinem Stickstoff durchgeführt. Die dabei benötigten Lösungsmittel wurden durch mehrstündiges Kochen mit Na/K-Legierung und anschließender Destillation unter Stickstoff getrocknet. Zur Darstellung von 1,2-Dichlortetramethyldisilan [8], 1,2,3-Trichlorpentamethyltrisilan [9], *Tris* (methyldichlorsilyl) methylsilan [10] und *Bis*(undecamethylcyclohexasilanyl) ether [19] sowie zur Synthese der funktionalisierten Cyclosilanderivate **6–14** [11–18] wurden aus der Literatur bekannte Methoden verwendet. Die Aufnahme der NMR Spektren erfolgte auf einem Bruker MSL 300 Spektrometer (Messfrequenzen: ^{29}Si: 59. 627 MHz, ^{1}H: 300.13 MHz, ^{13}C: 75.470 MHz) in $CDCl_3$ oder Toluol (D_2O-Kapillare als externer Lock) relativ zu externem *TMS*. Infrarotspektren wurden an einem Perkin-Elmer 883-Gerät als Nujolverreibungen gemessen. Für GC-Untersuchungen sowie für die Massenspektroskopie stand eine HP 5971/A/5890-II GC/MS Kombination mit einer HP 1 Kapillarsäule (Länge 25 m, Durchmesser 0.2 mm, Schichtdicke 0.33 μm Polydimethylsiloxan) zur Verfügung. Die Fluoreszenz-Emissions-Spektren wurden auf einem Photolumineszenzgerät zur Anregung der Emission, bestehend aus einer Xe-arc-Quelle und einem Doppelmomochromator, gemessen. Die Fluoreszenzintensität wurde durch Sammeln und Bündeln des Probensignals über ein Linsensystem und einen Monochromator auf einem gekühlten Photonendetektor bestimmt. Die PL-Spektren wurden nicht korrigiert, da das System in einem Bereich von 400 bis 800 nm ein konstantes Signal liefert. Flüssige Proben wurden in einer Infrasilküvette (0.5 cm Schichtdicke) vermessen, die festen Proben als Pulverschicht auf Papier. Präparation und Vermessung der Proben wurde unter Argon als Schutzgas durchgeführt.

Darstellung der offenkettigen Chlorsilane **3** *und* **4**

1,2,3,4-Tetraphenylhexamethyltetrasilan und 1,1,2,3,4,4-Hexaphenyltetramethyltetrasilan wurden in Analogie zur Literatur [20] durch Umsetzung von 1,2-Dichlor-1,2-diphenyldimethyldisilan mit 2 Äquivalenten $LiSiMe_2Ph$ bzw. $LiSiMePh_2$ hergestellt und durch $AlCl_3$-katalysierte Phenylabspaltung mit trockener HCl in Benzol in die chlorierten Derivate **3** und **4** überführt.

1,2,3,4-Tetraphenylhexamethyltetrasilan (Enantiomerengemisch)

Ausbeute: 62%; weiße Kristalle aus 2-Propanol; m.p.: 118–121°C; ^{29}Si-NMR (Toluol/D_2O, ext. *TMS*, ppm):– 17.81/–19.23 (Me_2PhSi-), –43.93/–44.16 (-SiMePh-); ^{1}H-NMR ($CDCl_3$, ext. *TMS*, ppm, rel. Int.): 7.287 (m, 20 H), 0.418/0.417 (s, 6 H), 0.241 (s, 12 H); ^{13}C-NMR ($CDCl_3$, ext. *TMS*, ppm): 139.14, 137.00, 134.93, 134.09, 128.51, 128.19, 127.86, 127.71, –3.69, –4.10, –7.02, –7.01; MS (70 eV): *m/z* (rel. Int.) = 510 (1.6, M^+), 375 (18.5, $Si_3Ph_3Me_4$), 298 (46.5, $Si_3Ph_2Me_4$), 197 (51.2, $SiPh_2Me$), 135 (100, $SiPhMe_2$), 105 (26.9, SiPh), 43 (10.8, SiMe); IR:ν = 1106 m, 834 m, 808 m, 780 v, 761 v, 731 v, 698 v, 665 m, 644 m, 466 w, 385 m, 341 w cm^{-1}.

1,1,2,3,4,4-Hexaphenyltetramethyltetrasilan (Enantiomerengemisch)

Ausbeute: 71%; weißes Pulver aus 2-Propanol; m.p.: 178–180°C; ^{29}Si-NMR (Toluol/D_2O, ext. *TMS*, ppm): –18.06/–19.18 ($MePh_2Si$-), –43.89/–44.36 (-SiMePh-); ^{1}H-NMR ($CDCl_3$, ext. *TMS*, ppm, rel. Int.): 7.2–7.0 (m, 60 H), 0.402 (s, 6 H), 0.361 (s, 6 H), 0.360 (s, 6 H), 0.316 (s, 6 H); ^{13}C-NMR ($CDCl_3$, ext. *TMS*, ppm): 137.15, 136.88, 136.57, 135.22, 135.17, 135.11, 129.71, 128.86, 128.78, 128.77, 128.69, 128.32, 127.96, 127.88, 127.82, 127.74, 127.63, –3.56, –3.62,–6.39, –6.42; MS (70 eV): *m/z* (rel. Int.) = 634 (0.25, M^+), 437 (8.2, $Si_3Ph_4Me_3$), 360 (11.8, $Si_3Ph_3Me_3$), 197 (100, $SiPh_2Me$), 135 (8.8, $SiPhMe_2$), 105 (6.4, SiPh), 43 (1.0, SiMe); IR: ν = 1105 m, 1099 m, 1087 m, 997 w, 809 w, 795 m, 786 m, 780 v, 763 v, 758 m, 736 v, 714 v, 697 v, 667 w, 505 m, 466 v, 461 m, 446 w cm^{-1}.

1,2,3,4-Tetrachlorhexamethyltetrasilan (Enantiomerengemisch) (**3**)

Ausbeute: 93%; farbloses Öl; ^{29}Si-NMR (Toluol/D_2O, ext. *TMS*, ppm): 19.76/19.60 ($ClMe_2Si$-), 2.73/1.86 (-SiMeCl-); ^{1}H-NMR ($CDCl_3$, ext. *TMS*, ppm, rel. Int.): 0.829/0.811 (s, jeweils 2 H), 0.709/0.701 (s, jeweils 4 H); ^{13}C-NMR ($CDCl_3$, ext. *TMS*, ppm): 2.63/2.14, –0.84/–1.04; MS

(70 eV): m/z (rel. Int.) = 344 (0.7, M^+), 329 (0.6, $Si_4Me_5Cl_4$), 309 (2.1, $Si_4Me_6Cl_3$), 116 (40.3, Si_2Me_4), 73 (100, $SiMe_3$), 43 (23.1, SiMe).

1,1,2,3,4,4-Hexachlortetramethyltetrasilan (Enantiomerengemisch) (**4**)

Ausbeute: 89%; farbloses Öl; ^{29}Si-NMR (Toluol/D_2O, ext. *TMS*, ppm): 24.81/24.67 (Cl_2MeSi-), 0.24/−0.78 (-SiMeCl-); ^{1}H-NMR ($CDCl_3$, ext. *TMS*, ppm, rel. Int.): 0.872/0.854 (s, jeweils 2H), 0.819/0.805 (s, jewels 2H); ^{13}C-NMR ($CDCl_3$, ext. *TMS*, ppm): 2.42/2.07, −0.95/−1.12.

Hydrolyse und Kondensation (allgemeine Arbeitsvorschrift)

Zur Hydrolyse der Halogensilane wurden 2 g der entsprechenden Substanz unter Schutzgasatmosphäre in 20–50 ml *THF* (je nach Löslichkeit) gelöst und 30 ml eines entgasten Gemisches aus *THF* und Wasser (1:1) bei 20°C zugetropft. Die Hydrolyse der Triflatverbindungen **9** und **10** erfolgte in Toluol, da die entstehende Trifluormethansulfonsäure *THF* unter Ringöffnung polymerisiert. Hatten sich nach Rühren über Nacht Feststoffe gebildet, wurden diese abfiltriert, ansonsten wurden Lösungsmittel und Wasser im Vakuum abgezogen. Nach mehrmaliger Extraktion der so erhaltenen Feststoffe mit *THF* zur Entfernung löslicher, niedermolekularer Fraktionen wurden die Produkte zur Vervollständigung der Kondensation bzw. zur Entfernung der letzten Wasser- und Lösungsmittelreste 5 h bei 150°C im Hochvakuum (0.02 mbar) getrocknet. Bei Annahme vollständiger Kondensation (erhaltene Summenformel $(Si_6R_6O_3)_n$) betragen die mittleren Ausbeuten der resultierenden unlöslichen Polymeren ungefähr 70%.

Dank

Die Autoren danken dem *Fonds zur Förderung der wissenschaftlichen Forschung* (Wien, Österreich) für finanzielle Unterstützung des Projektes im Rahmen des Forschungsschwerpunktes Siliciumchemie und der Firma *Wacker-Chemie* (Burghausen, Deutschland) für die Überlassung von Ausgangsmaterialien.

Literatur

[1] Dahn JR, Way BM, Fuller EW, Weydanz WJ, Tse JS, Klug DD, Van Buuren T, Tiedje T (1994) J Appl Phys **75**: 1946
[2] Brandt MS, Puchert T, Stutzmann M (1994) Siloxene: The Physics View. In: Corriu R, Jutzi P (eds) Tailor-Made Silicon-Oxygen Compounds. Vieweg, Wiesbaden, p 117
[3] Kautsky H (1921) Z Anorg Allg Chem **117**: 209
[4] Wöhler F (1863) Lieb Ann **127**: 257
[5] a) Dahn JR, Way BM, Fuller E (1993) Phys Rev B **48**: 17872; b) Dettlaf-Weglikowska U, Hönle W, Molassioti-Dohms A, Finkbeiner S, Weber J (1997) Phys Rev B **56**: 13132
[6] Deák P, Rosenbauer M, Stutzmann M, Weber J, Brandt MS (1992) Phys Rev Lett **69 (17)**: 2531
[7] Hengge E, Marketz H (1969) Monatsh Chem **100**: 898
[8] Sakurai H, Taminga K, Watanabe T, Kumada M (1966) Tetrahedron Lett **45**: 5493
[9] Schenzel K, Hassler K (1994) Spectrochim Acta **50A (1)**: 127
[10] Tandell RF, Urry G (1978) J Inorg Nucl Chem **40**: 1305
[11] Kovar D, Utvary K, Hengge E (1979) Monatsh Chem **110**: 1295
[12] Fossum E, Gordon-Wylie SW, Matyjaszewski K (1994) Organometallics **13**: 1695
[13] Hengge E, Kovar D (1977) J Organomet Chem **125**: C29
[14] Uhlig W (1997) Trends in Organomet Chem **2**: 1
[15] Hengge E, Kovar D (1979) Z Anorg Allg Chemie **459**: 123

[16] Hengge E, Kalchauer W, Schrank F (1986) Monatsh Chem **117**: 1399
[17] Spielberger A, Gspaltl P, Siegl H, Hengge E, Gruber K (1995) J Organomet Chem **499**: 241
[18] Eibl M, Katzenbeißer U, Hengge E (1993) J Organomet Chem **444**: 29
[19] Gspaltl P (1995) Dissertation, Technische Universität Graz, p 121
[20] Schenzel K, Hassler K (1994) Spectrochim Acta **50A (1)**: 127

Received June 30, 1998. Accepted (revised) August 4, 1998

Preparation and Characterization of Epitaxial $CaSi_2$ and Siloxene Layers on Silicon

Günther Vogg[1,*], **Nikta Zamanzadeh-Hanebuth**[1], **Martin S. Brandt**[1], **Martin Stutzmann**[1], and **Martin Albrecht**[2]

[1] Walter Schottky Institut, Technische Universität München, D-85748 Garching, Germany
[2] Institut für Werkstoffwissenschaften, Universität Erlangen-Nürnberg, D-91058 Erlangen, Germany

Summary. Epitaxial growth of thin $CaSi_2$ films on various silicon surfaces is described. Transforming the silicide into siloxene leads to high quality epitaxial siloxene films with different orientations. The structural quality of $CaSi_2$ and siloxene is investigated by transmission and scanning electron microscopy as well as X-ray diffraction analysis. Siloxene on (110)-Si enables direct observation of the anisotropic vibrational properties by IR transmission measurements. A diode structure was realized with siloxene by using contacts with different work functions as charge carrier injectors.

Keywords. Anisotropy of IR absorption; $CaSi_2$ epitaxy; Diode structure; Structural analysis.

Herstellung und Charakterisierung epitaktischer $CaSi_2$- und Siloxenschichten auf Silizium

Zusammenfassung. Das epitaktische Wachstum dünner $CaSi_2$-Filme auf verschiedenen Siliziumoberflächen wird beschrieben. Die Umwandlung des Silizids in Siloxen ergibt unterschiedlich orientierte epitaktische Siloxenfilme von hoher Schichtqualität. Die strukturelle Qualität von $CaSi_2$ und Siloxen wurde mit Hilfe von Transmissionselektronen- und Rastertunnelmikroskopie sowie Röntgendiffraktion untersucht. Siloxen auf (110)-Si ermöglicht einen direkten Nachweis der anisotropen Schwingungseigenschaften mittels IR-Transmissionsmessungen. Mit Siloxen und Kontakten mit unterschiedlicher Austrittsarbeit zur Ladungsträgerinjektion wurde eine Diodenstruktur hergestellt.

Introduction

Siloxene, or more correctly 2D-poly[1,3,5-trihydroxocyclohexasilane], has first been prepared by *Wöhler* in 1863 [1]. Later, *Kautsky* and *Hengge* have systematically studied the properties of this material [2, 3]. Whereas siloxene prepared according to the recipe of *Kautsky* [4] is difficult to analyze structurally, the structure and chemical composition of *Wöhler*'s siloxene has been identified by several groups [5, 6]. It consists of a backbone of two-dimensional Si layers nearly identical to the corrugated {111} double layers in crystalline Si. These layers are

* Corresponding author

terminated on both sides by H- and OH-groups and stacked upon each other with a layer distance of approximately 6 Å. Siloxene is synthesized from $CaSi_2$ which is the only known silicide already containing the described layered Si backbone structure. The calcium atoms are topochemically removed in concentrated HCl at low temperatures, leaving the Si layers intact. The two-dimensional polymer backbone is responsible for a variety of properties of this material which make it unique within Si based substances.

Studies concerning the chemical and physical properties of siloxene published so far have given rise to partly conflicting results. For example, the reported Si layer distances range from 5.5 Å [5] to 7 Å [6]. The only obvious conclusion is that small deviations in sample preparation and handling lead to structurally quite different compounds. To obtain a higher reproducibility it is necessary to carefully control the wet-chemical reaction of $CaSi_2$ to siloxene. Similarly important is the reproducibility, homogeneity, and quality of the starting material. In this paper we report the growth of thin epitaxial $CaSi_2$ layers on Si substrates. Chemical transformation of these epitaxial layers leads to highly ordered crystalline siloxene films with a good reproducibility. Such thin films are a prerequisite for more detailed studies of the notable anisotropy of the electrical, optical, and vibrational properties of siloxene which are otherwise difficult to perform. In addition, such films are necessary to evaluate the potential of siloxene for various device applications.

In this paper we briefly describe the preparation and characterization of thin $CaSi_2$ and siloxene films on Si. We first review the epitaxial growth of $CaSi_2$ on crystalline Si substrates with different orientations. Then, X-ray diffraction measurements on $CaSi_2$ and siloxene films are presented. For further structural characterization, the vibrational properties of siloxene are examined by infrared transmission measurements using polarized light. Finally, transport results obtained for a novel calcium-siloxene-Si diode structure are presented.

Results and Discussion

Epitaxial growth

The starting material for thin epitaxial siloxene films on Si are $CaSi_2$/Si heterostructures grown by reactive deposition epitaxy (RDE) [7] of calcium onto a heated Si substrate. As well known for transition metal silicides, the crystal quality of the epitaxial $CaSi_2$ layer depends strongly on the growth conditions, due to a sensitive equilibrium between nucleation and diffusion processes [8–10]. Our own studies, in agreement with previous work on other silicides, have identified Si to be the diffusing species in $CaSi_2$.

As expected, the Si layers in both $CaSi_2$ and siloxene correspond to the {111} layers in crystalline Si and therefore play a dominant role in the growth of the silicide. Most notably, the Si layers cause a fixed relation between the substrate and the silicide lattice. In addition, the anisotropic crystal structure leads to different growth conditions and morphology for various substrate orientations. In particular, we observed that the Si layers in $CaSi_2$ and siloxene grow parallel to the corresponding {111} layers of the Si substrate. Since there are four independent

(111) directions in crystalline Si, this can lead to inhomogeneous films with crystallites oriented in different directions. One can indeed observe this behaviour, again sensitively depending on the growth conditions. Using particular conditions, *e.g.* high Ca flow and low substrate temperature, one growth direction can be enhanced selectively, and therefore epitaxial films with a preferential orientation are obtained. Additionally, the use of different substrate orientations allows the growth of special and well defined layer orientations.

Epitaxial growth of $CaSi_2$ on (111)-Si by solid phase epitaxy (SPE) has already been reported [11, 12]. With RDE we are now able to prepare high quality epitaxial films on (111)-Si with a thickness of several hundred nm. Structural quality and lateral homogeneity is clearly shown by SEM and TEM micrographs (Fig. 1a) and, as discussed later, by XRD measurements. An increased Ca flow during growth leads to a formation of vertical crystallites which typically originate at the edges of the planar crystals. This behaviour has been observed earlier by *Braungart* and *Sigmund* [13]. The crystallites representing the other possible (111) directions finally coalesce and form a $CaSi_2$ film with a vertical layering. This is due to the anisotropic growth of $CaSi_2$: crystal growth parallel to the $CaSi_2$ layers is obviously much easier and therefore faster than perpendicular to them.

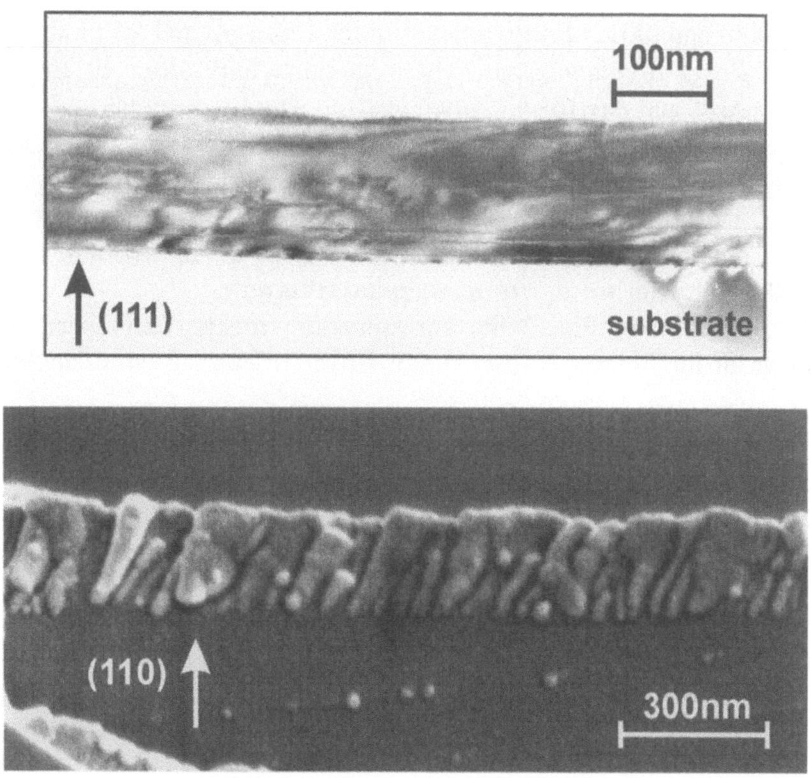

Fig. 1. a) Cross-sectional transmission electron micrograph of a 200 nm thick $CaSi_2$ layer on (111)-Si substrate; b) cross-sectional scanning electron micrograph of a 300 nm thick $CaSi_2$ layer on (110)-Si substrate

The same effect can be observed for (100) and (110) oriented substrate as well. In the case of (100)-Si the four (111) directions are equivalent, and the final layer orientation is therefore undefined. Macroscopically, a rectangular array of vertical crystallites is observed. (110) substrates, however, have a useful peculiarity. Two of the (111) directions are exactly perpendicular to the surface, causing a perpendicular orientation of the $CaSi_2$ layer as well. Figure 1b shows the structure of such a layer in a side view. It consists of parallel oriented crystallites with a tilt of about 10° relative to the exact vertical position. Tilt and probably a high concentration of dislocations are the consequences of the theoretical lattice mismatch of 19% between the layer spacing of $CaSi_2$ and the spacing of two next nearest Si-{111} layers in crystalline Si. Transforming the layer into siloxene should lead to structural destruction due to the increasing lattice constant. The presence of separated crystallites in the case of thin layers, however, enables various compensation processes so that a high crystalline quality is kept as well. We are therefore able to prepare samples with the *c*-axis of siloxene oriented parallel to the Si surface.

X-ray diffraction measurements

X-ray diffraction measurements were performed on both thin $CaSi_2$ and siloxene layers grown on (111)-Si substrates. The diffraction pattern of a 300 nm thick $CaSi_2$ layer (a similar layer is shown in Fig. 1a) is given in Fig. 2. On the right hand side, the (006) and (0012) *Bragg* reflexes of the $CaSi_2$ *tr3* and *tr6* modifications are seen in a *Bragg-Brentano* scan. These modifications differ in their stacking sequences (ABC and AABBCC) which leads to slightly different lattice constants. The observed values $c = 15.90$ Å (*tr3*) and $c = 30.63$ Å (*tr6*) are nearly identical to those of bulk $CaSi_2$ (15.98 Å and 30.6 Å, respectively) [14, 15]. The line widths of less than 0.1° show the high quality of the silicide layers. From the integrated diffraction intensity one can estimate the ratio of the volume fraction of approximately 1:1 for *tr3 vs. tr6* in an epitaxial layer.

In the left part of Fig. 2 the corresponding rocking curves are shown. A comparison of the 2Θ and Ω line widths of the *tr3* and *tr6* modifications of the

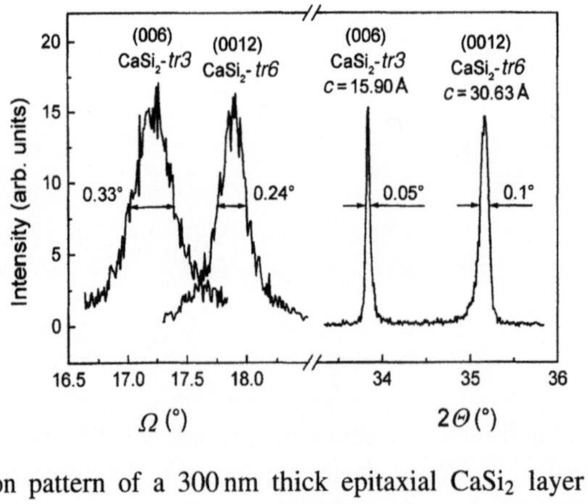

Fig. 2. X-ray diffraction pattern of a 300 nm thick epitaxial $CaSi_2$ layer on (111)-Si substrate; rocking (left) and *Bragg-Brentano* (right) curves are plotted together

epitaxial silicide shows remarkably different line widths for *tr3* and *tr6* which is probably caused by a phase separation. *Morar* and *Wittmer* have observed pure *tr6*-$CaSi_2$ in very thin epitaxial layers on (111)-Si substrates [11, 12]. We therefore assume that at first the *tr6* modification grows on the Si surface, whereas the *tr3* modification is formed above this *tr6* layer. This assumption is supported by the fact that the lattice mismatch between the *tr6* modification and Si (0.4%) is somewhat smaller than that for the *tr3* structure (−0.6%). It is therefore energetically more favorable to form *tr6*-$CaSi_2$ directly on the Si surface. The lattice mismatch causes a strain and a variation in the *c*-constant, thus leading to a broadening of the 2Θ curve. For the *tr6*-$CaSi_2$ located directly on the Si surface, the broadening is remarkably asymmetric towards higher *c*-values due to the compressive strain in the *a*-direction. On the other hand, the adjacent Si single crystal reduces the overall distortion so that the Ω-curve will be sharper. The *tr3* modification of $CaSi_2$ is further away from the Si surface and therefore less strained (towards lower *c*-values) but more distorted, thus explaining the opposite ratio of the line widths. The asymmetric shape of both 2Θ curves could also be evidence for a continuous phase mixing at the interface between the *tr3* and *tr6* silicide layers.

As expected, the crystalline quality decreases upon the topochemical reaction from silicide to siloxene. Figure 3 shows the (001) reflex of a 500 nm thick siloxene layer on a (111)-Si substrate. The 2Θ peak position at 14.25° corresponds to a layer spacing of 6.2 Å. The 2Θ line width of 1° must be taken as an upper limit because of the low resolution of 1° which was necessary due to the low signal intensity. Measurements with a better resolution showed a line width around 0.5°, which is nearly the same as observed for powder samples [16]. The asymmetric line shape indicates a distribution of layer spacings towards lower values of *c*. This may be caused by the beginning oxygen cross-linking of the siloxene layers during the measurement. On the other hand, the remarkable broadening of the rocking curve (line width 4°) is not surprising if one considers the ease of distortion in a layered compound with weak *van der Waals* interlayer bonding as in siloxene.

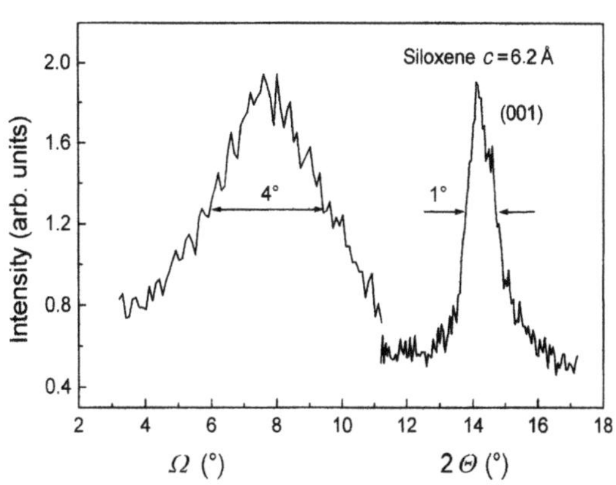

Fig. 3. X-ray diffraction pattern of a 500 nm thick siloxene layer on (111)-Si substrate; rocking (left) and *Bragg-Brentano* (right) curve are plotted together

Vibrational properties

Nearly all properties of siloxene will be anisotropic due to the layered structure of the material. In the case of optical absorption and refractive index, this has already been noted by *Kautsky* [17]. A quantitative analysis of such anisotropies, however, requires thin siloxene samples in which the *c*-axis lies in the plane of the thin crystal. Due to the weak interlayer forces in siloxene, such samples are difficult to prepare from bulk material. Here, the epitaxial layers on (110)-Si substrates can be used as a convenient alternative. As a first example of the kinds of experiments possible with these layers, we show here the results of a detailed analysis of the local vibrational modes with respect to the orientation of their dynamic dipole moment $\vec{P}$ in the crystal. Figure 4 shows the FTIR transmission spectra obtained for such a siloxene layer on a (110)-Si sample for the polarization of the infrared

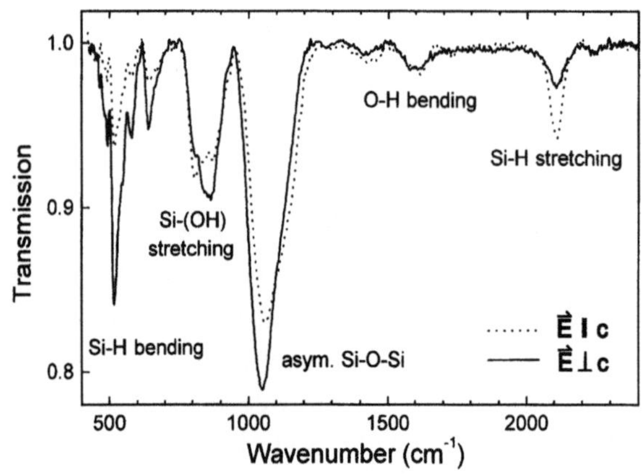

Fig. 4. FTIR transmission spectra of a 300 nm siloxene layer on (110)-Si substrate for different orientations of the polarization of the electric field vector

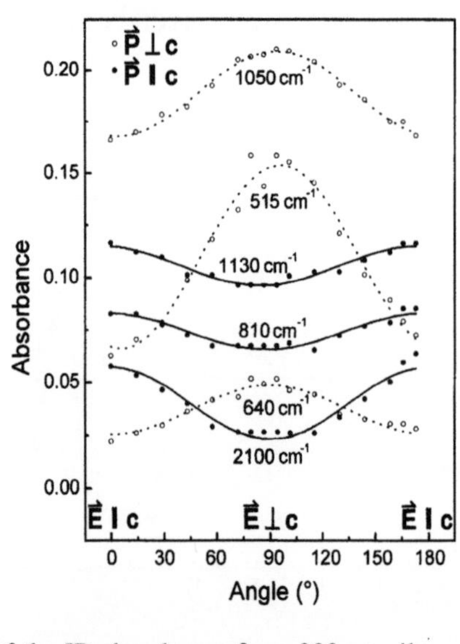

Fig. 5. Angular dependence of the IR absorbance for a 300 nm siloxene layer on (110)-Si substrate; the angle refers to the polarization of the electric field vector and the *c*-axis of the siloxene layers

electric field vector $\vec{E}$ parallel and perpendicular to the *c*-axis. Nearly all vibrations observed show a relative change in the transmission of about 50% upon change of polarization. This can be seen more quantitatively in Fig. 5 where the angular dependence of the absorbance is plotted as a function of the angle between $\vec{E}$ and c. Two sets of vibrations can be distinguished: 1130, 810, and 2100 cm^{-1} for $\vec{P} \| c$ (full circles in Fig. 5), and 1050, 515, and 640 cm^{-1} for $\vec{P} \perp c$ (open circles). Such a behaviour was certainly expected for the 2100 and 640 cm^{-1} modes which represent Si-H stretching and bending, respectively. However, the data also indicate that the 515 cm^{-1} mode linked to vibrations of the Si layers has an IR characteristic similar to the 640 cm^{-1} mode. This is due to a coupling of the Si-H bending modes and the Si phonon observed more clearly in *Raman* spectroscopy under H-D substitution. Finally, the asymmetric Si-O-Si stretching modes around 1050 and 1130 cm^{-1} indicate the presence of a significant concentration of impurity oxygen in the siloxene layer. Polarization dependent FTIR measurements show that oxygen is dominantly incorporated into the Si layer (1050 cm^{-1}); only a smaller fraction of the oxygen impurity atoms links adjacent layers (1130 cm^{-1}).

Electrical contacts

The strong photoluminescence and first indications of electroluminescence suggest that siloxene might be useful for active optical devices such as light emitting diodes. The particular structure of *Wöhler*'s siloxene prohibits conventional doping with group III or V elements as in crystalline Si. Since a pn-diode can therefore not be realized at present, we applied a technique which has been used successfully to fabricate organic LEDs [18]. The luminescent layer is packed between contact materials with highly different work functions. These contacts then work as electron and hole injectors into the thin luminescent layer. Due to the low carrier mobilities in siloxene, the siloxene layer has to be extremely thin to give the charge carriers injected at the opposite ends the possibility to undergo radiative recombination. The activation energy for dark conductivity in these structures is between 1.1 and 1.2 eV, as expected for an undoped semiconductor with a band gap energy of 2.5 eV. Similar values have been observed previously in siloxene crystals [19]. Figure 6 shows an *I vs. V* curve obtained from a 500 nm thick siloxene layer

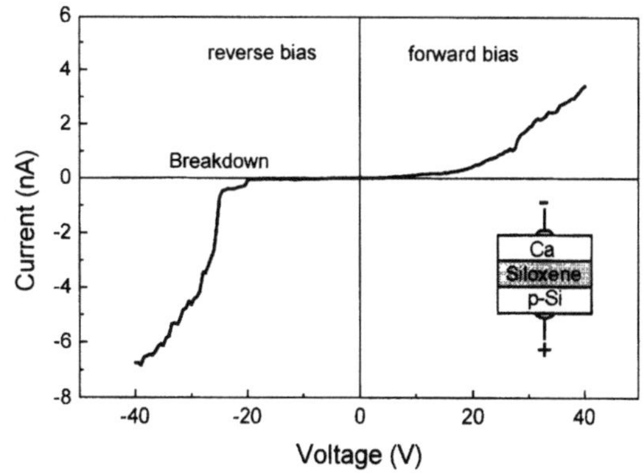

Fig. 6. I vs. V curve obtained from a 500 nm thick siloxene layer on Si; Ca and the p-type Si substrate were used as contact materials

on Si with Ca and the p-type Si substrate as contact materials. The *I–V* characteristics show a rectification typical for a diode. However, the currents flowing through these devices are still very small, and therefore the electroluminescence is barely detectable. This is in part due to the siloxene layer thickness which had to be increased to avoid pinholes short circuiting the diode. Using siloxene on (110)-Si substrates would further increase the current through the device, since the mobility parallel to the Si layers is about a factor of 1000 higher than perpendicular to them [19]. Work to realize such improved siloxene diodes is currently in progress.

Conclusions

The capability to fabricate epitaxial siloxene films with different orientations of the Si layers is an important step both with respect to further studies of the fundamental properties of this fascinating material as well as its applications. In particular, the anisotropic optical properties (polarization of photoluminescence, anisotropy of refractive index, optical activity) can now be determined quantitatively. However, both doping as well as stabilization of the material against photooxidation remain issues which have to be addressed prior to the successful realization of an efficient light emitting diode.

Experimental

Thin $CaSi_2$ films were prepared by thermal deposition of metallic Ca (Aldrich, purity >99.5%) onto heated Si substrates with different orientations ((111), (110), and (100)). The typical silicide layer thickness was several hundred nm. The layer morphology and especially the effect of different growth conditions were studied by transmission electron microscopy (TEM) and scanning electron microscopy (SEM). $CaSi_2$ was then transformed into siloxene using *Wöhler*'s method [1] (samples of 0.5 cm^2 were immersed in 5 ml 37% HCl at 0°C for 30 min). Both $CaSi_2$ and siloxene layers were examined by high resolution X-ray diffraction (XRD) for structural analysis (Philips Xpert triple axis diffractometer, Cu-K_α line). The vibrational properties were studied by IR transmission measurements using a *Fourier* transform infrared (FTIR) spectrometer (Bruker IFS 113). For transport measurements, Ca was used as the front contact on epitaxial siloxene films, whereas a p-type Si substrate was used as back electrode. To avoid the influence of ambient gases on the transport measurements, these were performed in a high vacuum.

Acknowledgements

The authors are grateful for financial support by the *Deutsche Forschungsgemeinschaft* within *Schwerpunktprogramm "Silicium-Chemie"* (Stu 134/4-2).

References

[1] Wöhler F (1863) Ann Chem Pharm **127**: 257
[2] Kautsky H, Blinoff G (1928) Z Phys Chem A **139**: 497
[3] Hengge E (1967) Fortschr Chem Forsch **9**: 145
[4] Kautsky H, Herzberg G (1924) Z Anorg All Chem **139**: 135
[5] Dahn JR, Way BM, Fuller E, Tse JS (1993) Phys Rev B **48**: 17872
[6] Weiss A, Beil G, Meyer H (1979) Z Naturforsch **34b**: 25

[7] Tung RT (1994) Epitaxial Silicides. In: Mahajan S (ed) Handbook on Semiconductors, vol 3. Elsevier, Amsterdam
[8] Tu KN, Mayer JW (1978) Silicide Formation. In: Poate JM, Tu KN, Mayer JW (eds) Thin Films – Interdiffusion and Reactions. Wiley, New York, p 393
[9] Nicolet MA, Lau SS (1983) Formation and Characterization of Transition Metal Silicides. In: Einspruch NG (ed) VLSI Electronics, vol 6. Academic Press, New York
[10] Murarka SP (1983) Silicides for VLSI Applications. Academic Press, New York
[11] Morar JF, Wittmer M (1988) J Vac Sci Technol A **6**: 1340
[12] Morar JF, Wittmer M (1988) Phys Rev B **37**: 2618
[13] Braungart R, Sigmund H (1980) Z Naturforsch **35a**: 1268
[14] Janzon KH, Schäfer H, Weiss A (1968) Z Naturforsch **23b**: 1544
[15] Evers J (1979) J Sol Stat Chem **28**: 369
[16] Dettlaff-Weglikowska U, Hönle W, Molassioti-Dohms A, Finkbeiner S, Weber J (1997) Phys Rev B **56**: 13132
[17] Kautsky H, Zocher H (1922) Z Phys **9**: 267
[18] Brown AR, Bradley DDC, Burroughes JH, Friend RH, Greenham NC, Burn PL, Holmes AB, Kraft A (1992) Appl Phys Lett **61**: 2793
[19] Brandt MS, Puchert T, Stutzmann M (1997) Solid State Commun. **102**: 365

Received June 12, 1998. Accepted (revised) July 24, 1998

Synthesis of Alkoxysilyl Substituted Cyclophosphazenes and their Properties in the Sol-Gel Process

Andreas Schneider, Stefan Kairies, and **Klaus Rose***

Fraunhofer Institute for Silicate Research, D-97082 Würzburg, Germany

Summary. Cyclophosphazenes with different alkoxysilyl substituents were synthesized as new starting compounds for material synthesis *via* the sol-gel process. Their hydrolysis was investigated by FT *Raman* spectroscopy, and the thermal stability of the hydrolyzed and condensed new cyclophosphazenes was determined *via* thermogravimetric analysis. The organopolysiloxane of the type $(N{=}P(CH_2SiO_{3/2})_2)_5$ is thermally stable up to 430°C in air. Cyclophosphazenes bearing 3-aminopropyltriethoxysilane (*APTES*) substituents were cocondensed with methacryloxypropyltrimethoxysilane, and some properties of coatings obtained from this composition were compared with the properties of a coating made from a mixture of tetraethoxysilane (*TEOS*) and methacryloxypropyltrimethoxysilane.

Keywords. Alkoxysilyl substituted cyclophosphazenes; Hydrolysis; Thermal stability; Coatings.

Synthese von alkoxysilylsubstituierten Cyclophosphazenen und ihre Eigenschaften im Sol-Gel-Prozeß

Zusammenfassung. Cyclophosphazene mit unterschiedlichen Alkoxysilylsubstituenten wurden als neue Ausgangsverbindungen für die Materialsynthese im Sol-Gel-Prozeß hergestellt. An den neuen Cyclophosphazenverbindungen wurde die Hydrolysereaktion mittels FT-*Raman*-Spektroskopie untersucht. Die thermische Stabilität der hydrolysierten und kondensierten Cyclophosphazene wurde mittels Thermogravimetrie bestimmt. Das Polysiloxan mit der Zusammensetzung $(N{=}P(CH_2SiO_{3/2})_{25}$ ist an Luft bis 430°C stabil. Cyclophosphazene mit 3-Aminopropyltriethoxysilan (*APTES*) als Substituent wurden mit Methacryloxypropyltrimethoxysilan kokondensiert. Die aus dem resultierenden Sol hergestellten Schichten wurden mit Schichten aus Tetraethoxysilan (*TEOS*) und Methacryloxypropyltrimethoxysilan verglichen.

Introduction

During the last years, sol-gel derived coatings have been well established for various applications. The sol-gel route, originally directed towards the synthesis of purely inorganic materials, is increasingly being extended to the preparation of inorganic/organic copolymers [1]. Starting from hydrolyzable molecular com-

* Corresponding author

pounds *e.g.* $RSi(OEt)_3$, formation of a siloxane network is induced by reaction with water. Polymerizable or special functional groups *R* influence the chemical and physical properties of the resulting materials to a high degree [2]. In order to increase the field of applications for this type of material, new compositions are investigated world-wide with the aim to improve and tailor the desired material properties. Since the sol-gel process requires monomeric hydrolyzable alkoxysilyl compounds, new alkoxysilyl substituted cyclophosphazenes were synthesized with the goal to incorporate these thermally stable ring systems into organically modified polysiloxanes *via* sol-gel processing [3].

Results and Discussion

Synthesis of the precursor molecules

The incorporation of cyclophospazenes into organopolysiloxanes *via* sol-gel processing affords the synthesis of monomeric hydrolyzable alkoxysilyl substituted cyclophosphazenes as precursor molecules. The easiest way to introduce an alkoxysilyl group is the reaction between hexachlorocylotriphosphazene and 3-aminopropyltriethoxysilane (*APTES*). According to a general procedure developed by *Allcock* for the synthesis of aminosubstituted cyclophosphazenes [4], 3-aminopropyltriethoxysilane was reacted with hexachlorocyclotriphosphazene in toluene.

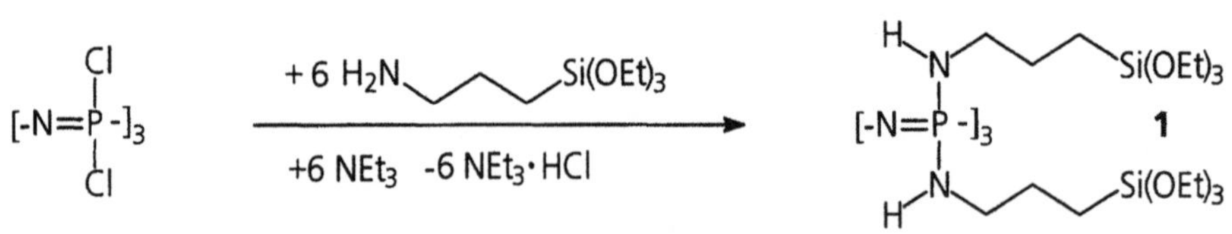

Scheme 1

This reaction provided **1** in 82% yield. In an analogous reaction using 6 equivalents of butylamine in the first step and 3 equivalents of 3-aminopropyltriethoxysilane as well as an excess of triethylamine in the second step, compound **2** was isolated as a mixture of geminal and non-geminal isomers.

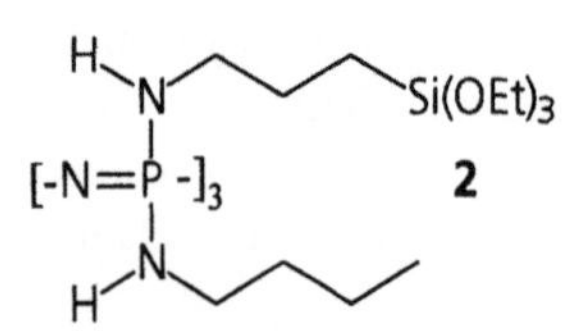

Scheme 2

Van den Grampel et al. introduced allylaryl substituents into cyclotriphosphazene for the purpose of hydrosilylation [5]. Following this route, three chlorine atoms in hexachlorocyclotriphosphazene were substituted by potassium-2-methoxy-4-allylphenolate, leading to the non-geminal product **3**. In the second step, the remaining chlorine atoms were replaced by a stoichiometric amount of sodium phenolate. Compound **4** was isolated in a yield of 49% calculated from the amount of hexachlorocyclotriphosphazene used. The substitution of all chlorine atoms is necessary due to the instability of the phosphorus-chlorine bond towards hydrolysis.

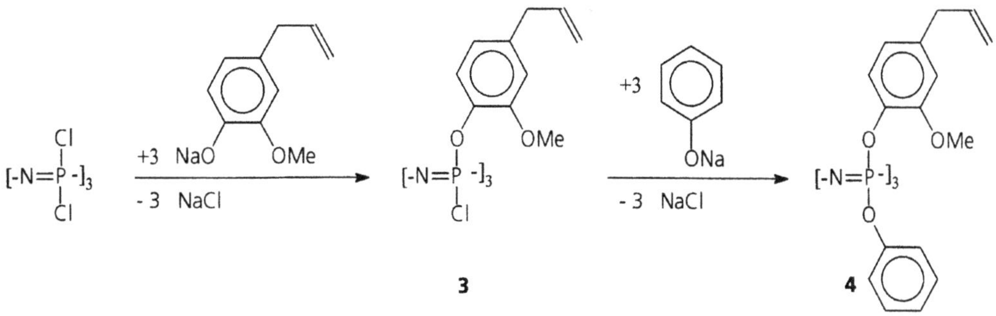

Scheme 3

Hydrosilylation of **4** with triethoxysilane under catalysis of $[H_2PtCl_6]$/2-propanol in *THF* finally afforded the silylated compound **5** in 78% yield.

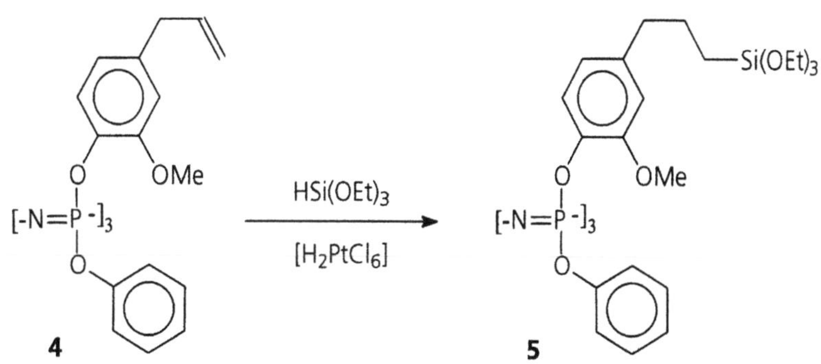

Scheme 4

In order to obtain cyclophosphazenes containing a direct phosphorus-carbon linkage without nitrogen or oxygen as heteroatom in the bridging unit between the phosphazene ring and the alkoxysilyl moiety, a different approach was necessary. The cyclophosphazene ring is formed by a thermal decomposition reaction of $(H_3C)_2BrP{=}NSi(CH_3)_3$. A modification of the pyrolysis reaction of the N-silyl-P-bromophosphinimine established by *Neilson et al.* [6] leads to decamethylcyclopentaphosphazene in a yield of 84%.

Complete or partial lithiation of decamethylcyclopentaphosphazene followed by the addition of chlorotrimethoxysilane results in the formation of the silylated compounds **6** and **7**.

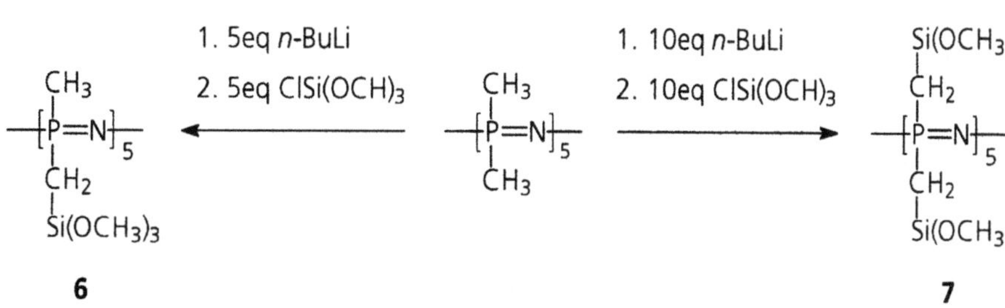

Scheme 5

FT Raman spectroscopic investigations

Raman spectroscopy has proven to be a powerful tool for the investigation of the sol-gel process [7]. *Raman* data provide information about the hydrolysis reaction of the alkoxysilanes leading to the silsesquioxane network. In Fig. 1, the *Raman* spectra of **1** during the reaction with water are shown. After starting the hydrolysis using a stoichiometric amount of water with respect to the number of hydrolyzable alkoxysilyl groups and NH_4F as catalyst, the band at $650\,cm^{-1}$ diminishes. This band can be assigned to the $\nu(SiO_3)$ stretchting vibration of the $-Si(OEt)_3$ moiety [7] and shows the decrease of the unhydrolyzed compound **1** during formation of the siloxane network. The band at $884\,cm^{-1}$ assigned to the CO-stretching vibration of ethanol [8] appears during the reaction indicating the evolving of ethanol due to hydrolysis of **1**.

For the evaluation of the reaction kinetics, the intensities of the *Raman* bands at $650\,cm^{-1}$ and $884\,cm^{-1}$ were determined and plotted against the reaction time (Fig. 2).

Generally, the decrease of the $\nu(SiO_3)$ stretchting vibration goes along with the increase of the CO-stretching of ethanol as expected. The curves show that the decrease of the $-Si(OEt)_3$ moiety is complete after 120 min, whereas EtOH formation is nearly finished after about 200 min. This phenomenon is explained by the formation of partially hydrolyzed intermediate species (RSi(OH)(OEt)$_2$ and RSi(OH)$_2$OEt) which could not be identified due to their low intensities in this *Raman* experiment, but which also eliberate ethanol.

It turned out that the hydrolysis of **1** depends strongly on the catalyst used. As can be seen from Fig. 3, only NH_4F is significantly effective. The same behaviour is observed for **2**. Therefore, NH_4F is used as a catalyst in the following experiments with **1** and **2**.

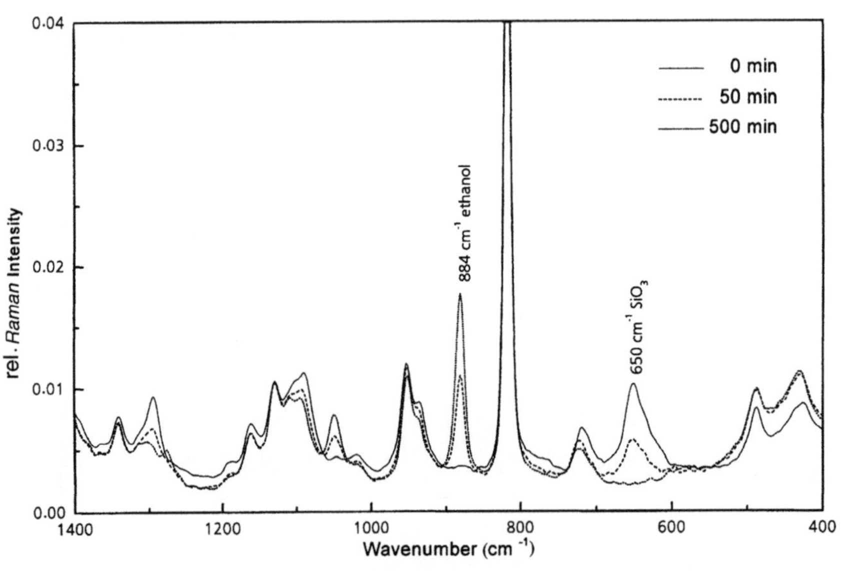

Fig. 1. *Raman* spectra of **1** during hydrolysis at 20°C with 0.1 M NH_4F as catalyst; the characteristic *Raman* bands are indicated in the plot; the plots are recorded 0 min (solid), 50 min (dashed), and 500 min (dotted) after start of the hydrolysis reaction

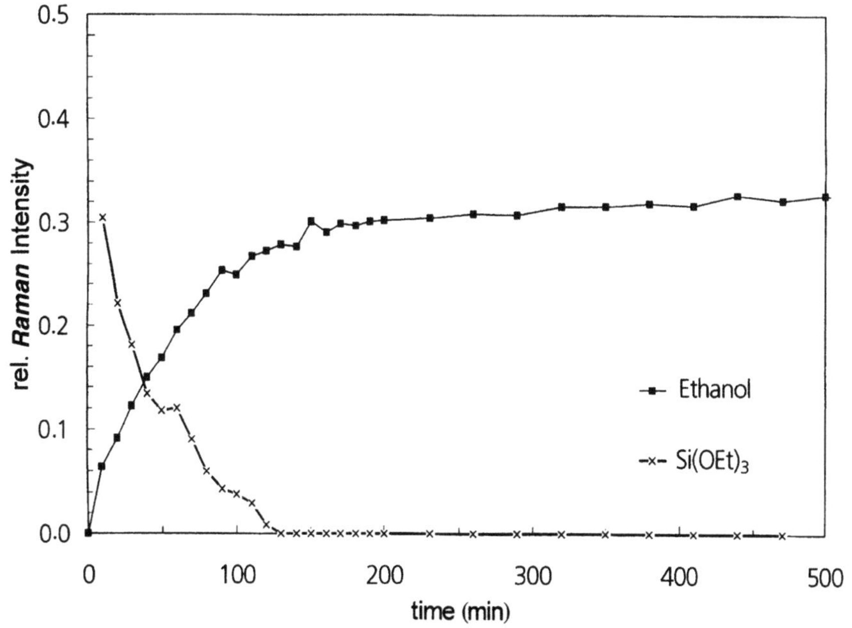

Fig. 2. Kinetic plot of the hydrolysis of 1; the curve indicated by squares represents the evolving of ethanol, the curve indicated by crosses represents the decrease of unhydrolyzed **1**

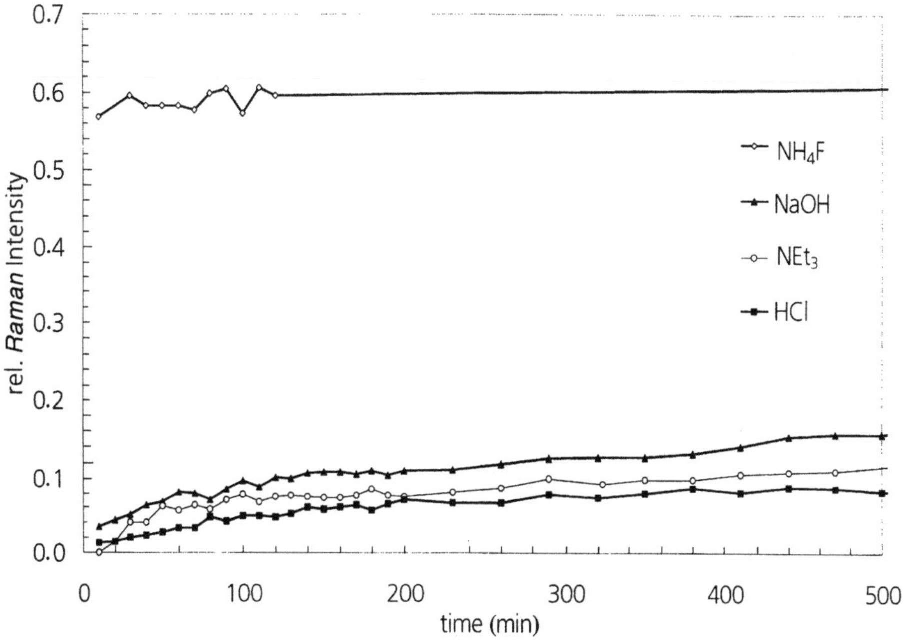

Fig. 3. Kinetic plot of EtOH formation from **1** (filled squares: catalysis with 0.1 *M* HCl, circles: 0.1 *M* NEt_3, triangles: 0.1 *M* NaOH, unfilled squares 0.1 *M* NH_4F)

A completely different behaviour was observed with **6** and **7**. The hydrolysis rate is four times faster than for **1** and **2** and independent from the catalyst used. In the present case, even the reaction without any catalyst proceeds in the same time as the catalyzed reaction. The effect is shown for **6** in Fig. 4. This behaviour is

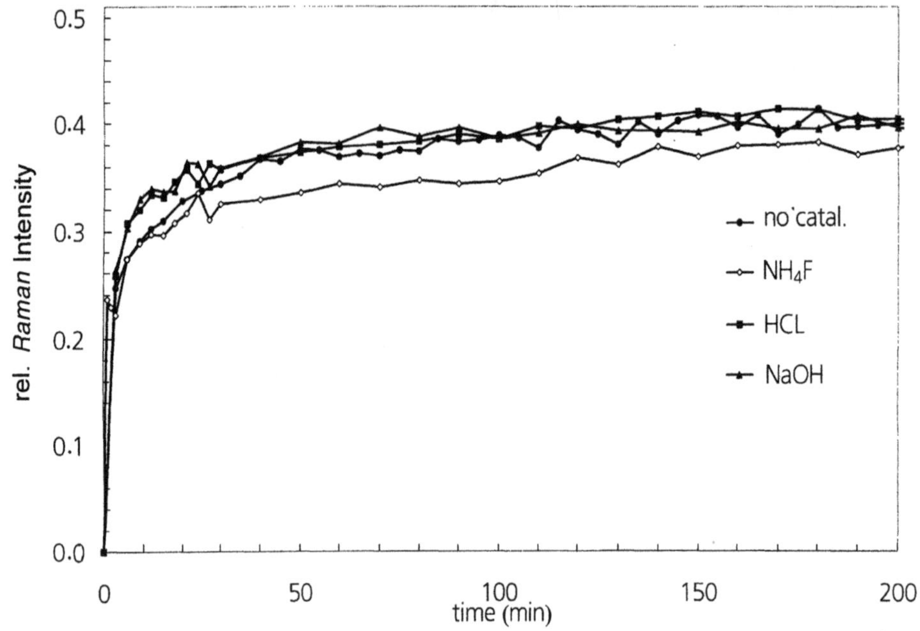

Fig. 4. Kinetic plot of EtOH formation from **6** (filled squares: 0.1 *M* HCl catalysis, circle: without catalysis, triangles: 0.1 *M* NaOH catalysis, unfilled squares: 0.1 *M* NH_4F catalysis)

partly attributed to the more reactive methoxy groups. It is also possible that a neighbouring group mechanism is operating and that the nitrogen atoms of the phosphazene ring act as a donor and a pentavalent intermediate is formed that stretches and weakens the Si–OMe bonds.

Thermal stability of cyclophosphazene modified organopolysiloxanes

The samples obtained from the hydrolysis experiments were dried for two hours at 120°C in an oven and submitted to TGA. The decomposition temperature measured in air was defined as the temperature where a weight loss of 5% had occurred [9]. The hydrolysis products derived from **1** and **2** are fairly instable and start to decompose around 220°C. Both the stabilizing influence of the phenyl group and the presence of an oxygen link to the phosphazene ring in the siloxane derivative of **5** increase significantly the stability up to 320°C. The highest thermal stability was measured for the hydrolysis products of **6** and **7** which exhibit decomposition temperatures of 345°C and 430°C, respectively. On the one hand, the direct P–C-linkage in **6** and **7** increases the stability of the resulting phosphazene siloxanes compared to derivatives with an oxygen or a nitrogen as bridging atom. On the other hand, the higher number of alkoxysilyl groups in **7** compared to **6** faciliates inorganic silicate-like crosslinking also leading to a higher thermal stability.

Coating properties

An important feature of the polysiloxane coating materials developed at the Fraunhofer Institute for Silicate Research known as ORMOCER®s (Trademark of the *Fraunhofer-Gesellschaft zur Förderung der angewandten Forschung e.V.*) is their hardness. Therefore, they are widely used as scratch resistant coatings on various substrates [10, 11]. In order to elucidate the properties of polysiloxanes containing cyclophosphazenes, the new alkoxysilanes **1** and **2** were cocondensed with methacryloxy-propyltrimethoxysilane (*MEMO*). The obtained sols were applied on glass and polycarbonate substrates and cured by UV induced polymerization of the methacrylic group using Irgacure 500® as photoinitiator. The resulting coatings were compared to a UV cured coating derived from tetraethoxysilane (*TEOS*) and *MEMO*. In preliminary investigations it turned out that the hardness of the coatings prepared from **1** (*MEMO*:**1** = 16:0.9) and **2** (*MEMO*:**2** = 16:1.8) was 256 N/mm^2 and 276 N/mm^2, respectively. Layers prepared from *MEMO* and TEOS (ratio:16:4) exhibit only 208 N/mm^2. This significant increase of hardness is remarkable since the amount of hydrolyzable and crosslinkable alkoxy groups in the sols derived from *TEOS*, **1** and **2** were equal. The higher hardness of the layers made from **1** and **2** can be attributed to the higher degree of three-dimensional inorganic crosslinking of the large silyl substituted cyclophosphazene. The higher hardness of the layer derived from **2** compared to that from **1** can be explained by the increased steric hindrance in **1** which possesses 18 alkoxy groups. This high steric hindrance will interfere with the hydrolysis reaction and on a larger scale with the condensation reaction. Thus, the formation of the silicate-like network is limited. This assumption is manifested by solid state ^{29}Si NMR experiments. The ^{29}Si CP/MAS spectra of the hydrolysis and condensation products derived from **1** show a smaller intensity of the T^3-peak than the spectra of the product derived from **2**. Therefore, the inorganic crosslinking in products derived from **2** is higher than in products derived from **1**.

Experimental

^1H, ^{13}C, and ^{31}P NMR spectra were recorded using a Bruker AMX 400 NMR spectrometer. Solid state ^{29}Si NMR spectra were obtained a Bruker DSX 400 instrument. IR spectra were measured using a Nicolet FTIR spectrometer Magna 750. *Raman* spectra were recorded in a tempered *Raman* cell attached to a Bruker RFS100 *Raman* spectrometer. Mass spectra were performed on a Finnigan MAT 8200. All reactions for the synthesis of phosphazene monomers were carried out under an atmosphere of dry argon in oven dried glassware. Hydrolysis reactions were conducted in ambient atmosphere. To the synthesized polysiloxanes 1% (w/w) Irgacure 500® (Ciba Geigy) was added, and the coatings solutions were applicated *via* knife coating on a glass or polycarbonate substrate and UV cured within 20 seconds. The universal hardness was determined by microindentation with a Fischerscope H100.

$(NP[NHCH_2CH_2CH_2Si(OEt)_3]_2)_3$ (**1**; $C_{54}H_{132}N_9O_{18}P_3Si_6$)

6.64 g (30 mmol) 3-aminopropyltriethoxysilane were added dropwise at 0°C to a solution of 1.78 g (5.0 mmol) hexachlorocyclotriphosphazene and 3.34 g (33 mmol) triethylamine in 100 ml toluene. After the addition, the reaction mixture was heated to reflux for 24 h. Filtration under argon followed by solvent removal left 5.97 g (82%) **1** as colourless viscous liquid.

^1H NMR ($CDCl_3$, δ, 400 MHz): 0.73 (t, CH_2Si, 12H), 1.14 (t, OCH_2CH_3, 54H), 1.75 (m, $CH_2CH_2CH_2$, 12H), 3.07 (t, NCH_2, 12H), 3.75 (q, OCH_2CH_3, 36H) ppm; ^{13}C NMR ($CDCl_3$, δ, 100 MHz): 8.09 (CH_2Si), 18.5 (OCH_2CH_3), 26.6 ($CH_2CH_2CH_2$), 44.7 (NCH_2, 12H), 58.4

(OCH_2CH_3) ppm; ^{31}P NMR ($CDCl_3$, δ, 162 MHz): 17.9 ppm; IR (Film): $\nu = 1183–1195\,cm^{-1}$ (P=N).

(NP[NHCH₂CH₂CH₂Si(OEt₃)][NHCH₂CH₂ CH₂CH₃])₃ (**2**; $C_{39}H_{96}N_9O_9P_3Si_3$)

6.32 g (86.4 mmol) *n*-butylamine were added dropwise at 0°C to a solution of 5.0 g (14.4 mmol) hexachlorocyclotriphosphazene in 100 ml toluene. After the addition, the reaction mixture was heated to reflux for 24 h. The solution was decanted from the precipitate. 9.56 g (43.2 mmol) 3-aminopropyltriethoxysilane were added to the obtained solution. The resulting mixture was refluxed for 30 h. Filtration under argon followed by solvent removal left 12.7 g (82%) **2** as a mixture of isomers. **2** was a colourless highly viscous liquid.

^{1}H NMR ($CDCl_3$, δ, 400 MHz): 0.73 (t, CH_2Si, 6H), 0.93 (t, $CH_2CH_2CH_2CH_3$, 9H), 1.14 (t, OCH_2CH_3, 27H), 1.63 (m, $CH_2CH_2CH_2CH_3$, 6H), 1.71 (m, $CH_2CH_2CH_2CH_3$, 6H), 1.75 (m, $CH_2CH_2CH_2$, 6H), 3.07 (t, NCH_2, 5H), 3.11–3.19 (m, (($NHCH_2$)($NHCH_2CH_2CH_2CH_3$)), 12H), 3.26 (t, $NHCH_2CH_2CH_2CH_3$), 3.75 (q, OCH_2CH_3, 18H) ppm; ^{13}C NMR ($CDCl_3$, δ, 100 MHz): 8.09 (CH_2Si), 10.2 ($CH_2CH_2CH_2CH_3$), 18.9 (OCH_2CH_3), 22.5 ($CH_2CH_2CH_2CH_3$), 25.7 ($CH_2CH_2CH_2CH_3$), 26.6 ($CH_2CH_2CH_2$),.44.7 (NCH_2), 58.4 (OCH_2CH_3) ppm; ^{31}P NMR ($CDCl_3$, δ, 162 MHz): 14.2 (d, P(NHBu)$_2$, $J_{PNP} = 22$ Hz), 16.3 (dd, P(NHBu)(*AMEO*), $J_{PNP} = 17$ Hz, 22 Hz), 18.4 (dd, P((*AMEO*)$_2$, $J_{PNP} = 17$ Hz, 22 Hz) ppm; IR (Film): $\nu = 1192–1196\,cm^{-1}$ (P=N).

(NP[O(C₆H₃)(OMe)CH₂CHCH₃]Cl)₃ (**3**; $C_{30}H_{33}Cl_3N_3O_6P_3$)

A solution prepared from 9.85 g (60 mmol) 4-allyl-2-methoxyphenol and 1.44 g (60 mmol) sodium hydride in 100 ml *THF* was added dropwise at 0°C to 6.96 g (20.0 mmol) hexachlorocyclotriphosphazene in 60 ml *THF*. The reaction mixture was stirred for 12 h at 20°C and subsequently heated to reflux for 4 h. The solvent was removed *in vacuo*. The obtained liquid was dissolved in 50 ml diethylether, washed with water, dried over Na_2SO_4, and the solvent was removed *in vacuo*. Traces of 4-allyl-2-methoxyphenol were removed at 120°C (0.4 mbar), and 10.1 g (13.8 mmol) of **3** was isolated as a slightly yellow solid.

^{1}H NMR ($CDCl_3$, δ, 400 MHz): 3.36 (d, $ArCH_2$, 6H), 3.87 (s, OCH_3, 9H), 5.11 (m, $=CH_2$, 6H), 5.85 (m, CH_2CHCH_2, 3H), 7.03 (m, ArH, 9H) ppm; ^{31}P NMR ($CDCl_3$, δ, 162 MHz): 13.3 (P(OAr)(Cl)) ppm; IR (CCl_4): $\nu = 1235–1242$ (P=N) cm^{-1}.

(NP[O(C₆H₃)(OMe)CH₂CHCH₂]OPh)₃ (**4**; $C_{45}H_{42}N_3O_6P_3$)

A solution prepared from 3.78 g (40.2 mmol) phenol and 0.96 g (40.2 mmol) sodium hydride in 50 ml *THF* was added to a solution of 8.60 g **3** in 50 ml THF. The mixture was heated to reflux for 24 h. *THF* was removed *in vacuo*. 30 ml water and 1 ml 0.1 *M* HCl were added to the crude product. The aqueous solution was extracted with $CHCl_3$. The combined organic layers were dried over Na_2SO_4, and the solvent was removed *in vacuo*. The liquid was dissolved in 20 ml pentane and purified *via* flash chromatography (silica gel 50, 30–70 mesh, petroleum 40–60:ethylacetate = 4:1). After evaporation of the solvent *in vacuo*, 7.74 g (71%) of **4** were obtained as a colourless viscous liquid.

^{1}H NMR ($CDCl_3$, δ, 400 MHz): 3.34 (d, $ArCH_2$, 6H), 4.97 (m, $=CH_2$, 6H), 5.72 (m, CH_2CHCH_2, 3H), 7.22–7.28 (m, ArH, 27H) ppm; ^{31}P NMR ($CDCl_3$, δ, 162 MHz): 8.82 (P(OAr)(OPh)) ppm; IR (CCl_4): $\tilde{\nu} = 1235–1242, 1254–1272$ (P=N) cm^{-1}.

(NP[O(C₆H₃)(OMe)CH₂CH₂CH₂Si(OEt)₃]OPh)₃ (**5**; $C_{63}H_{90}N_3O_{15}Si_3$)

4.53 g (5.00 mmol) **4** were dissolved in 15 ml *THF* and 0.5 ml of a 10^{-6} *M* solution of H_2PtCl_6 in 2-propanol were added. After stirrig for 1 h at 20°C, 2.4 g (15.0 mmol) triethoxysilane were added and heated to reflux for 24 h. Volatile substances were removed *in vacuo*. The crude yellow product was dispersed in 30 ml heptane and filtered over a 3 cm layer of silica gel 60. The filtrate was concentrated to provide 5.45 g (78%) **5** as a colourless liquid.

^{1}H NMR ($CDCl_3$, δ, 400 MHz): 0.59 (t, CH_2Si, 6H), 1.13 (t, OCH_2CH_3, 27H), 1.53 (m, $CH_2CH_2CH_2$, 6H), 3.32 (t, $ArCH_2$, 6H), 3.81 (q, OCH_2CH_3, 18H), 3.89 (s, OCH_3, 9H) 7.22–7.28 (m, ArH, 24H) ppm; ^{31}P NMR ($CDCl_3$, δ, 162 MHz): 8.93 (P(OAr)(OPh)) ppm; IR (CCl_4): $\tilde{\nu}$ = 1235–1242, 1262–1274 (P=N) cm^{-1}.

$[NP(CH_3)_2]_5$ ($C_{10}H_{30}N_5P_5$)

32.7 g (143 mmol) P-bromo-P,P-dimethyl-N-silylphosphinimine were heated for 2 h to 160°C in a round bottom flask equipped with a reflux condenser and a liquid separator. The trimethylbromosilane formed during the reaction was continuously removed *via* the liquid seperator. To complete the reaction, the temperature was slowly rised to 200°C. After 5 h the reaction was completed. Sublimation of the obtained white needles yielded 9.02 g (84%) $[NP(CH_3)_2]_5$.

^{1}H NMR ($CDCl_3$, δ, 400 MHz): 1.45 ($P(CH_3)_2$, $^3J_{PH}$ = 12 Hz) ppm; ^{31}P NMR ($CDCl_3$, δ, 162 MHz): 19.25 ($P(CH_3)_2$) ppm; El-MS: *m/z* = 375.3 (M^+, 19%), 360.1 (M^+-CH_3, 57%), 300.0 (M^+-$NP(CH_3)_2$, 58%), 285.1 (M^+-$NP(CH_3)_2$-CH_3, 100%), 225.1 (M^+-2 $NP(CH_3)_2$, 5%), 210.1 (M^+-2 $NP(CH_3)_2$-CH_3, 57%).

$(NP[CH_3][CH_2Si(OMe)_3])_5$ (**6**; $C_{25}H_{70}N_5O_{15}P_5Si_5$)

40 ml of a 2 *M* solution of *n*-BuLi in heptane were added dropwise to an ice-cooled solution of 3.00 g (8.00 mmol) decamethylcyclopentaphosphazene in 80 ml *THF*. After stirring for 12 h, this reaction mixture was added to a vigorously stirred solution of 8.50 g (42.0 mmol) chlorotrimethoxysilane in 100 ml *THF* at a temperature of −78°C. The mixture was warmed up to room temperature, stirred for a period of 12 h, and then filtered. The solvent and unreacted chlorotrimethoxysilane were removed *in vacuo*. The crude product was dissolved in 80 ml dichloromethane and filtered over a 2 cm layer of silica gel 60. After evaporation of the solvent, 11.5 g (72%) **6** were obtained as a colourless viscous liquid.

^{1}H NMR ($CDCl_3$, δ, 400 MHz): 0.96 (s, PCH_3, 21H), 1.27 (br, PCH_2Si, 6H), 3.38 (s, OCH_3, 18H) ppm; ^{13}C NMR ($CDCl_3$, δ, 100 MHz): 6.31 (PCH_3) 9.09 (CH_2Si), 56.2 (OCH_3) ppm; ^{31}P NMR ($CDCl_3$, δ, 162 MHz): 25.2 (PCH_3CH_2Si) ppm; IR (KBr): $\tilde{\nu}$ = 1187 (P=N) cm^{-1}.

$(NP[CH_2Si(OMe)_3]_2)_5$ (**7**; $C_{40}H_{105}N_5O_{30}P_5Si_{10}$)

7 was prepared analogous to **6** but using twofold amounts of *n*-BuLi and chlorotrimethyoxysilane.

Yield: 72%; ^{1}H NMR ($CDCl_3$, δ, 400 MHz) 1.42 (br, PCH_2Si, 20H), 3.23 (s, OCH_3, 90H) ppm; ^{13}C NMR ($CDCl_3$, δ, 100 MHz): 9.09 (CH_2Si), 56.2 (OCH_3)ppm; ^{31}P NMR ($CDCl_3$, δ, 162 MHz): 25.2 ($P(CH_2Si)_2$) ppm; IR (KBr): $\tilde{\nu}$ = 1187 (P=N) cm^{-1}.

Hydrolysis experiments for Raman investigations

General procedure

If not mentioned otherwise, 1 g of the alkoxysilyl substituted cyclophosphazene was dissolved in 1 g of solvent (ethylacetate or 2-propanol). A stoichiometric amount of water (calculated by the number of alkoxy groups) or instead a 0.1 *M* or 1 *M* aqueous solution of a catalyst was added, and the mixture was stirred in a cell inside the *Raman* spectrometer.

Hydrolysis of **5**

3.41 g (2.44 mmol) **5** were dissolved in 20 ml ethyl acetate; 0.40 g 0.1 *M* HCl were added, and the mixture was stirred for 12 h at 20°C.

Synthesis of the coating sols

Sol 1 prepared from *MEMO* and *TEOS*:

1.16 g (64.5 mmol) 0.1 *M* aqueous solution of NH_4F were added to an ice-cooled solution of 4 g (16.1 mmol) methacryloxypropyltrimethoxysilane (*MEMO*) and 0.84 g (4.03 mmol) *TEOS* in 14.4 g ethyl acetate. After 10 min the ice-bath was removed and the mixture was stirred for another 2 h.

Sol 2 prepared from MEMO and **1**:

1.16 g (64.5 mmol) 0.1 *M* aqueous solution of NH_4F were added to an ice cooled solution of 4 g (16.1 mmol) methacryloxypropyltrimethoxysilane (*MEMO*) and 1.30 g (0.89 mmol) **1** in 14.4 g ethyl acetate. After 10 min the ice-bath was removed and the mixture was stirred for another 2 h.

Sol 3 prepared from MEMO and **2**:

To an ice cooled of 4 g (16.1 mmol) methacryloxypropyltrimethoxysilane (*MEMO*) and 1.81 g (1.79 mmol) **2** in 7.2 g ethyl acetate and 7.2 g ethanol, 1.16 g (64.5 mmol) 0.1 *M* aqueous solution of NH_4F were added. After 10 min the ice-bath was removed and the mixture was stirred for another 2 h.

Finally, photoinitiator was added to the sols, and the sols were applied on the substrates as described above.

Acknowledgements

Financial support from the *Deutsche Forschungsgemeinschaft* is gratefully acknowledged.

References

[1] Novak BM (1993) Adv Mat **5**: 422
[2] Schubert U, Hüsing N, Lorenz A (1995) Chem Mater **7**: 2010
[3] Kairies S, Rose K (1998) In: Auner N, Weis J (eds) Organosilicon Chemistry III. Wiley-VCH, Weinheim, p 543
[4] Allcock HR (1970) Phosphorus-Nitrogen-Compounds, Academic Press; Allcock HR, Kugel HR (1966) Inorg Chem **5**: 1716
[5] Puyenbroek R, Jekel AP, van den Grampel JC (1991) J Inorg Organomet Polym **1**: 105
[6] Wisian-Neilson P, Neilson RH (1980) Inorg Chem **19**: 1875
[7] Posset U, Lankers M, Kiefer W, Steins H, Schottner G (1993) Appl Spectrosc **47**: 1600
[8] Dollish FR, Fateley WG, Bentley FF (1974) Characteristic Raman Frequencies of Organic Compounds. Wiley, New York
[9] Cassidy PE (1979) Thermally Stable Polymers. Dekker, New York, p 21
[10] Amberg-Schwab S, Arpac E, Glaubitt W, Rose K, Schottner G, Schubert U (1991) In: Vincenzini P (ed) High Performance Films and Coatings. Elsevier, Amsterdam, p 203
[11] Greiwe K, Glaubitt W, Amberg-Schwab S, Piana K (1992) Mat Res Soc Symp Proc **271**: 725

Received June 12, 1998. Accepted (revised) September 30, 1998

A Zwitterionic λ^5Si-Silicate with an Almost Ideal Square-Pyramidal Si Coordination Polyhedron: Synthesis and Crystal Structure Analysis

Reinhold Tacke*, **Joachim Heermann**, **Melanie Pülm**, and **Elmar Gottfried**

Institut für Anorganische Chemie, Universität Würzburg, D-97074 Würzburg, Germany

Summary. The pentacoordinate silicon compound ((morpholinio)methyl)*bis*(3,4,5,6-tetrachlorobenzene-1,2-diolato(2−))silicate–acetonitrile ($\mathbf{4}\cdot CH_3CN$), a zwitterionic spirocyclic λ^5Si-silicate, has been synthesized and structurally characterized by single-crystal X-ray diffraction analysis. Compound $\mathbf{4}\cdot CH_3CN$ was prepared by reaction of the silanes $(MeO)_3SiCH_2NR_2$, $Ph(MeO)_2$-$SiCH_2NR$, or $Me(MeO)_2SiCH_2NR_2$ (NR_2 = morpholino) with two mol equivalents of 3,4,5,6-tetrachlorobenzene-1,2-diol in acetonitrile at room temperature. The Si coordination polyhedron of $\mathbf{4}\cdot CH_3CN$ in the crystal is a nearly ideal square pyramid, the basal positions being occupied by the four oxygen atoms of the two diolato(2−) ligands. Significant differences between the isotropic ^{29}Si chemical shifts of $\mathbf{4}\cdot CH_3CN$ in the crystal ($\delta = -85.1$) and in solution ($\delta = -131.9$, *DMSO*-d_6) indicate special structural features in solution.

Keywords. Pentacoordinate silicon; Zwitterionic λ^5Si-silicate; Crystal structure analysis; Solid-state ^{29}Si NMR; Solution-state ^{29}Si NMR.

Ein zwitterionisches λ^5Si-Silicat mit einem nahezu idealen quadratisch-planaren Si-Koordinationspolyeder: Synthese und Kristallstrukturanalyse

Zusammenfassung. Die pentakoordinierte Siliciumverbindung ((Morpholinio)methyl)*bis*(3,4,5,6-tetrachlorbenzol-1,2-diolato(2−))silicat–Acetonitril ($\mathbf{4}\cdot CH_3CN$), ein zwitterionisches spirocyclisches λ^5Si-Silicat, wurde synthetisiert und mittels Einkristallröntgenstrukturanalyse strukturell charakterisiert. Verbindung $\mathbf{4}\cdot CH_3CN$ wurde durch Reaktion der Silane $(MeO)_3SiCH_2NR_2$, $Ph(MeO)_2SiCH_2NR_2$ oder $Me(MeO)_2SiCH_2NR_2$ (NR_2 = Morpholino) mit zwei Mol-Äquivalenten 3,4,5,6-Tetrachlorbenzol-1,2-diol in Acetonitril bei Raumtemperatur dargestellt. Das Si-Koordinationspolyeder von $\mathbf{4}\cdot CH_3CN$ im Kristall ist eine nahezu ideale quadratische Pyramide, wobei die vier Sauerstoffatome der beiden Diolato(2−)-Liganden die basalen Positionen besetzen. Drastische Unterschiede in den isotropen ^{29}Si-chemischen Verschiebungen von $\mathbf{4}\cdot CH_3CN$ im Kristall ($\delta = -85.1$) und in Lösung ($\delta = -131.9$, *DMSO*-d_6) weisen auf Besonderheiten der Struktur in Lösung hin.

* Corresponding author

Introduction

In a series of recent papers, we have reported on the syntheses and crystal structures of some zwitterionic (ammoniomethyl)*bis*(benzene-1,2-diolato(2−))silicates [1], such as compounds **1** [1e,i] and **2** [1e,i]. Although most of the Si coordination polyhedra of pentacoordinate silicon compounds of this particular formula type have shown to be more or less distorted trigonal bipyramids, several intermediate structures between the ideal trigonal bipyramid and the ideal square pyramid have been observed [1].

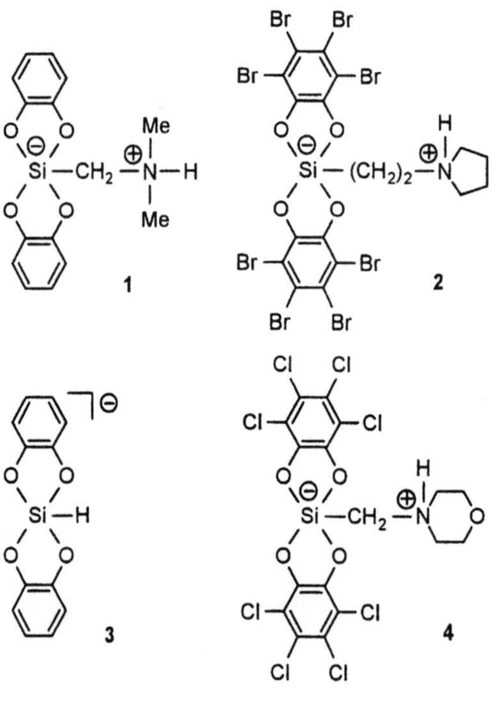

1 **2** **3** **4**

This is in good accordance with the results of *ab initio* studies with the anionic model species **3**. As shown by SCF/SVP geometry optimizations, the trigonal-bipyramidal structure of **3** (C_2 symmetry) is only 5.82 kJ mol^{-1} more stable than the alternative square-pyramidal structure (C_s symmetry) [2]. Thus, it is likely to assume that the geometries of the Si coordination polyhedra of the above-mentioned zwitterionic λ^5*Si*-silicates in the crystal are mainly governed by the crystal packing, including hydrogen bonds.

We have now succeeded in preparing a zwitterionic λ^5*Si*-silicate with an almost ideal square-pyramidal Si coordination polyhedron. We report here on the synthesis and crystal structure of ((morpholinio)methyl)*bis*(3,4,5,6-tetrachlorobenzene-1,2-diolato(2−))silicate–acetonitrile (**4** · CH_3CN). The studies presented here were carried out as a part of our systematic investigations on zwitterionic λ^5*Si*-silicates (in this context, see Ref. [3]; for reviews dealing with pentacoordinate silicon compounds, see Ref. [4]).

Results and Discussion

Syntheses

The zwitterionic λ^5Si-silicate **4** was synthesized by reaction of the silanes **5** [1h], **6** [1h], or **7** [5] with two mol equivalents of 3,4,5,6-tetrachlorobenzene-1,2-diol (Scheme 1). All syntheses were carried out in acetonitrile at room temperature. The first method (**5**→**4**) involves cleavage of three Si–O bonds (formation of three mol equivalents of methanol), whereas the second (**6**→**4**) and third method (**7**→**4**) are based on the cleavage of two Si–O bonds and one Si–C bond (formation of two mol equivalents of methanol and one mol equivalent of benzene and methane, respectively).

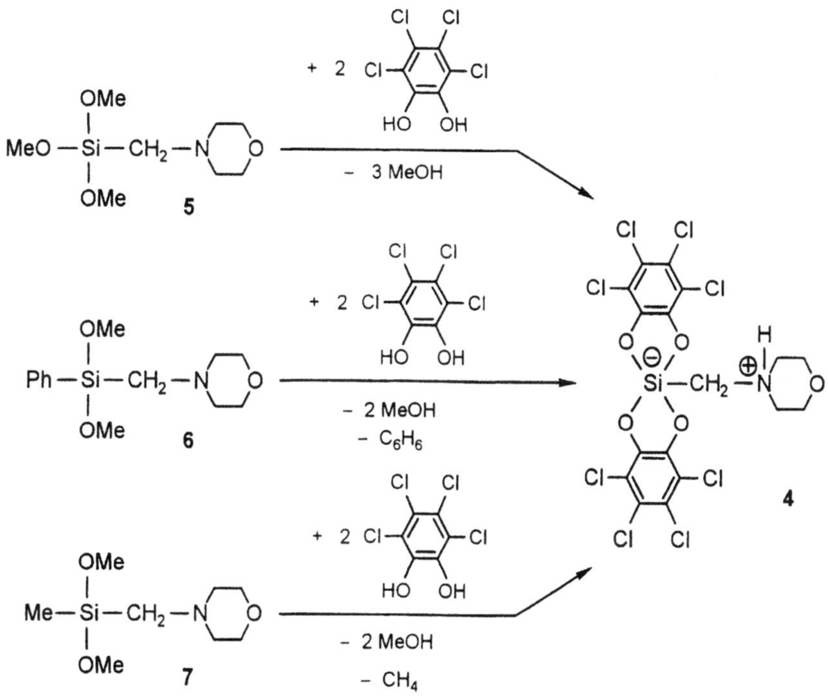

Scheme 1

The zwitterionic λ^5Si-silicate **4** was isolated as the crystalline acetonitrile solvate **4**·CH_3CN (yield 77% (method 1), 33% (method 2), 60% (method 3)). Its identity was established by elemental analyses (C, H, N), solution-state (^1H, ^{13}C, ^{29}Si) and solid-state (^{29}Si CP/MAS) NMR experiments, and mass spectrometric studies. In addition, compound **4**·CH_3CN was structurally characterized by single-crystal X-ray diffraction.

Crystal structure analysis

Compound **4**·CH_3CN crystallizes in the space group $P\bar{1}$, the unit cell containing two zwitterions **4** and two acetonitrile molecules. The crystal data and experimental parameters used for this study are given in Table 1 (for further

Table 1. Crystal data and experimental parameters for the crystal structure analysis of **4** · CH_3CN

Empirical formula	$C_{19}H_{14}Cl_8N_2O_5Si$
Formula mass ($g \cdot mol^{-1}$)	662.01
Collection temperature (K)	133(2)
$\lambda(MoK_\alpha)$ (Å)	0.71073 Å
Crystal system	triclinic
Space group (no.)	$P\bar{1}$(2)
a (Å)	10.0594(1)
b (Å)	11.1881(1)
c (Å)	11.7878(1)
α (deg)	108.755(1)
β (deg)	95.931(1)
γ (deg)	96.621(1)
V (Å^3)	1233.86(2)
Z	2
D(calcd.) ($g \cdot cm^{-3}$)	1.782
μ (mm^{-1})	0.999
F(000)	664
Crystal dimensions (mm)	0.8×0.5×0.5
2θ range (deg)	3.70–56.62
Index ranges	$-13 \leq h \leq 13$, $-14 \leq k \leq 14$, $0 \leq l \leq 15$
Reflections collected	18088
Independent reflections	5917
R_{int}	0.0210
Max/min transmission	0.6350/0.5021
Reflections used	5917
Parameters	322
S^a	1.048
Weight parameters a/b^b	0.0383/0.5661
$R1^c(I > 2\sigma(I))$	0.0267
$wR2^d$ (all data)	0.0716
Extinction coefficient	0.0082(8)
Max/min residual electron density (e · Å^{-3})	+0.462/−0.322

$^a S = (\Sigma(w(F_o^2 - F_c^2)^2)/(n-p))^{0.5}$, n = no. of reflections, p = no. of parameters; $^b w^{-1} = \sigma^2(F_o^2) + (aP)^2 + bP, P = (F_o^2 + 2F_c^2)/3$; $^c R1 = \Sigma(||F_o| - |F_c||)/\Sigma|F_o|$; $^d wR2 = (\Sigma(w(F_o^2 - F_c^2)^2)/\Sigma(w(F_o^2)^2))^{0.5}$

Table 2. Selected interatomic distances (Å) and angles (deg)

Si–O1	1.7513(10)	Si–O4	1.7421(10)
Si–O2	1.7309(10)	Si–C1	1.898(2)
Si–O3	1.7455(10)		
O1–Si–O2	88.93(5)	O2–Si–O4	152.54(5)
O1–Si–O3	157.76(5)	O2–Si–C1	105.43(5)
O1–Si–O4	86.43(5)	O3–Si–O4	89.27(5)
O1–Si–C1	105.49(6)	O3–Si–C1	96.75(6)
O2–Si–O3	84.88(5)	O4–Si–C1	101.89(5)

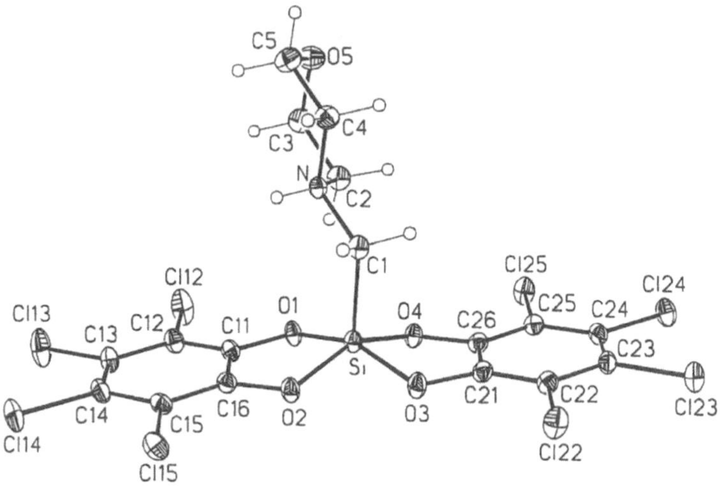

Fig. 1. Molecular structure of **4** in the crystal of **4**$\cdot CH_3CN$ (probability level of displacement ellipsoids 50%), showing the atomic numbering scheme

details, see Experimental). The molecular structure of **4** in the crystal of the acetonitrile solvate **4**$\cdot CH_3CN$ is depicted in Fig. 1. Selected interatomic distances and angles are listed in Table 2.

The coordination polyhedron around the silicon atom of **4** can be described as a nearly ideal square pyramid, the basal positions being occupied by the four oxygen atoms of the two 3,4,5,6-tetrachlorobenzene-1,2-diolato(2−) ligands. The two chelating ligands are almost coplanar, with the central silicon atom deviating from this plane by 0.38 Å.

As shown in Fig. 2 and 3, the zwitterions and acetonitrile molecules are connected by intermolecular N–H···N hydrogen bonds (N–H, 0.82(2) Å; H···N1A, 2.10(2) Å; N···N1A, 2.879(2) Å; N–H···N1A, 157(2)°). Obviously, the crystal packing of **4**$\cdot CH_3CN$ is mainly governed by intermolecular interactions between the almost coplanar aromatic 3,4,5,6-tetrachlorobenzene-1,2-diolato(2−) ligands of the zwitterions (mean distance between the two planes 3.37 Å, maximum deviation 0.04 Å), leading to a layer structure (Fig. 2). The acetonitrile molecules are also arranged almost coplanar to these layers. As can be seen from Fig. 3, the intermolecular interactions between the diolato(2−) ligands of two zwitterions leads to the formation of centrosymmetric dimeric building blocks in the crystal. It is likely to assume that this particular intermolecular interaction is mainly responsible for the existence of the energetically less favorable square-pyramidal Si coordination polyhedron of **4** in the crystal of **4**$\cdot CH_3CN$.

NMR studies

The solution-state ^{1}H and ^{13}C NMR data of **4**$\cdot CH_3CN$ (solvent: *DMSO*-d_6) are compatible with the existence of the zwitterion **4** in solution. However, the results of the solid-state and solution-state ^{29}Si NMR studies are indicative of special structural features in solution. The isotropic ^{29}Si chemical shifts observed in the

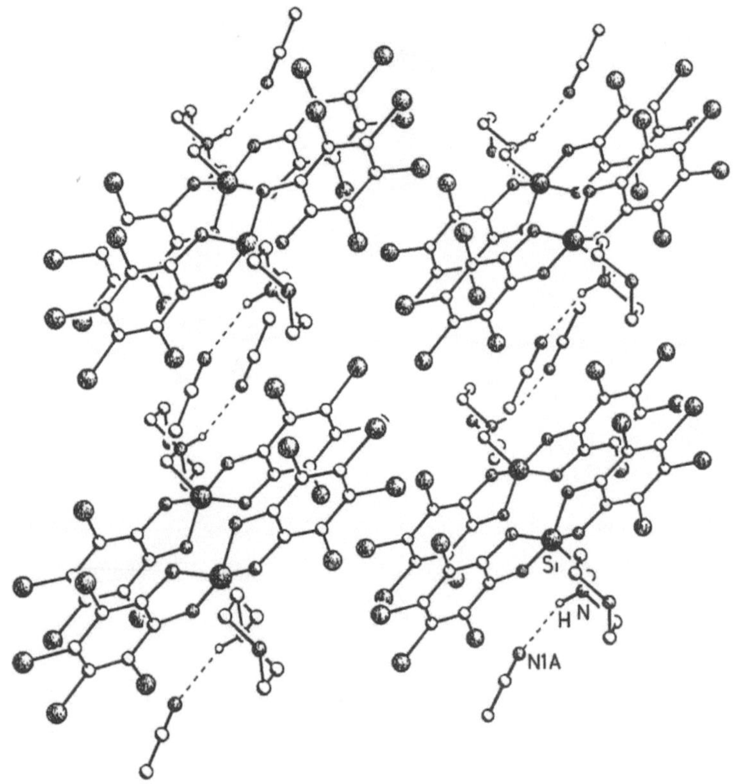

Fig. 2. Cell plot for **4** · CH_3CN in view along [001], showing the layer structure in the crystal lattice; hydrogen atoms (except NH) are omitted for clarity

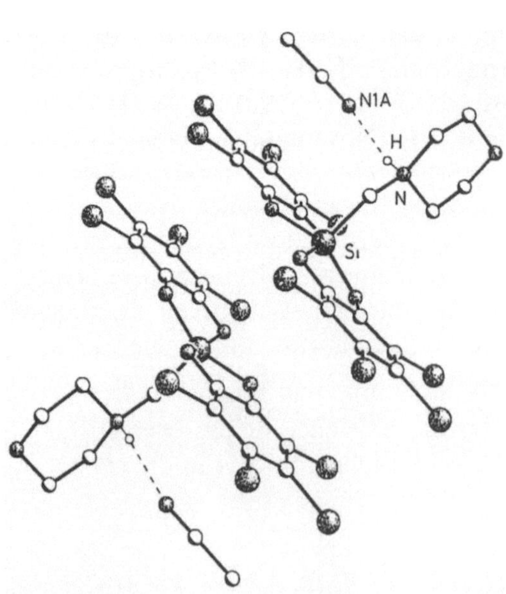

Fig. 3. Centrosymmetric dimeric building block in the crystal of **4** · CH_3CN; the mean distance between the two planes formed by the two almost coplanar diolato(2–) ligands of each zwitterion amounts to 3.37 Å

crystal ($\delta = -85.1$) and in solution ($\delta = -131.9$) differ significantly from each other, whereas the respective solid-state and solution-state ^{29}Si NMR data for compounds **1** and **2** are very similar. This remarkable discrepancy might be explained in terms of hexacoordination of the silicon atom of **4** in solution (*e.g.* intermolecular interactions of **4** with *DMSO*-d_6). However, further studies have to be performed to check this (rather speculative) hypothesis and to find an explanation for the quite different behavior of **1**/**2** and **4** in solution.

Experimental

Syntheses

All syntheses were carried out under dry nitrogen. The organic solvents used were dried and purified according to standard procedures and stored under nitrogen. Melting points were determined with a Leitz Biomed microscope equipped with a heater (Leitz, Model M 350). The ^{1}H, ^{13}C, and ^{29}Si solution-state NMR spectra were recorded at room temperature on a Bruker DRX-300 NMR spectrometer (^{1}H, 300.1 MHz; ^{13}C, 75.5 MHz; ^{29}Si, 59.6 MHz). *DMSO*-d_6 was used as solvent. Chemical shifts (ppm) were determined relative to internal *DMSO*-d_5 (^{1}H, $\delta = 2.49$), *DMSO*-d_6 (^{13}C, $\delta = 39.5$), and external *TMS* (^{29}Si, $\delta = 0$). Assignment of the ^{13}C NMR data was supported by DEPT 135 experiments. The ^{29}Si CP/MAS NMR spectrum was recorded at room temperature on a Bruker DSX-500 NMR spectrometer with double air bearing rotors (ZrO_2, diameter 7 mm) containing *ca.* 200 mg of sample (99.4 MHz, *TMS* as external standard ($\delta = 0$), spinning rate 3980 Hz, contact time 7 ms, 90° ^{1}H transmitter pulse length 5.5 μs, repetition time 25 s). Mass spectra were obtained with a Finnigan MAT-8430 mass spectrometer (FAB MS, 3-nitrobenzyl alcohol as liquid matrix, xenon as FAB source).

((Morpholinio)methyl)bis(3,4,5,6-tetrachlorobenzene-1,2-diolato(2−))silicate–Acetonitrile (**4**·CH_3CN)

Method 1. Compound **5** (2.00 g, 9.04 mmol) was added dropwise at room temperature within 2 min to a solution of 3,4,5,6-tetrachlorobenzene-1,2-diol (4.48 g, 18.1 mmol) in acetonitrile (35 ml; formation of a precipitate after *ca.* 2 min). The reaction mixture was stirred for 2 min and then kept undisturbed at room temperature for 3 d. The precipitate was filtered off, washed with cold acetonitrile (2×10 ml), and then recrystallized from acetonitrile. The product was isolated by filtration and dried *in vacuo* (0.001 torr, 20°C, 5 h) to give 4.59 g of colorless crystals.

Yield: 77%; m.p.: 278°C (dec); ^{1}H NMR: $\delta = 2.06$ (s, 3H, CH_3CN), 2.35 (d, $^{3}J_{HH} = 5.1$ Hz, 2H, $SiCH_2N$), 2.83–3.03, 3.41–3.58, 3.59–3.80, and 3.82–4.00 (m, 8H, CCH_2N, CCH_2O), 8.3 (br s, 1H, NH) ppm; ^{13}C NMR: $\delta = 1.2$ (CH_3CN), 54.4 ($SiCH_2N$), 54.9 (NCH_2C), 63.5 (CCH_2O), 113.0 (C-4/C-5, $C_6Cl_4O_2$), 118.1 (C-3/C-6, $C_6Cl_4O_2$), 147.5 (C-1/C-2, $C_6Cl_4O_2$) ppm, CH_3CN not localized; ^{29}Si NMR: $\delta = -131.9$ ppm; ^{29}Si CP/MAS NMR: $\delta = -85.1$ ppm; FAB MS (positive ions): $m/z = 622$ (12) $(M(^{12}C_{17}{}^{1}H_{11}{}^{35}Cl_6{}^{37}Cl_2{}^{14}N^{16}O_5{}^{28}Si)_{Zwitterion}+H^+)$, 154 (100) (matrix); FAB MS (negative ions): $m/z = 620$ (12) $(M(^{12}C_{17}{}^{1}H_{11}{}^{35}Cl_6{}^{37}Cl_2{}^{14}N^{16}O_5{}^{28}Si)_{Zwitterion}-H^+)$, 153 (100) (matrix); $C_{19}H_{14}Cl_8N_2O_5Si$ (662.0); calcd.: C 34.47, H 2.13, N 4.23; found: C 34.4, H 2.1, N 4.2.

Method 2. Compound **6** (1.00 g, 3.74 mmol) was added dropwise at room temperature within 3 min to a solution of 3,4,5,6-tetrachlorobenzene-1,2-diol (1.85 g, 7.46 mmol) in acetonitrile (15 ml; formation of a precipitate after *ca.* 20 min). The reaction mixture was stirred for 80 min and then kept undisturbed at room temperature for 8 d. The precipitate was filtered off and washed with cold acetonitrile (2×10 ml). The product was isolated by filtration and then dried *in vacuo* (0.001 torr, 20°C, 5 h) to give 815 mg (33%) of colorless crystals. M.p.: 278°C (dec); for analytical data, see above.

Method 3. Compounds **7** (1.00 g, 4.87 mmol) was added dropwise at room temperature within 3 min to a solution of 3,4,5,6-tetrachlorobenzene-1,2-diol (2.42 g, 9.76 mmol) in acetonitrile (35 ml; formation of a precipitate after *ca.* 1 d). The reaction mixture was stirred for 3 min and then kept undisturbed at room temperature for 5 d. The precipitate was filtered off, washed with cold acetonitrile (2×5 ml), and then recrystallized from acetonitrile. The product was isolated by filtration and then dried *in vacuo* (0.001 torr, 20°C, 5 h) to give 1.92 g (60%) of colorless crystals. M.p.: 278°C (dec); for analytical data, see above.

Trimethoxy((morpholino)methyl)silane (**5**): Synthesis according to Ref. [1h].

Dimethoxy((morpholino)methyl)phenylsilane (**6**): Synthesis according to Ref. [1h].

Dimethoxy(methyl)((morpholino)methyl)silane (**7**): Synthesis according to Ref. [5].

Crystal structure analysis

A suitable single crystal (obtained by recrystallization from acetonitrile at room temperature) was mounted in inert oil (Riedel-de Haën, RS 3000) on a glass fiber and then transferred to the cold gas stream of a Stoe-Huber-Siemens four-circle diffractometer fitted with a Siemens CCD detector (monochromated MoK_α radiation ($\lambda = 0.71073$ Å); for the low-temperature attachment, see Ref. [6]). A semiempirical absorption correction with all data was applied. The structure was solved by direct methods [7]. All non-hydrogen atoms were refined anisotropically [8]. A riding model was employed in the refinement of the CH hydrogen atom positions. The NH hydrogen atom was localized in difference *Fourier* syntheses and refined freely.

Additional material to the structure determination may be ordered from Fachinformationszentrum Karlsruhe, Gesellschaft für wissenschaftlich-technische Information mbH, D-76344 Eggenstein-Leopoldshafen, Germany, referring to the deposition number CSD-408919, the names of the authors, and the citation of the present paper.

Acknowledgements

This work was supported by the *Deutsche Forschungsgemeinschaft* and the *Fonds der Chemischen Industrie*. In addition, we thank Prof. *G. M. Sheldrick* (University of Göttingen) for providing diffractometer facilities for collecting the X-ray diffraction data.

References

[1] a) Strohmann C, Tacke R, Mattern G, Kuhs WF (1991) J Organomet Chem **403**: 63; b) Tacke R, Sperlich J, Strohmann C, Mattern G (1991) Chem Ber **124**: 1491; c) Tacke R, Sperlich J, Strohmann C, Frank B, Mattern G (1992) Z Kristallogr **199**: 91; d) Tacke R, Wiesenberger F, Lopez-Mras A, Sperlich J, Mattern G (1992) Z Naturforsch, Part B **47**: 1370; e) Tacke R, Lopez-Mras A, Sperlich J, Strohmann C, Kuhs WF, Mattern G, Sebald A (1993) Chem Ber **126**: 851; f) Tacke R, Lopez-Mras A, Sheldrick WS, Sebald A (1993) Z Anorg Allg Chem **619**: 347; g) Tacke R, Becht J, Lopez-Mras A, Sperlich J (1993) J Organomet Chem **446**: 1; h) Sperlich J, Becht J, Mühleisen M, Wagner SA, Mattern G, Tacke R (1993) Z Naturforsch, Part B **48**: 1693; i) Tacke R, Mütileisen M, Lopez-Mras A, Sheldrick WS (1995) Z Anorg Allg Chem **621**: 779

[2] Dannappel O, Tacke R (1996) Intramolecular Ligand Exchange of Zwitterionic Spirocyclic *Bis*(1,2-benzenediolato(2−))organosilicates: Ab initio Studies of the Bis[1,2-benzenediolato-

(2−)]hydridosilicate(1−) Ion. In: Auner N, Weis J (eds) Organosilicon Chemistry II. Verlag Chemie, Weinheim, p 453

[3] Tacke R, Pfrommer B, Lunkenheimer K, Hirte R (1998) Organometallics **17**: 3670 and references cited therein

[4] a) Tandura SN, Voronkov MG, Alekseev NV (1986) Top Curr Chem **131**: 99; b) Sheldrick WS (1989) Structural Chemistry of Organic Silicon Compounds. In: Patai S, Rappoport Z (eds) The Chemistry of Organic Silicon Compounds, Part 1. Wiley, Chichester, UK, p 227; c) Holmes RR (1990) Chem Rev **90**: 17; d) Chuit C, Corriu RJP, Reye C, Young JC (1993) Chem Rev **93**: 1371; e) Wong CY, Woollins JD (1994) Coord Chem Rev **130**: 175; f) Lukevics E, Pudova OA (1996) Chem Heterocycl Compd **353**: 1605; g) Holmes RR (1996) Chem Rev **96**: 927

[5] Becht J (1994) Thesis, University of Karlsruhe, Germany

[6] Kottke T, Lagow RJ, Stalke D (1996) J Appl Crystallogr **29**: 465

[7] Sheldrick GM SHELXS-96 (1996) University of Göttingen, Germany; Sheldrick GM (1990) Acta Crystallogr A **46**: 467

[8] Sheldrick GM SHELXL-96 (1996) University of Göttingen, Germany

Received July 3, 1998. Accepted (revised) September 21, 1998

Preparation and Structure of 1-(Trifluoromethyl)silatrane

Reint Eujen*, **Achim Roth,** and **David J. Brauer**

Fachbereich 9 – Anorganische Chemie, Bergische Universität – GH Wuppertal, D-42097 Wuppertal, Germany

Summary. 1-(Trifluoromethyl)silatrane, $CF_3Si(OCH_2CH_2)_3N$, was prepared from $CF_3Si(OCH_3)_3$ and triethanolamine and characterized by NMR, vibrational, and mass spectroscopy. The compound crystallizes in the monoclinic space group $P2_1/n$ with $a = 777.2(1)$, $b = 1236.9(2)$, $c = 1002.8(1)$ pm, $\beta = 91.33(1)°$, $Z = 4$. The equatorial Si–O bond lengths of the almost perfect trigonal bipyramid average 165.3(2) pm, the axial Si–C bond length is 94.6(1) pm. The transannular Si–N contact, 202.4(1) pm, is considerably shorter than in other organylsilatranes.

Keywords. Silicon; Silatrane; Trifluoromethyl compounds.

Synthese und Struktur von 1-(Trifluormethyl)silatran

Zusammenfassung. 1-(Trifluoromethyl)silatran, $CF_3Si(OCH_2CH_2)_3N$, wurde aus $CF_3Si(OCH_3)_3$ und Triethanolamin gewonnen und durch NMR-, Schwingungs- und Massenspektroskopie charakterisiert. Die Verbindung kristallisiert in der monoklinen Raumgruppe $P2_1/n$ mit $a = 777.2(1)$, $b = 1236.9(2)$, $c = 1002.8(1)$ pm, $\beta = 91.33(1)°$, $Z = 4$. Die equatorialen Si–O-Abstände der nahezu perfekten trigonalen Bipyramide betragen im Schnitt 165.3(2) pm, der axiale Si–C-Abstand 194.6(1) pm. Der transanulare Si–N-Kontakt, 202.4(1) pm, ist beträchtlich kürzer als in anderen Organylsilatranen.

Introduction

Silatranes and related compounds have attracted great attention over the last decades not only because of their unique structure with pentacoordinated silicon and a transannular N→Si bond but also due to their biological activity. Several review articles covering both structural and biological aspects are available [1–7]. The nature of the Si–N interaction has been the subject of recent theoretical studies [8–11] which agree on the statement that only a very small amount of energy is required to shorten or lengthen the Si–N distance. Thus, the Si–N distances of 1-methylsilatrane and 1-fluorosilatrane have been found to be *ca.* 28 pm shorter in the solid state [12, 13] than in the gas phase [14, 15], and this shortening has been attributed to packing forces [10]. On the other hand, the Si–N bond distance is also

* Corresponding author

sensitive to the *Lewis* acidity of the silicon atom and thus reflects the electronic effect of a substituent bonded to silicon. For example, in the solid state the shortest Si–N distance has been found in the 1-(dimethyloxonio)silatranium cation with 196.5 pm, whereas those in the isothiocyanato-[16], fluoro-[13] or chlorosilatrane [17], which slightly exceed 200 pm, are significantly shorter than in organylsilatranes. It is expected that substitution of silicon with the electronegative CF_3 group will result in a similarly short Si–N contact. Earlier attempts to produce the CF_3 substituted silatrane have failed [18]. In the following we report on the synthesis, spectroscopic properties, and structure of 1-(trifluoromethyl)silatrane.

Results and Discussion

1-(Trifluoromethyl)silatrane is readily accessible from $CF_3Si(OCH_3)_3$ and triethanolamine as a white crystalline substance which is stable in air and insoluble in non-polar solvents, but readily dissolves in polar solvents like dimethylformamide, dimethylsulfoxide, or N-methylpyrrolidone.

$$CF_3Si(OCH_3)_3 + N(CH_2CH_2OH)_3 \xrightarrow[-CH_3OH]{THF/RT} CF_3Si(OCH_2CH_2)_3N \qquad (1)$$

Combined calorimetric and thermogravimetric analyses show that the compound decomposes in air at 201°C without melting and without any noticeable thermal effect, but with loss of one third of its weight (33%). Strongly exothermic degradation follows at 285°C with loss of another 18% of weight.

Spectra

The silatrane was characterized by NMR spectroscopy using *DMSO*-d_6 as solvent (^{1}H: $\delta(OCH_2)$ 3.81 ppm, $\delta(NCH_2)$ 3.09 ppm, $^3J(HH)$ 6.0 Hz; ^{13}C: $\delta(OCH_2)$ 56.6 ppm, $\delta(NCH_2)$ 50.5 ppm, $\delta(CF_3)$ 131.2 ppm, $^1J(CF)$ 325.8 Hz; ^{19}F: $\delta(CF_3)$ −65.1 ppm; ^{29}Si: δ −95.4 ppm, $^2J(SiF)$ 40.8 Hz). The increase of $^1J(CF)$ with simultaneous decrease of $^2J(SiF)$ with respect to the corresponding values of $CF_3Si(OCH_3)_3$ (315.5 and 53.6 Hz, respectively) follows the pattern observed upon increase of the coordination number for other CF_3 substituted silicon compounds [19].

The mass spectrum shows the characteristics of two different fragmentation pathways – both loss of the exocyclic substituent and skeleton fragmentation. Besides the molecular ion ($m/z = 243$, 9%), loss of CF_2 ($m/z = 193$, 68%) and of CF_3 to the silatranyl ion ($m/z = 174$, 100%) dominates – the latter constituting the base peak of most organylsilatranes [20]. In addition, an intense peak is observed at $m/z = 150$ (85%) which is indeed the base fragment ion of 1-fluorosilatrane, $[C_4H_9FNO_2Si]^+$ [13].

The infrared and *Raman* spectra listed in Table 1 show the typical pattern of most silatranes. For the 15 non-hydrogen skeleton atoms $10a_1$, $3a_2$, and $12e$ modes are expected, both a_1 and e being infrared and *Raman* active. The assignment of the a_1 modes is based on a normal coordinate analysis of silatrane [21] and on *Raman* polarization data obtained for molten methylsilatrane [22] which are included in Table 1 for comparison. There have been lots of discussion and attempts to assign

Table 1. Vibrational spectra of polycrystalline 1-(trifluoromethyl)silatrane ($R = CF_3$) and related silatranes (cm^{-1})

Raman $R = CF_3$	IR	*Raman* $R = H/CH_3$ [22]	Assignment
203 m		184 m/182 m	
233 m-s		(–/242 w)	$e(\rho\ CF_3)$
297 m-s		(–/288 m)	$a_1(\nu\ Si\text{-}CF_3)$
362 m		348 w/354 m,p	a_1[a]
	442 s	–/455 w	e[a]
497 m	492 m	490 w/471 m,p	a_1[a]
526 m			$e(\delta_{as}\ CF_3)$
596 s	596 s	593 s/580 vs,p	a_1[a]
		–/619 vs,p	$a_1\ (\nu\ Si–CH_3)$
625 w	625 w		
648 s	648/652 s	630 m/696 m	$a_1(\nu_s\ SiO_3)$
716 m	716 m	–/(718 w)	$a_1(\delta_s\ CF_3)$
767 w	767 w		$\rho\ CH_2$
813 w	805/811 vs	760 m/769 m	$e\ (\nu_{as}\ SiO_3)$
877 w	871/877 w	864 w/873 w	e[a]
916 m	913 m	911 s/908 m,p	a_1[a]
944 m	938/946 m	942 s/944 m,p	a_1[a]
1028 m	1029 vs		$e(\nu_{as}CF_3)$
1043 m	1048 m	1043 w/1050 m	
1079 m	1086 vs		$a_1(\nu_s\ CF_3)$
1110 m	1115/1118 s	1104 w/1113 m	a_1[a]
1170 m	1167 w	1165 w/1171 m	e[a]
1228 w	1225 m		
1248 m	1245 w	1240 m/1237 w	} CH_2
1274 s	1277 m	1267 m/1278 s	
1353 w		–/1350 w	twisting and
1390 w		–/1381 w	wagging
1460 m-s	1455 m	1455 m/1452 s,p	} $a_1(\delta_s\ CH_2)$
1490 s	1489 m	1486 m/1487 s,p	
2898 s	2891 s	2876 s/2875	} $\nu_s\ CH_2$
2904 s	2897 s		
2963 s	2958 m	2928 s/2948	$\nu_{as}CH_2$
3002 s	2998 m	2967 m/2975	$\nu_{as}CH_2$

[a] Skeleton mode

the Si–N stretching mode although it is evident from the kinetics of the a_1 cage modes that there is no unique Si–N stretch and that an assignment of frequencies to individual internal coordinates is unreasonable. Additionally, the flatness of the calculated potential curves indicates that the contribution of the Si–N stretching force constant to the total potential energy distribution of the a_1 cage modes will be low. In another approximation, the Si–N stretch may be described as a combination of the symmetric SiO_3 and NC_3 deformations. In the CF_3 derivative, these low-energy modes will be further coupled to the Si–CF_3 stretch which appears at 386 cm^{-1} in $CF_3Si(CH_3)_3$ [23] or 402 cm^{-1} in CF_3SiH_3 [24]. The longer Si–C bond in the silatrane implies that the Si–CF_3 stretch should occur at even lower wavenumbers, and an assignment to the *Raman* band at 297 cm^{-1} seems reasonable. Similarly, the $\nu_s(CF_3)$ and $\delta_s(CF_3)$ modes are considerably red-shifted with respect to tetracoordinated CF_3Si compounds. The general increase of almost all other frequencies and especially of those between 350 and 700 cm^{-1} is in accordance with a more rigid skeleton and stronger Si–N interaction.

Crystal structure

No crystallographic symmetry is imposed on the molecule; the structure closely complies with C_3 symmetry with the CF_3 group being rotated about the Si–C bond by 9.4(2)° from an exactly staggered conformation with respect to the SiO_3 entity. The structure is shown in Fig. 1, and selected bond distances and angles are given in Table 2. As found in numerous silatranes [25], the three heterocyclic five-membered rings in this compound adopt an envelope conformation with the C(2), C(4), or C(6) atoms deviating on the average by 51(1) pm from the corresponding plane defined by the other four ring atoms.

A comparison with the structure of the methyl analog [12] shows that fluorination of the methyl group lengthens the Si–C bond by 7.6(6) pm, at the same time leading to a slight but significant shortening (1.3(3) pm) of the average Si–O bond length and to a large contraction (15.1(4) pm) of the transannular Si–N contact. The SiO_3 pyramid is flatter, the average values of both the O–Si–C and O–Si–N angles being closer (4.4(6))° to the perfectly flattened 90° value than are those of the methyl compound.

Whereas the Si–CF_3 bond is significantly longer than the Si–CH_3 bond in the methyl analog, it is not much longer than those found in tetravalent

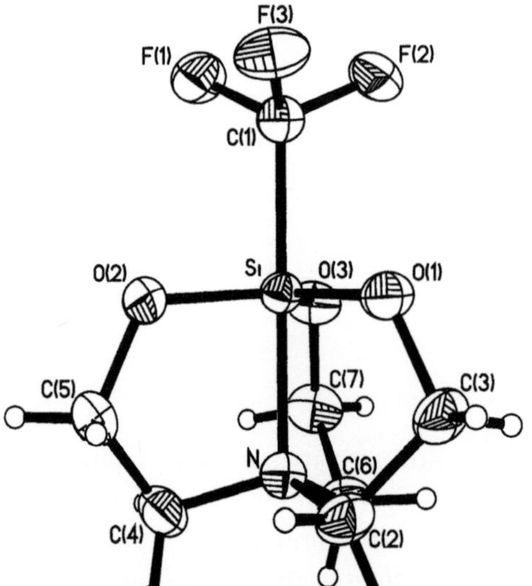

Fig. 1. Perspectivic drawing of 1-(trifluoromethyl)silatrane with 50% probability thermal ellipsoids for the non-hydrogen atoms

Table 2. Selected or average bond lengths (pm) and angles (°) for 1-(trifluoromethyl)silatrane

Si–O(1)	165.17(8)	C(1)–F_{av}	135.9(7)
Si–O(2)	165.29(8)	O–C_{av}	142.7(2)
Si–O(3)	165.42(8)	N–C_{av}	148.1(3)
Si–C(1)	194.61(10)	C–C_{av}	151.4(4)
Si–N	202.42(9)		
C(1)–Si–N	179.54(4)	O–Si–N_{av}	87.1(2)
O(1)–Si–O(2)	119.96(5)	Si–O–C_{av}	119.4(2)
O(1)–Si–O(3)	120.44(5)	C–N–C_{av}	113.5(5)
O(2)–Si–O(3)	118.85(5)	Si–N–C_{av}	105.1(1)
C(1)–Si–O(1)	92.49(4)	F–C–F_{av}	104.3(3)
C(1)–Si–O(2)	93.16(4)	Si–C–F_{av}	114.3(9)
C(1)–Si–O(3)	92.94(4)	N–C–C_{av}	105.9(2)
		O–C–C_{av}	108.1(3)

trifluoromethylsilanes [24, 26–28]. The short transannular Si–N distance, 202.4 pm, is readily associated with the high *Lewis* acidity of the silicon atom induced by the high electronegativity of the CF_3 moiety. The similarity to the values reported for the chloro-and fluorosilatrane (202.3 and 204.2 pm, respectively) clearly demonstrates the pseudohalide character of the CF_3 group despite severe differences in the nature of the Si–CF_3 and Si–X bonding.

Experimental

Chemicals were obtained from commercial sources and used without further purification. (Trifluoromethyl)trimethoxysilane was prepared according to Ref. [29]. *Raman* spectra were obtained on a Cary 82 spectrometer with Kr^+ excitation at 647.1 nm. Infrared spectra were recorded with a Bruker IFS 25 spectrometer as KBr pellets. Combined calorimetric and thermogravimetric analyses (DSC/TGA) were made with a simultaneous DSC/TG instrument, Netzsch STA 409. NMR spectra were recorded with a Bruker AC 250 instrument (^{1}H, 250.13 MHz; ^{19}F, 235.36 MHz; ^{13}C, 62.90 MHz; ^{29}Si, 49.69 MHz) and referenced to internal *DMSO*-d_6 at 2.58 ppm (^{1}H) and 39.50 ppm (^{13}C), external *TMS* (^{29}Si) and $CFCl_3$ (^{19}F), corrections being made for the shift of the lock substance.

1-(Trifluoromethyl)silatrane

To 1.90 g (10 mmol) of $CF_3Si(OCH_3)_3$ in 8 ml of *THF*, 1.49 g of $N(CH_2CH_2OH)_3$ (10 mmol) in 3 ml of *THF* were added dropwise. The reaction mixture was stirred for 2 h at ambient temperature. The white precipitate was filtered off and recrystallized from methanol yielding 1.08 g (44%) of 1-(trifluoromethyl)silatrane.

X-ray structural investigation

Crystals of (trifluoromethyl)silatrane were grown from a solution in methanol, and a specimen with the dimensions 0.23 × 0.32 × 0.50 mm was glued to a glass fiber. X-ray measurements were at 22°C

Table 3. Atomic coordinates and equivalent isotropic displacement parameters (pm^2) for 1-(trifluoromethyl)silatrane

	x	y	z	$U(eq)$[a]
Si	0.05804(3)	0.39859(2)	0.25484(2)	240(1)
F(1)	−0.09707(10)	0.19504(6)	0.23071(8)	450(2)
F(2)	−0.25944(9)	0.32395(6)	0.15832(8)	453(2)
F(3)	−0.22548(10)	0.29782(7)	0.36871(7)	479(2)
O(1)	−0.05849(9)	0.49992(6)	0.31069(8)	352(2)
O(2)	0.17164(10)	0.32319(6)	0.36066(8)	347(2)
O(3)	0.08771(10)	0.38377(7)	0.09304(7)	355(2)
N	0.25984(11)	0.50208(7)	0.25561(8)	279(2)
C(1)	−0.13722(12)	0.30007(8)	0.25372(10)	301(2)
C(2)	0.20777(15)	0.59314(9)	0.34182(13)	370(2)
C(3)	0.0154(2)	0.60531(9)	0.32015(14)	398(2)
C(4)	0.40655(13)	0.43935(10)	0.31259(12)	372(2)
C(5)	0.33100(15)	0.36335(11)	0.41387(13)	394(2)
C(6)	0.2810(2)	0.53248(10)	0.11428(11)	372(2)
C(7)	0.2339(2)	0.43257(11)	0.03384(11)	388(2)

[a] $U(eq)$ is defined as one third of the trace of the orthogonalized U_{ij} tensor

with a Siemens P3 diffractometer equipped with a graphite monochromator and employing MoK_α radiation. The compound crystallizes in the monoclinic space group $P2_1/n$ with $a = 777.2(1)$, $b = 1236.9(2)$, $c = 1002.8(1)$ pm, $\beta = 91.33(1)°$, $Z = 4$, and $D_{calc} = 1.677\ g \cdot cm^{-3}$. The hemisphere of data with $l \geq 0$ and $2.6° \leq \theta \leq 35.1°$ was collected by the θ–2θ scan technique. The intensities were derived by profile analysis, corrected for the small fluctuations ($\pm 1\%$) of the three periodically monitored standards and by integration for absorption (transmission: 0.9221–0.9457), and converted into structure factor amplitudes. Of the 8861 intensities measured, 4274 are unique; 3204 of them have $I > 2\sigma(I)$. The structure was solved by direct methods and refined conventionally with anisotropic and isotropic temperature factors for the non-hydrogen and hydrogen atoms, respectively. The refinement on F^2 of the 185 parameters, which included an extinction correction, converged with a conventional residual $R_1 = 0.034$ for the observed reflections and a weighted residual (F^2) $wR_2 = 0.095$ for all reflections. The final difference density map contains features between 376 and $-236\ e \cdot nm^{-3}$ and confirms the structure. Coordinates of the non-hydrogen atoms are listed in Table 3 [30]. Direct methods, least-squares refinement, and the ORTEP drawing were made with the SHELXTL program package (Version 5.03).

Acknowledgements

Financial support by the *Deutsche Forschungsgemeinschaft* (*Schwerpunktprogramm Siliciumchemie*) and the *Fonds der Chemischen Industrie* is gratefully acknowledged.

References

[1] Voronkov MG (1966) Pure Appl Chem **13**: 35
[2] Voronkov MG (1979) Top Curr Chem **84**: 77
[3] Tandura SN, Voronkov MG, Alekseev NV (1986) Top Curr Chem **131**: 99

[4] Lukevics E, Pudova O, Sturkovich R (1989) Molecular Structure of Organosilicon Compounds. Ellis Horwood, Chichester
[5] Chuit C, Corriu RJP, Reye C, Young JC (1993) Chem Rev **93**: 1371
[6] Greenberg A, Wu G (1990) Struct Chem **1**: 79
[7] Hencsei P (1991) Struct Chem·**2**: 21
[8] Schmidt MW, Windus TL, Gordon MS (1995) J Amer Chem Soc **117**: 7480
[9] Dahl T, Skancke PN (1996) Int J Quant Chem **60**: 567
[10] Csonka GI, Hencsei P (1996) J Comput Chem **17**: 767
[11] Csonka GI, Hencsei P (1996) J Mol Struct **362**: 199
[12] Párkányi L, Bihátsi L, Hencsei P (1978) Cryst Struct Commun **7**: 435
[13] Párkányi L, Hencsei P, Bihátsi L, Müller T (1984) J Organomet Chem **269**: 1
[14] Shen Q, Hiderbrandt RL (1980) J Mol Struct **64**: 257
[15] Forgacs G, Kolonits M, Hargittai I (1990) Struct Chem **1**: 245
[16] Narula SP, Shankar R, Kumar M, Chadha RK, Janaik C (1997) Inorg Chem **36**: 1268
[17] Kemme AA, Bleidelis YaYa, Pestunovich VY, Baryshok VP, Voronkov MG (1978) Dokl Akad Nauk SSSR **243**: 688
[18] Voronkov MG, Baryshok VP, Petukhov LP, Rakhlin VI, Mirskov RG, Pestunovich VA (1988) J Organomet Chem **358**: 39
[19] Eujen R, Roth A, (to be published)
[20] Bochkarev VN, Chernyshev AE, Vitkovskii VYu, Voronkov MG (1985) J Gen Chem USSR **55**: 1209
[21] Imbenotte M, Palavit G, Legrand P, Huvenne JP, Fleury G (1983) J Mol Spectrosc **102**: 40
[22] Imbenotte M, Palavit G, Legrand P (1983) J Raman Spectrosc **14**: 135
[23] Eujen R (1987) Spectrochim Acta **43A**: 1165
[24] Beckers H, Bürger H, Eujen R, Rempfer B, Oberhammer H (1986) J Mol Struct **140**: 281
[25] Voronkov MG, Dyakov VM, Kirpichenko SV (1982) J Organometal Chem **233**: 1
[26] Beckers H, Brauer DJ, Bürger H, Wilke CJ (1988) J Organometal Chem **356**: 31
[27] Rempfer B, Pfafferott G, Oberhammer H, Beckers H, Bürger H, Eujen R, Boggs JE (1986) Rev Chim Miner **23**: 551
[28] Durig JR, Attia G, Groner P, Beckers H, Bürger H (1998) J Chem Phys **88**: 545
[29] Beckers H, Bürger H, Bursch P, Ruppert I (1986) J Organomet Chem **316**: 41
[30] Further crystallographic details may be obtained from Fachinformationszentrum Karlsruhe, 76344 Eggenstein-Leopoldshafen, Germany, by quoting the deposit number CSD-410073, the authors and the literature reference

Received June 29, 1998. Accepted (revised) August 11, 1998

Hexacoordinate Silicon-Azomethine Complexes: Synthesis, Characterization, and Properties

Frank Mucha, Jörg Haberecht, Uwe Böhme, and **Gerhard Roewer***

Technische Universität Bergakademie Freiberg, Institut für Anorganische Chemie, D-09596 Freiberg, Bundesrepublik Deutschland

Summary. N,N′-Ethylene-*bis*(2-hydroxyacetophenoneimine) = *salen**H_2 and N,N′-ethylene-*bis*(3,5-di-*tert*-butyl-salicylideneimine) = *salen*$^{\ddagger}H_2$ react both with $SiCl_4$ under formation of hexacoordinate silicon compounds ((*salen*′)$SiCl_2$, *salen*′ = *salen** or *salen*‡). The analogous fluoro derivatives (*salen*′)SiF_2 have been prepared by reaction of (*salen*′)$SiCl_2$ with ZnF_2. X-ray structure analysis of (*salen**)SiF_2 clearly demonstrates the octahedral coordination of silicon. *Salen** acts as a tetradentate chelating ligand, two halogen atoms remaining at the silicon atom. The reduction of (*salen**)$SiCl_2$ by alkaline metal affords polysilanes containing main chain hexacoordinate silicon. Coupling with acetylides results in polycarbosilanes with a Si–C≡C–Si backbone.

Keywords. *Salen* complexes; Hexacoordinate silicon; Silicon complexes.

Hexakoordinierte Silicium-Azomethin-Komplexe: Synthese, Charakterisierung und Eigenschaften

Zusammenfassung. N,N′-Ethylen-*bis*(2-hydroxyacetophenonimin) = *salen**H_2 und N,N′-Ethylen-*bis*(3,5-di-*tert*-butyl-salicylidenimin) = *salen*$^{\ddagger}H_2$ reagieren mit $SiCl_4$ unter Bildung hexakoordinierter Siliciumverbindungen ((*salen*′)$SiCl_2$, *salen*′ = *salen** bzw. *salen*‡). Die analogen Fluorderivate (*salen*′)SiF_2 wurden durch Reaktion von (*salen*′)$SiCl_2$ mit ZnF_2 hergestellt. Die Röntgenstrukturanalyse von (*salen**)SiF_2 zeigt eindeutig eine oktaedrische Koordination des Siliciumatoms. *Salen** wirkt als vierzähniger Chelatligand, zwei Halogenatome verbleiben am Silicium. Die Reduktion von (*salen**)$SiCl_2$ mittels Alkalimetallen führt zu Polysilanen mit hexakoordiniertem Silicium in der Hauptkette des Polymers. Kupplung mit Acetyliden ergibt Polycarbosilane mit einem Si–C≡C–Si-Skelett.

Introduction

Compounds with hypervalent silicon atoms attract interest from both the structural and reactivity point of view [1]. Anions of type **1** are established as tetradentate ligands for transition metal ions [2]. The free acids *salen*′H_2 (*salen*′ = different substituents R^1 and R^2) are easily available by condensation of 1,2-diamines with aromatic α-hydroxyaldehydes or benzoylacetone.

* Corresponding author

The structures as well as the magnetic and electronic properties of *salen* metal complexes can be tuned by different substituents at the *salen* ligands [3]. Transition metal *salen* complexes have been investigated as model compounds for biologically important porphyrin and corrin complexes, *e.g.* vitamin B_{12} coenzyme [4, 5]. The tetracoordinating *salen* ligand causes similar electronic properties of the resulting metal complexes as the macrocyclic porphyrin or corrin ligands. Thus, similar supernucleophilic properties like in coenzyme B_{12} have been shown for [Co(*salen*)]. It can be reduced to $[Co(salen)]^-$ which reacts with alkyl halides in an oxidative addition to form [R-Co(*salen*)] [6]. Some of these compounds coordinate molecular oxygen [7], and they are efficient catalysts for stereoselective oxygen transfer reactions like the *Jacobsen* epoxidation [8–10].

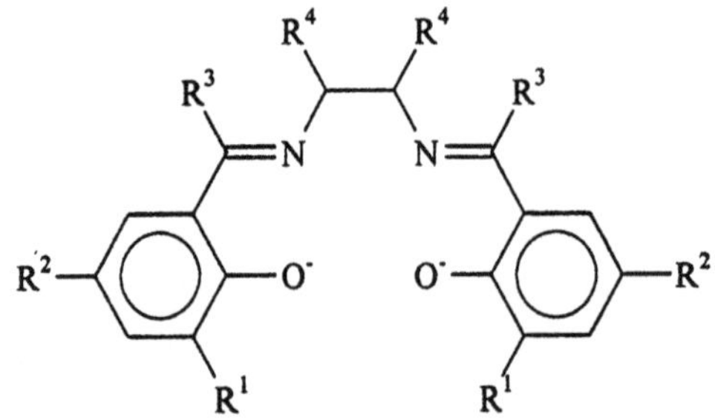

1

Azomethine ligands of type **1** should be able to coordinate also the silicon atom leading to square planar complexes. Further substituents at the silicon atom should create hypercoordinated silicon derivatives. The preparation of hypercoordinated silicon complexes was the target of our investigations. Till now, only some rare examples of *salen* silicon compounds are known from the literature [11], but characterization of these compounds seems to be inconclusive [12]. Structural aspects are uncertain due to the lack of crystal structural data.

Results and Discussion

Owing to expected differences in chemical reactivity and solubility, we used two different *salen* ligands: N,N′-ethylene-*bis*(2-oxyacetophenoneimine) = *salen**H_2 ($R^1 = R^2 = R^4 = H$, $R^3 = CH_3$) and N,N′-ethylene-*bis*(3,5-di-*tert*-butyl-salicylideneimine) = *salen*‡H_2 ($R^1 = R^2 =$ *tert*-butyl, $R^3 = R^4 = H$). *Salen**H_2 contains methyl groups at the azomethine carbon atom instead of hydrogen. This prevents undesired side reactions caused by the azomethine protons. *Salen*‡H_2 is substituted with two *tert*-butyl groups per phenyl group and should therefore give better soluble compounds. We were able to perform X-ray structure analyses of both ligands as free acids (*salen**H_2(**1a**,) and *salen*‡H_2(**1b**)).

The crystal structure of **1a** is shown in Fig. 1; atomic coordinates, bond lengths, and angles are summarized in Tables 1 and 2, respectively. Crystallographic data and data collection parameters are listed in Table 8.

The crystal structure of **1b** is demonstrated in Fig. 2; atomic coordinates, bond lengths, and angles are listed in Tables 3 and 4, respectively. Crystallographic data and data collection parameters are given in Table 8.

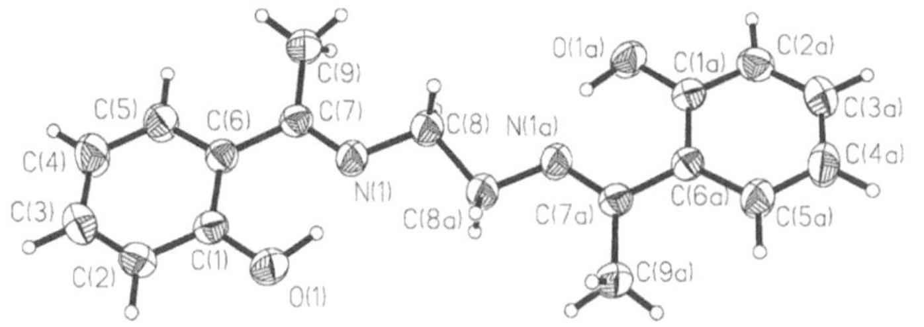

Fig. 1. ORTEP drawing (50% probability ellipsoids) of **1a**

Table 1. Atomic coordinates ($\times 10^4$) and equivalent isotropic displacement parameters ($\mathring{A}^2 \times 10^3$) for **1a**; *U(eq)* is defined as one third of the trace of the orthogonalized U_{ij} tensor

	x	*y*	*z*	*U(eq)*
N(1)	2490(2)	4479(1)	6409(1)	40(1)
O(1)	5443(2)	3932(1)	4792(1)	52(1)
C(1)	6341(2)	3636(1)	6576(2)	39(1)
C(2)	8268(2)	3207(1)	6741(2)	47(1)
C(3)	9190(2)	2875(1)	8511(2)	54(1)
C(4)	8262(3)	2969(1)	10191(2)	57(1)
C(5)	6404(2)	3404(1)	10069(2)	48(1)
C(6)	5397(2)	3750(1)	8278(2)	37(1)
C(7)	3414(2)	4218(1)	8149(2)	37(1)
C(8)	514(2)	4946(1)	6123(2)	43(1)
C(9)	2587(3)	4378(1)	10013(2)	51(1)

Table 2. Bond lengths (Å) and angles (°) for **1a**

Bond	Distance	Structural fragment	Angle
N(1)–C(7)	1.287(1)	C(7)–N(1)–C(8)	121.57(9)
N(1)–C(8)	1.462(2)	O(1)–C(1)–C(2)	118.4(1)
O(1)–C(1)	1.340(1)	O(1)–C(1)–C(6)	121.7(1)
C(1)–C(2)	1.397(2)	C(2)–C(1)–C(6)	119.8(1)
C(1)–C(6)	1.416(2)	C(3)–C(2)–C(1)	120.5(1)
C(2)–C(3)	1.370(2)	C(2)–C(3)–C(4)	120.7(1)
C(3)–C(4)	1.391(2)	C(5)–C(4)–C(3)	119.4(1)
C(4)–C(5)	1.380(2)	C(4)–C(5)–C(6)	121.6(1)
C(5)–C(6)	1.402(2)	C(5)–C(6)–C(1)	117.8(1)
C(6)–C(7)	1.476(2)	C(5)–C(6)–C(7)	121.6(1)
C(7)–C(9)	1.498(2)	C(1)–C(6)–C(7)	120.5(1)
C(8)–C(8)[a]	1.511(2)	N(1)–C(7)–C(6)	117.4(1)
		N(1)–C(7)–C(9)	122.8(1)
		C(6)–C(7)–C(9)	119.7(1)
		N(1)–C(8)–C(8)[a]	109.0(1)

[a] Symmetry transformations used to generate equivalent atoms: −x, −y+1, −z+1

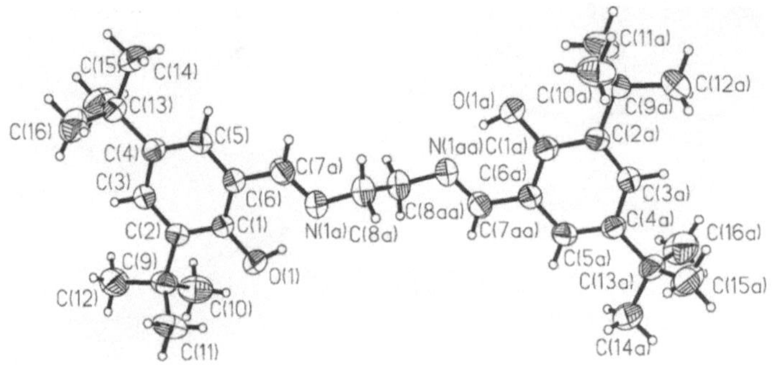

Fig. 2. ORTEP drawing (50% probability ellipsoids) of **1b**

Table 3. Atomic coordinates ($\times 10^4$) and equivalent isotropic displacement parameters ($Å^2 \times 10^3$) for **1b**; $U(eq)$ is defined as one third of the trace of the orthogonalized U_{ij} tensor

	x	*y*	*z*	*U(eq)*
O(1)	9583(1)	178(2)	3878(1)	62(1)
N(1A)	9642(1)	2404(3)	4690(1)	53(1)
C(7A)	9005(1)	1670(4)	4667(1)	51(1)
C(8A)	9990(1)	3869(3)	5100(1)	56(1)
N(1B)	9453(5)	3235(15)	4490(4)	57(2)
C(7B)	8853(7)	2351(20)	4494(4)	52(3)
C(8B)	9689(7)	5246(19)	4781(4)	70(3)
C(1)	8910(1)	−537(2)	3888(1)	46(1)
C(6)	8598(1)	222(2)	4261(1)	50(1)
C(5)	7908(1)	−534(2)	4277(1)	50(1)
C(4)	7526(1)	−2038(2)	3937(1)	47(1)
C(3)	7857(1)	−2766(2)	3572(1)	49(1)
C(2)	8534(1)	−2067(2)	3530(1)	45(1)
C(9)	8858(1)	−2919(2)	3113(1)	56(1)
C(10)	8969(1)	−1045(3)	2780(1)	79(1)
C(11)	9599(1)	−4022(3)	3351(1)	77(1)
C(12)	8345(1)	−4559(4)	2772(1)	95(1)
C(13)	6774(1)	−2944(2)	3947(1)	55(1)
C(14)	6482(1)	−1824(3)	4348(1)	76(1)
C(15)	6207(1)	−2556(4)	3433(1)	83(1)
C(16)	6847(1)	−5327(3)	4067(1)	97(1)

The bond lengths and angles of the phenyl groups and the ethylene bridges connecting the azomethine units are in the normal range for **1a** and **1b**. Both molecules are located at crystallographic inversion centres. This implies a transoid conformation in the solid state with a torsion angle N(1)–C(8)–C(8a)–N(1a) of 180.0°. The ethylenediamine bridge connecting the aromatic rings in **1b** is disordered. The disorder was resolved with site occupation factors of 0.84 for C(7a)–N(1a)–C(8a) and 0.16 for C(7b)–N(1b)–C(8b). The small site occupation of 16% results in large standard deviations for bonds and angels at C(7b), N(1b), and C(8b).

Table 4. Bond lengths (Å) and angles (°) for **1b**

Bond	Distance	Structural fragment	Angle
O(1)–C(1)	1.355(2)	C(7A)–N(1A)–C(8A)	118.6(2)
N(1A)–C(7A)	1.273(3)	N(1A)–C(7A)–C(6)	122.8(2)
N(1A)–C(8A)	1.451(2)	N(1A)–C(8A)–C(8A)[a]	110.6(2)
C(7A)–C(6)	1.468(3)	C(7B)–N(1B)–C(8B)	119.7(9)
C(8A)–C(8A)[a]	1.512(4)	N(1B)–C(7B)–C(6)	124.4(9)
N(1B)–C(7B)	1.26(1)	C(8B)[a]–C(8B)–N(1B)	109(1)
N(1B)–C(8B)	1.48(1)	O(1)–C(1)–C(6)	120.0(1)
C(7B)–C(6)	1.49(1)	O(1)–C(1)–C(2)	119.8(1)
C(8B)–C(8B)[a]	1.46(2)	C(6)–C(1)–C(2)	120.3(1)
C(1)–C(6)	1.402(2)	C(5)–C(6)–C(1)	120.1(1)
C(1)–C(2)	1.407(2)	C(5)–C(6)–C(7A)	118.3(1)
C(6)–C(5)	1.399(2)	C(1)–C(6)–C(7A)	121.4(1)
C(5)–C(4)	1.372(2)	C(5)–C(6)–C(7B)	118.4(5)
C(4)–C(3)	1.401(2)	C(1)–C(6)–C(7B)	117.3(5)
C(4)–C(13)	1.536(2)	C(4)–C(5)–C(6)	121.5(1)
C(3)–C(2)	1.389(2)	C(5)–C(4)–C(3)	116.8(1)
C(2)–C(9)	1.537(2)	C(5)–C(4)–C(13)	123.3(1)
C(9)–C(11)	1.529(2)	C(3)–C(4)–C(13)	120.0(1)
C(9)–C(12)	1.529(2)	C(2)–C(3)–C(4)	124.8(1)
C(9)–C(10)	1.533(2)	C(3)–C(2)–C(1)	116.6(1)
C(13)–C(16)	1.513(2)	C(3)–C(2)–C(9)	121.8(1)
C(13)–C(15)	1.530(2)	C(1)–C(2)–C(9)	121.6(1)
C(13)–C(14)	1.535(2)	C(11)–C(9)–C(12)	108.0(2)
		C(11)–C(9)–C(10)	109.3(1)
		C(12)–C(9)–C(10)	108.0(1)
		C(11)–C(9)–C(2)	110.4(1)
		C(12)–C(9)–C(2)	111.5(1)
		C(10)–C(9)–C(2)	109.6(1)
		C(16)–C(13)–C(15)	110.5(2)
		C(16)–C(13)–C(14)	108.4(2)
		C(15)–C(13)–C(14)	106.9(1)
		C(16)–C(13)–C(4)	109.6(1)
		C(15)–C(13)–C(4)	109.6(1)
		C(14)–C(13)–C(4)	111.8(1)

[a] Symmetry transformations used to generate equivalent atoms: −x+2, −y+1, −z+1

The *ortho*-hydroxy-azomethine unit is stabilized in both compounds by intramolecular hydrogen bridges (O–H····N) with an H····N distance of 1.547 Å in **1a** and 1.766 to 1.796 Å in **1b**. Intermolecular hydrogen bridges are found in the structure of **1a**, with a distance of 2.492 Å for 01···H(9c) which is less than the sum of the *van der Waals* radii of oxygen and hydrogen (1.2 Å+1.4 Å = 2.6 Å) [13]. The hydrogen bridges between the molecules are situated in a plane leading to a layered structure. Similar intermolecular hydrogen bridges are prevented in **1b** by the bulky *tert*-butyl groups; therefore, only intramolecular hydrogen bridges are found in the structure of **1b** between O(1) and two hydrogen atoms of the direct neighbouring *tert*-butyl group (O(1)···H(10a) = 2.380 Å, O(1)···H(11a) = 2.332 Å).

The molecules of **1a** are nearly planar in the solid state with only 0.54 Å distance between the planes of the phenyl rings. The planes of the two phenyl rings are 0.769 Å apart from each other in **1b**. The phenyl rings surrounded by bulky *tert*-butyl groups are packed in the crystal lattice of **1b** in a way leaving enough space for the linking *bis*(azomethine)ethylene bridge to be disordered. The transoid conformations for **1a** and **1b** are not retained in solution. Both molecules are able to rotate freely around the bridge N(1)–C(8)–C(8a)–N(1a) in solution, as indicated by the fact that only one set of signals is found in their NMR spectra.

The key compound (*salen**)$SiCl_2$ (**2**) can be prepared *via* three different routes (a, b and c, Scheme 1). Elemental analysis indicates the formation of a complex of the composition (*salen**)$SiCl_2$. Preparation of **2** *via* reaction of *salen**H_2 with $SiCl_4$ (a) yields a product with two ^{29}Si NMR signals and 18 peaks in the ^{13}C NMR spectrum. This points to the formation of two different isomers (**2**-*trans* and **2**-*cis*) in a ratio of 1:1 according to the intensity of the signals. It demonstrates that the $SiCl_2$ unit is not exclusively linear in azomethine complexes as has been pretended in the literature [11]. The same type of isomerism has been observed in tin *salen* compounds and confirmed by *Mössbauer* spectroscopy [14].

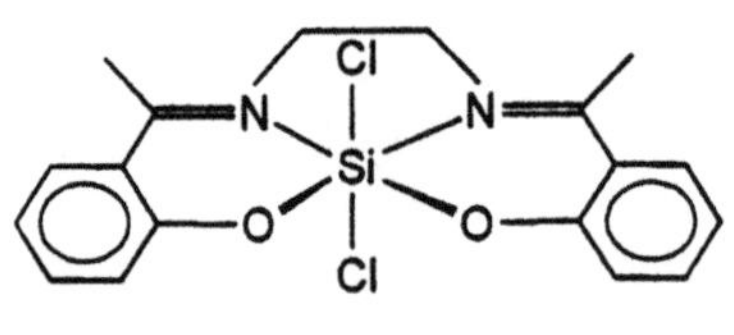

2-*trans*

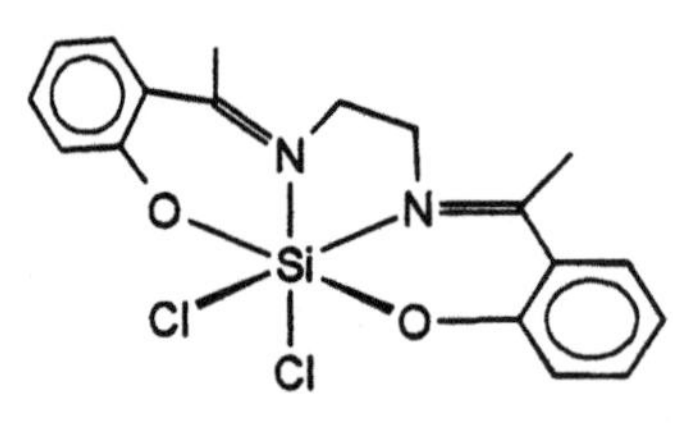

2-*cis*

The large high field shift of the ^{29}Si NMR signals for the isomers of **2** (−186.1, −188.0 ppm) indicates the presence of hexacoordinate silicon atoms. Reaction of Si_2Cl_6 with *salen**H_2 (b in Scheme 1) gives only one isomer with 9 signals in the ^{13}C NMR spectrum and one ^{29}Si NMR signal at −188 ppm. Both hydrogen chloride and hydrogen are evolved in this reaction.

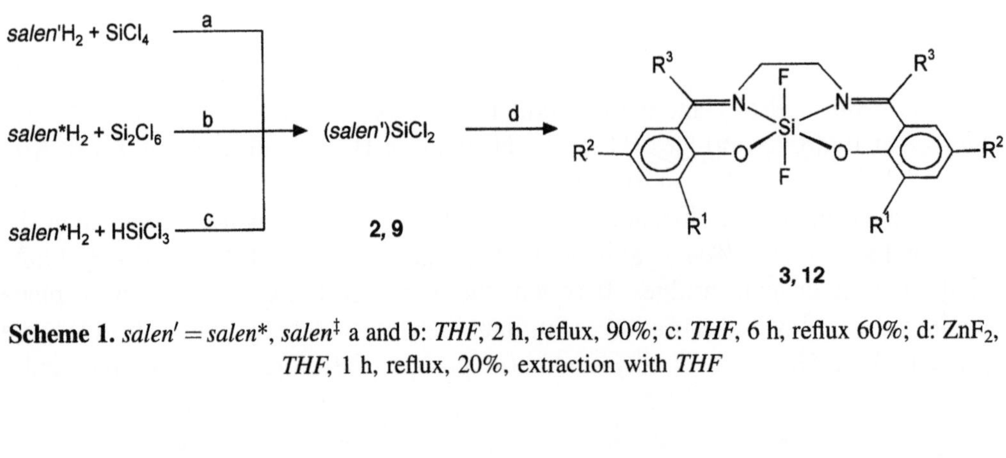

Scheme 1. *salen'* = *salen**, *salen*‡ a and b: *THF*, 2 h, reflux, 90%; c: *THF*, 6 h, reflux 60%; d: ZnF_2, *THF*, 1 h, reflux, 20%, extraction with *THF*

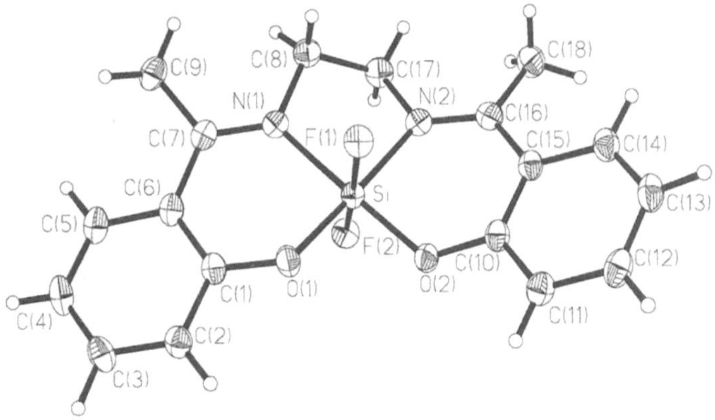

Fig. 3. ORTEP drawing (50% probability ellipsoids) of **3**

The chlorine atoms in **2** and **9** can be substituted by fluorine treating **2** or **9** with ZnF_2 in *THF*. For the preparation of the fluoro derivative **3**, the mixture of isomers **2**-*trans* and **2**-*cis* was used. The resulting product **3** represents only one isomer as concluded from NMR data. This information suggests a rearrangement of the coordination sites of the *salen** ligand during nucleophilic substitution of chlorine by fluorine. The mechanism of this reaction as well as the energy differences between **2**-*trans* and **2**-*cis* seem to be quite interesting and will be the subject of further investigations. Complex **3** is much more soluble in organic solvents than the chloro derivative. Therefore, we were able to obtain single crystals of **3** by recrystallization from acetonitrile. The X-ray crystal structure analysis of **3** provides the structure shown in Fig. 3. Atomic coordinates, bond lengths, and angles are listed in Tables 5, 6, and 7. Crystallographic data and data collection parameters are given in Table 8.

3 represents the first structurally characterized hexacoordinate silicon compound with a tetradentate azomethine ligand. The structure was solved and refined in space group $P2_1/a$ with four molecules **3** and four molecules acetonitrile per unit cell. There is a distorted octahedral coordination geometry around silicon. The fluorine atoms are situated at axial positions (*trans* isomer), whereas the *salen** ligand occupies the four equatorial positions.

There are a number of crystal structures of *bis*-chelate compounds with hexacoordinate silicon. Most of these have essentially a tetrahedral arrangement around silicon with the coordinated nitrogen donor atoms capping the tetrahedra at relatively large distances (N–Si between 2.5–3.0 Å). [15] The Si–F distance in **3** corresponds well with bond lengths found in other hypervalent silicon compounds (1.60–1.73 Å) [15, 16]. The distances Si–O and Si–N are remarkably short compared to other hexacoordinate silicon compounds [15]. The distortion of the octahedral coordination environment around silicon probably originates from the conformation of the chelating *salen** ligand. There is a considerable torsion along the atoms N(1)–C(8)–C(17)–N(2) (about 46°). This causes one azomethine unit to fold under and the other one above the plane N(1)–Si–N(2).

Very promising are experiments to obtain stacked polymers **4** by reaction of **2** or **3** with alkaline metals or magnesium (Scheme 2). These polymers are only

Table 5. Atomic coordinates ($\times 10^4$) and equivalent isotropic displacement parameters ($Å^2 \times 10^3$) for **3**; $U(eq)$ is defined as one third of the trace of the orthogonalized U_{ij} tensor

	x	y	z	$U(eq)$
Si	3589(1)	1306(1)	2583(1)	19(1)
F(1)	4444(1)	134(1)	2469(1)	24(1)
F(2)	2885(1)	2582(1)	2749(1)	27(1)
N(1)	4581(1)	1773(1)	3745(1)	21(1)
O(1)	2805(1)	398(1)	3224(1)	27(1)
C(1)	2717(1)	407(1)	4164(1)	22(1)
C(2)	1815(2)	−230(2)	4432(1)	26(1)
C(3)	1665(2)	−284(2)	5397(1)	31(1)
C(4)	2395(2)	292(2)	6119(1)	34(1)
C(5)	3291(2)	899(2)	5868(1)	30(1)
C(6)	3481(1)	970(1)	4888(1)	23(1)
C(7)	4473(1)	1568(1)	4647(1)	22(1)
C(8)	5554(2)	2408(2)	3498(1)	27(1)
C(9)	5356(2)	1919(2)	5474(1)	29(1)
N(2)	4511(1)	2315(1)	1899(1)	22(1)
O(2)	2735(1)	958(1)	1514(1)	25(1)
C(10)	3054(1)	744(2)	651(1)	23(1)
C(11)	2452(2)	−107(2)	22(1)	27(1)
C(12)	2750(2)	−382(2)	−875(1)	30(1)
C(13)	3652(2)	192(2)	−1171(1)	30(1)
C(14)	4222(2)	1066(2)	−575(1)	28(1)
C(15)	3942(1)	1369(2)	343(1)	23(1)
C(16)	4572(1)	2284(2)	976(1)	24(1)
C(17)	5199(2)	3149(2)	2573(1)	27(1)
C(18)	5283(2)	3179(2)	529(1)	33(1)
N(3)	5148(3)	3569(3)	7748(2)	72(1)
C(19)	5933(2)	3026(2)	7867(2)	48(1)
C(20)	6935(2)	2319(2)	8012(2)	60(1)

slightly soluble in organic solvents and more stable against moisture than the monomers. They have been characterized by elemental analysis, IR, and NMR . Furthermore, it is possible to react **2** with dilithiumacetylide or the corresponding *Grignard* reagent to obtain polymers of type **5** which are linked by Si−C≡C−Si units (Scheme 1). Polymers with similar backbone but different chelating ligands have been described recently [17].

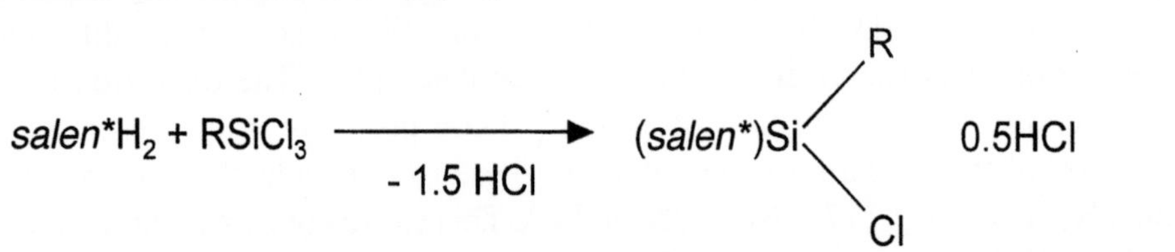

(1)

R=Me (**6**·0.5HCl), Ph (**7**·0.5HCl)

Table 6. Bond lengths (Å) for **3**

Bond	Distance	Bond	Distance
Si–F(2)	1.670(1)	C(7)–C(9)	1.506(2)
Si–F(1)	1.677(1)	C(8)–C(17)	1.519(2)
Si–O(1)	1.721(1)	N(2)–C(16)	1.294(2)
Si–O(2)	1.724(1)	N(2)–C(17)	1.473(2)
Si–N(2)	1.931(2)	O(2)–C(10)	1.336(2)
Si–N(1)	1.937(1)	C(10)–C(11)	1.402(2)
N(1)–C(7)	1.296(2)	C(10)–C(15)	1.412(2)
N(1)–C(8)	1.472(2)	C(11)–C(12)	1.382(2)
O(1)–C(1)	1.324(2)	C(12)–C(13)	1.394(3)
C(1)–C(6)	1.408(2)	C(13)–C(14)	1.378(3)
C(1)–C(2)	1.410(2)	C(14)–C(15)	1.407(2)
C(2)–C(3)	1.380(2)	C(15)–C(16)	1.466(2)
C(3)–C(4)	1.389(3)	C(16)–C(18)	1.506(2)
C(4)–C(5)	1.380(3)	N(3)–C(19)	1.127(4)
C(5)–C(6)	1.415(2)	C(19)–C(20)	1.446(4)
C(6)–C(7)	1.472(2)		

The reaction between *salen**H_2 and organofunctionalized trichlorosilanes gives adducts containing 0.5 equivalents of hydrogen chloride (Eq. (1)). Thermogravimetric analysis shows that the adducts **6** · 0.5HCl and **7** · 0.5HCl lose hydrogen chloride at higher temperatures (**6** · 0.5HCl from 140 to 270°C, **7** · 0.5HCl from 100 to 260°C). The association of hydrogen chloride with complexes **6** and **7** is partially reversed in solution as concluded from the ^{29}Si NMR spectra of these adducts which are identical with those of (*salen**)Si(Cl)*R*.

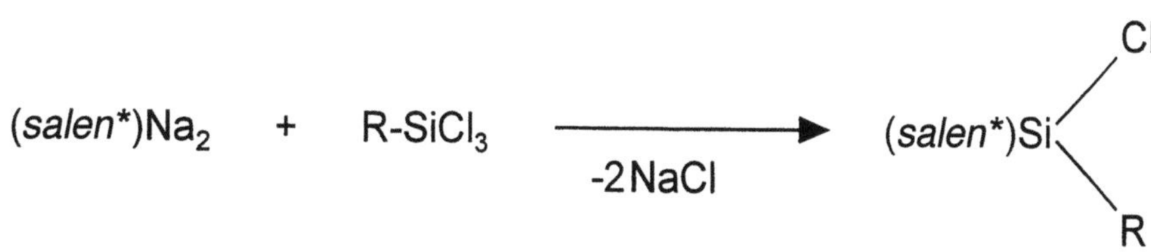

(2)

R=Me(**6**), Ph(**7**)

The complexes (*salen**)Si(Cl)*R* are available by reaction of (*salen**)Na_2 with $RSiCl_3$ (R = Me, Ph; Eq. (2)). Only one isomer is formed in both cases as can be seen from the NMR data. At the moment it is not possible to decide whether these are the *cis*- or *trans*-isomers. *Bis*(organo)silicon-*salen* complexes (*salen**)SiR_2 are accessible by reaction of lithium alkyl compounds Li*R* with (*salen**)$SiCl_2$ (**2**); for instance, (*salen**)Si(n-C_4H_9)$_2$ (**8**) results from the interaction between **2** and n-C_4H_9Li in *n*-hexane. These organo silicon compounds are better soluble in organic solvents than **2**.

We were successful in preparing (*salen*‡)$SiCl_2$ (**9**), (*salen*‡)SiF_2 (**12**), and (*salen*‡)Si(Cl)*R* (**10** and **11**) by methods analogous to those described for the *salen** derivatives. The nucleophilic substitution of the two chloro substituents in

Table 7. Angles (°) for **3**

Structural fragment	Angle	Structural fragment	Angle
F(2)–Si–F(1)	172.40(5)	C(1)–C(6)–C(5)	117.9(2)
F(2)–Si–O(1)	93.17(6)	C(1)–C(6)–C(7)	121.5(1)
F(1)–Si–O(1)	91.68(6)	C(5)–C(6)–C(7)	120.5(2)
F(2)–Si–O(2)	92.00(6)	N(1)–C(7)–C(6)	120.5(2)
F(1)–Si–O(2)	93.85(6)	N(1)–C(7)–C(9)	121.3(2)
O(1)–Si–O(2)	89.87(6)	C(6)–C(7)–C(9)	118.2(2)
F(2)–Si–N(2)	87.60(6)	N(1)–C(8)–C(17)	107.6(1)
F(1)–Si–N(2)	87.36(6)	C(16)–N(2)–C(17)	121.8(1)
O(1)–Si–N(2)	178.00(6)	C(16)–N(2)–Si	126.8(1)
O(2)–Si–N(2)	91.94(6)	C(17)–N(2)–Si	111.4(1)
F(2)–Si–N(1)	86.9(5)	C(10)–O(2)–Si	125.4(1)
F(1)–Si–N(1)	86.93(5)	O(2)–C(10)–C(11)	117.9(2)
O(1)–Si–N(1)	93.27(6)	O(2)–C(10)–C(15)	122.5(2)
O(2)–Si–N(1)	176.75(6)	C(11)–C(10)–C(15)	119.5(2)
N(2)–Si–N(1)	84.93(6)	C(12)–C(11)–C(10)	120.7(2)
C(7)–N(1)–C(8)	120.9(1)	C(11)–C(12)–C(13)	120.3(2)
C(7)–N(1)–Si	127.6(1)	C(14)–C(13)–C(12)	119.3(2)
C(8)–N(1)–Si	111.5(1)	C(13)–C(14)–C(15)	121.8(2)
C(1)–O(1)–Si	129.9(1)	C(14)–C(15)–C(10)	118.2(2)
O(1)–C(1)–C(6)	123.8(2)	C(14)–C(15)–C(16)	121.0(2)
O(1)–C(1)–C(2)	116.3(2)	C(10)–C(15)–C(16)	120.8(2)
C(6)–C(1)–C(2)	119.8(2)	N(2)–C(16)–C(15)	120.2(2)
C(3)–C(2)–C(1)	120.4(2)	N(2)–C(16)–C(18)	121.1(2)
C(2)–C(3)–C(4)	120.6(2)	C(15)–C(16)–C(18)	118.8(2)
C(5)–C(4)–C(3)	119.5(2)	N(2)–C(17)–C(8)	105.9(1)
C(4)–C(5)–C(6)	121.7(2)	N(3)–C(19)–C(20)	179.4(3)

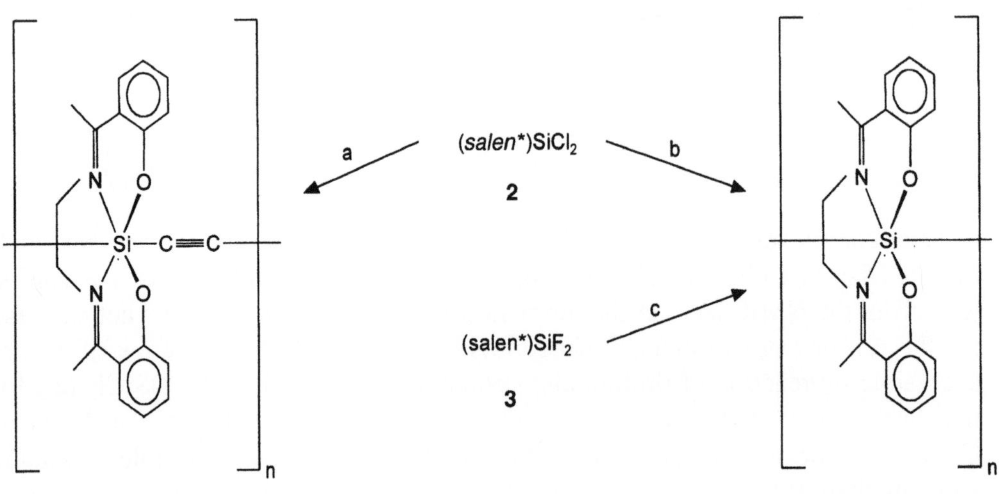

Scheme 2. a: LiCCLi or BrMgCCMgBr, *THF*/*n*-hexane, 6 h, reflux; b: (*salen**)$SiCl_2$ and K (1:2), toluene, 6 h, reflux; c: (*salen**)SiF_2 and Li, *THF*, 2 h, room temperature

(*salen*‡)$SiCl_2$ (**9**) should be possible under mild conditions. However, all attempts to perform this substitution with *Grignard* reagents, lithium organyls, or silyl lithium compounds failed so far, leading to decomposition of (*salen*‡)$SiCl_2$ in each case. (*Salen*‡)$SiCl_2$ (**9**) for instance reacted with MeMgI under extrusion of **1b**. We assume that the nucleophilic reagents attack the azomethine proton of the ligand. This initializes the decomposition of the *salen*‡ silicon complex.

The UV VIS spectra of the *salen** complexes in acetonitrile are very similar to each other. The long wavelength maxima ($n \rightarrow \pi^*$) are located between 310 and 330 nm. The formation of the (*salen*‡)SiX_2 complexes (X = Cl, F) gives rise to significant additional UV absorption in comparison with the free ligand *salen*$^‡H_2$. The band of the ligand at 265 nm is shifted bathochromically in both cases: (*salen*‡)$SiCl_2$ (**9**) to 280 nm, (*salen*‡)SiF_2 (**12**) to 284 nm. A similar effect occurs in the region above 300 nm. The absorption maximum of *salen*$^‡H_2$(332 nm) is moved to longer wavelengths for both complexes (356 nm). These facts lead to the conclusion that the lone pairs at the donor atoms of the ligand are less involved in bonds in the silicon compound than in the free ligand *salen*$^‡H_2$.

The compounds derived from the ligand *salen*‡ are considerably better soluble in common organic solvents than the analogous *salen** complexes due to the four *tert*-butyl groups at the ligand. The inaccessiability of new derivatives by nucleophilic substitution represents a major drawback of this class of compounds. Therefore, we will prepare modified ligand systems with protective alkyl groups at the azomethine carbon atom and additional alkyl groups at the phenyl rings for better solubility.

Experimental

IR spectra were recorded with a Nicolet 510 A spectrometer. Bruker DPX 400 (liquid) and MSL 300 (solid) spectrometers were used for measuring NMR spectra; chemical shifts are referenced to *TMS*. All solvents were distilled from sodium/benzophenone before use. The operations for preparing the silicon complexes were carried out under an atmosphere of Argon. N,N′-Ethylene-*bis*(2-hydroxy-acetophenoneimine) (*salen**H_2, **1a**) [18] and N,N′-ethylene-*bis*(3,5-di-*tert*-butyl-salicylideneimine) (*salen*$^‡H_2$, **1b**) [19] were synthesized as reported previously.

Dichloro-(N,N′-ethylene-bis(2-hydroxyacetophenoneiminato))-silane ((salen)*$SiCl_2$, **2**-*trans*/**2**-*cis*)

2.87 g (16.9 mmol) $SiCl_4$ or 2.27 g (8.45 mmol) Si_2Cl_6 were added dropwise to a solution of 5 g (16.9 mmol) (*salen**)H_2 in *THF*. The resulting mixture was refluxed for 2 h and cooled. The precipitated white powder was filtered off with suction, washed with *THF*, and dried under vacuum (5.7 g, 90%).

$C_{18}H_{18}N_2O_2SiCl_2$; calcd.: C 55.0, H 4.6, N 7.1, Si 7.1, Cl 18.1; found: C 54.8, H 5.4, N 7.2, Si 7.1, Cl 19.1; ^{13}C NMR (*DMSO*-d_6, mixture of isomers, δ in ppm): C_1 157.3/158.1, C_2 118.8/119.4, C_3 135.1/136.9, C_4 120.6/120.7, C_5 129.9/131.2, C_6 120.2/120.5, C_7 176.4/171.6, C_8 46.1/46.7, C_9 18.6/20.7; ^{29}Si NMR (DMSO-d_6, mixture of isomers, δ in ppm): −186.1/−188.0; IR: ν(C=N) 1630 cm^{-1}, ν(Si–Cl) 521 cm^{-1}, δ(Si–O–C) 1104 cm^{-1}.

From Si_2Cl_6: ^{13}C-NMR (*DMSO*-d_6, δ, ppm): C_1 157.3, C_2 119.4, C_3 136.9, C_4 120.7, C_5 131.2, C_6 120.5, C_7 176.4, C_8 46.7, C_9 20.7; ^{1}H NMR (DMSO-d_6, δ, ppm): H_2–H_5 6.8–8.1, H_8 4.2, H_9 1.5; ^{29}Si NMR (*DMSO*-d_6, δ, ppm): −188.0.

N,N′-Ethylene-bis(2-hydroxyacetophenoneiminato)-difluorosilane (*(salen**)SIF_2, **3**)

2.6 g (25.4 mmol) ZnF_2 were added to a stirred suspension of 5 g (12.7 mmol) *(salen*)$SiCl_2$ in THF* during 30 min. The mixture was heated for one hour under reflux and then cooled. The precipitated solid material was extracted with *THF*. After extraction, the filtered white compound was washed with *THF* and dried under vacuum (0.9 g, 20%).

$C_{18}H_{18}N_2O_2SiF_2$; calcd.: C 60.0, H 5.0, N 7.8; found: C 54.5, H 4.9, N 7.6[a]; ^{13}C NMR ($CDCl_3$, δ, ppm): C_1 160.3, C_2–C_5 134.4–116.9, C_6 121.9, C_7 170.7, C_8 46.4, C_9 18.2; ^1H NMR ($CDCl_3$, δ, ppm): H_2–H_5 6.7–8.1, H_8 4.1, H_9 2.3; ^{29}Si NMR ($CDCl_3$, δ, ppm): −187.9, t, ^2J(Si–F) = 179.2 Hz; IR: ν(C=N) 1610 cm^{-1}, ν(Si–F) 954 cm^{-1}.

Poly-(N,N′-ethylene-bis(2-hydroxyacetophenoneiminato))-silane((-salen)*Si*-)n*, **4**)

1 g K was added to a stirred suspension of 5 g (*salen**)$SiCl_2$ in toluene and heated for 6 h under reflux. After cooling the reaction mixture was filtered. From the dark clear filtrate the product was obtained by addition of *n*-hexane (0.4 g, 10%).

($C_{18}H_{18}N_2O_2Si$)*n*; calcd.: C 67.1, H 5.6, N 8.7, Si 8.7; found: C 62.7, H 6.0, N 7.5, Si 8.2[a]; ^{13}C NMR ($CDCl_3$, δ, ppm): $C_{1/7}$ 156 (broad); $C_{2\text{-}6}$ 137–116, C_8 47, C_9 22; ^1H NMR ($CDCl_3$, δ, ppm): $H_{2\text{-}5}$ 6.8–7.5, H_8 3.9, H_9 1.5; ^{29}Si NMR (solid state, δ, ppm): −130 (broad); IR: ν (C=N) 1613 cm^{-1}, δ(Si–O–C) 1100 cm^{-1}.

Poly-(μ-ethinyl-(N,N′-ethylene-bis(2-hydroxyacetophenoneiminato)))-silane((-salen)*SiCC*-)n*, **5**)

16 mmol LiCCLi or 16 mmol BrMgCCMgBr were slowly added to a stirred suspension of 5 g (12.7 mmol) *salen**$SiCl_2$ in *THF*. The mixture was heated under reflux for 6 h. The light yellow precipitate was filtered off with suction. It can be purified by extraction with *THF* (0.5 g, 10%).

($C_{20}H_{18}N_2O_2Si$)*n*; calcd.: C 69.4, H 5.2, N 8.1, Si 8.1; found: C 62.2, H 5.9, N 7.9, Si 7.4; ^{13}C NMR (*DMSO*-d_6, δ, ppm): C_1 168.4, $C_{2\text{-}6}$ 132.6–117, C_7 169.1, C_8 51.4, C_9 22.2; ^1H NMR (*DMSO*-d_6, δ, ppm): $H_{2\text{-}5}$ 6.2–7.6, H_8 3.6, H_9 0.86; ^{29}Si NMR (solid state, δ, ppm): −88 (broad); IR: ν(C=N) 1607 cm^{-1}, ν(Si–C) 840 cm^{-1}; *Raman*: ν(C≡C) 2260 cm^{-1}.

Chloro-(N,N′-ethylene-bis(2-hydroxyacetophenoneiminato))-methylsilane hemi hydrogen chloride (*(salen**)Si(Cl)Me · 0.5HCl, **6** · 0.5HCl)

2.5 g (16.9 mmol) $MeSiCl_3$ were added to a stirred solution of 5 g (16.9 mmol) (*salen**)H_2 in *THF*; the resulting mixture was refluxed for 4 h. After cooling to room temperature, the precipitated white product was filtered off with suction, washed with *THF* and dried under vacuum (5.9 g, 90%).

$C_{19}H_{21}N_2O_2SiCl \cdot 0.5HCl$; Calcd.: C 58.3, H 5.5, N 7.2, Si 7.2, Cl 13.6; found: C 59.1, H 6.0, N 7.3, Si 6.8, Cl 14.8[a]; ^{13}C NMR(*DMSO*-d_6, δ, ppm): C_1 159.1/154.0, $C_{2\text{-}6}$ 136.2–116, C_7 185.4/181.4, C_8 46.2, C_9 23.7, C_{14} 6.5; ^1H NMR (*DMSO*-d_6, δ, ppm): $H_{2\text{-}5}$ 6.9–7.7, H_8 4.3, H_9 1.75; ^{29}Si NMR (*DMSO*-d_6, δ, ppm): −150.7; ^{29}Si NMR (solid state, δ, ppm): −105.3; IR: ν(C=N) 1629 cm^{-1}, ν(Si–Cl) 509/526 cm^{-1}, ν(Si–C) 799 cm^{-1}, δ(Si–O–C) 1110 cm^{-1}.

Chloro-(N,N′-ethylene-bis(2-hydroxyacetophenoneiminato))-phenylsilane hemi hydrogen chloride (*(salen**)Si(Cl)Ph · 0.5HCl, **7** · 0.5HCl)

7 · 0.5 HCl was prepared as described for **6** · 0.5HCl (6.9 g, 90%).

$C_{24}H_{23}N_2O_2SiCl \cdot 0.5HCl$; ^{13}C NMR(*DMSO*-d_6, δ, ppm): C_1 158.6/154.0, $C_{2\text{-}6}$/$C_{14a\text{-}d}$ 136.0–116.7, C_7 185.5/182.1, C_8 45.9, C_9 19.4; ^1H NMR (*DMSO*-d_6, δ, ppm): $H_{2\text{-}5}$ 6.7–7.9, H_8 4.0, H_9 2.4;

[a] Determined carbon content too low owing to the formation of silicon carbide during analysis

^{29}Si NMR(*DMSO*-d$_6$, δ, ppm): −173.1; IR: ν(C=N) 1615 cm^{-1}, ν(Si–Cl)514 cm^{-1}, ν(Si–C)840 cm^{-1}, δ(Si–O–C)1110 cm^{-1}.

Chloro-(N,N′-ethylene-bis(2-hydroxyacetophenoneiminato))-methylsilane ((salen)*Si(Cl)Me, **6***)*

33.8 mmol NaOMe were added to a stirred mixture of 5 g (16.9 mmol) *salen**H$_2$ in *THF*/MeOH. After 30 min the disodium salt of the ligand was precipitated by addition of hexane. It was filtered off and dried under vacuum. A mixture of 1.7 g (11.1 mmol) MeSiCl$_3$ and 4 g (11.1 mmol) (*salen**)Na$_2$ was stirred for 30 min in *THF* at room temperature; then the precipitated product suspension was filtered. Finally, the solid was extracted with *THF*/toluene (0.83 g, 20%).

$C_{19}H_{21}N_2O_2SiCl$; calcd.: C 61.1, H 5.6, N 7.5, Si 7.5, Cl 9.5; found: C 56.6, H 5.2, N 7.4, Si 7.1, Cl 10.6[a]; ^{13}C NMR (*DMSO*-d$_6$, δ, ppm): C$_1$ 158.4, C$_{2-6}$ 136.2–116.0, C$_7$ 174.8, C$_8$ 46.0, C$_9$ 20.4, C$_{14}$ 6.6; ^{1}H NMR (*DMSO*-d$_6$, δ, ppm): H$_{2-5}$ 6.7–7.9, H$_8$ 4.2, H$_9$ 1.75, H$_{14}$ 0.2; ^{29}Si NMR (*DMSO*-d$_6$, δ, ppm): −150.7; ^{29}Si NMR (solid state, δ, ppm): −168.7; IR: ν(C=N) 1629 cm^{-1}, ν(Si–Cl) 509 cm^{-1}, ν(Si–C) 798 cm^{-1}.

Chloro-(N, N′-ethylene-bis(2-hydroxyacetophenoneiminato))-phenylsilane ((salen)*Si(Cl)Ph,**7***)*

7 was prepared as described for **6** (1.45 g, 30%).

$C_{24}H_{23}N_2O_2SiCl$; calcd.: C 66.3, H 5.3, N 6.4, Si 6.4, Cl 8.2; found: C 64.7, H 4.9, N 6.7, Si 5.8, Cl 9.1[a]; ^{13}C NMR (*DMSO*-d$_6$, δ, ppm): C$_1$ 159.2, C$_{2-6}$/C$_{14a-d}$ 136.0–116.0, C$_7$ 174.6, C$_8$ 46.1, C$_9$ 19.1; ^{1}H NMR (*DMSO*-d$_6$, δ, ppm): H$_{2-5}$/H$_{14b-d}$ 6.7–7.7, H$_8$ 4.2, H$_9$ 1.75; ^{29}Si NMR (*DMSO*-d$_6$, δ, ppm): −173.1; IR: ν(C=N) 1612 cm^{-1}, ν(Si–Cl) 488 cm^{-1}, ν(Si–C) 840 cm^{-1}.

Dibutyl-(N,N′-ethylene-bis(2-hydroxyacetophenoneiminato))-salen ((salen)*Si(*n*-C$_4$H$_9$)$_2$, **8***)*

A solution of 33.8 mmol *n*-BuLi in *n*-hexane was added dropwise to an ice-cooled stirred suspension of 5 g (16.9 mmol) *salen**SiCl$_2$. After stirring for 2 days the mixture was concentrated, and the formed yellow precipitate was extracted with *THF* (1.40 g, 30%).

$C_{26}H_{36}N_2O_2Si$; ^{13}C NMR (*DMSO*-d$_6$, δ, ppm): C$_1$ 168.7, C$_7$ 169.2, C$_8$ 44.9, C$_9$ 18.0, Si-Bu; ^{29}Si NMR (*DMSO*-d$_6$, δ, ppm): −215.5; IR: ν(C=N) 1612 cm^{-1}, ν(Si–C) 882 cm^{-1}, δ(Si–O–C) 1050–1110 cm^{-1}.

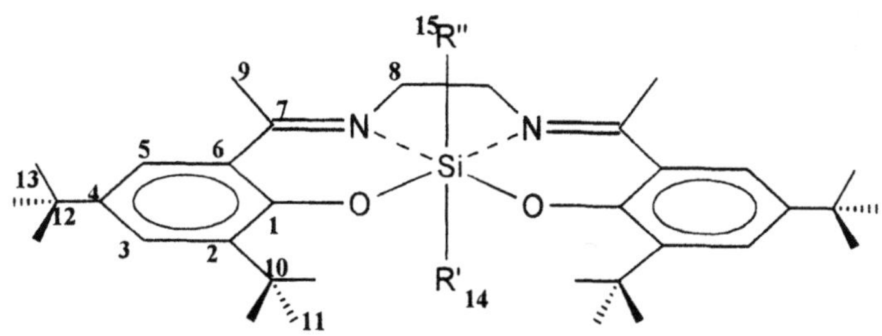

Numbering scheme for the NMR data of compounds **9–12**

*Dichloro-(N,N′-ethylene-bis(3,5-di-tert-butyl-salicylideneiminato))-silane ((salen‡)*SiCl$_2$, **9***)*

A 100 ml *Schlenk* flask equipped with a magnetic stirring bar was charged with N,N′-ethylene-*bis*(3,5-di-*tert*-butyl-salicylideneimine) (1.6 g, 3 mmol) in 50 ml *THF* and silicon tetrachloride (0.52 g, 0.35 ml, 3 mmol). After 15 min, the mixture was heated under reflux for 3 h; then it was cooled to the room temperature. The mixture volume was diminished to 10 ml by evaporation and 40 ml hexane were added. The yellow product was filtered off and dried under vacuum (yield: 90%).

$C_{32}H_{46}Cl_2N_2O_2Si$ (589.8); calcd.: C 65.16, H 7.86, N 4.75; found: C 64.25, H 7.95, N 6.08; IR (KBr disk): $\nu = 2958$, 2910, 2871 (CH), 1654, 1619 (C=N), 1556, 1475, 1463, 1428 (C=CH) cm^{-1}; ^1H NMR ($CDCl_3$, δ, ppm): 8.07 (2H, N–CH), 7.64 and 7.07 (4H, phenyl), 4.25 (4H, CH_2), 1.56 and 1.30 (36H, C–CH_3); ^{13}C NMR ($CDCl_3$, δ, ppm): C_7 165.4, C_1 157.8, C_6 116.3, C_8 51.9; ^{29}Si NMR ($CDCl_3$, δ, ppm): −187.6.

The organo(chloro) derivatives (*salen*‡)Si(Cl)*R* (**10** and **11**) were obtained by the same procedure.

Chloro-(N,N′-ethylene-bis(3,5-di-tert-butyl-salicylideneiminato))-methylsilane ((salen‡)Si(Cl)Me, **10***)*

Yield: 94%; $C_{33}H_{49}ClN_2O_2Si$; IR (KBr disk): $\nu = 2960$, 2909, 2871 (CH), 1652, 1619 (C=N), 1568, 1476, 1445, 1427 (C=CH) cm^{-1}; ^1H NMR ($CDCl_3$, δ, ppm): 9.37 (2H, N=CH), 7.75 and 7.53 (4H, phenyl), 4.50 and 4.64 (4H, CH_2), 1.50 and 1.34 (36H, C–CH_3), 0.45 (3H, Si–CH_3); ^{13}C NMR ($CDCl_3$, δ, ppm): C_7 173.2, C_1 156.0, C_6 117.1, C_8 53.9, C_{14} 2.4; ^{29}Si NMR ($CDCl_3$, δ, ppm): −104.5.

Chloro-(N,N′-ethylene-bis(3,5-di-tert-butyl-salicylideneiminato))-phenylsilane (salen‡*)*Si(Cl)Ph, **11***)*

Yield: 83%; $C_{38}H_{51}ClN_2O_2Si$; IR (KBr disk): $\nu = 2964$, 2907, 2871 (CH), 1648, 1617 (C=N), 1571, 1472, 1442, 1431 (C=CH) cm^{-1}; ^1H NMR ($CDCl_3$, δ, ppm): 9.66 (2H, N=CH), 7.76 and 7.56 (4H, phenyl), 4.56 and 4.33 (4H, CH_2), 1.38 and 1.35 (36H, C–CH_3), 7.16–7.28 (5H, Si-Ph); ^{13}C NMR ($CDCl_3$, δ, ppm): C_7 174.1, C_1 157.7, C_6 117.3, C_8 52.8, $C_{14a\text{–}d}$ 137.1, 130.8, 130.5, 128.2; ^{29}Si NMR ($CDCl_3$, δ, ppm): −118.5.

N,N′-Ethylene-bis(3,5-di-tert-butyl-salicylideneiminato)-difluorosilane ((salen‡*)*SiF_2, **12***)*

A 100 ml *Schlenk* flask equipped with a magnetic stirring bar was charged with dichloro-(N,N′-ethylene-*bis*(3,5-di-*tert*-butyl-salicylideneiminato))-silane (2 g, 3.4 mmol) in 50 ml *THF* and ZnF_2 (1 g, 9.7 mmol). The mixture was heated under reflux for 8 h and then cooled. After filtration, the solution volume was diminished to 10 ml by evaporation of *THF*; then, 40 ml hexane were added. The white product was filtered and dried under vacuum (yield: 94%). The solid is stable on air and moisture. Single crystals for X-ray structure analysis were obtained from acetic ester/cyclohexane.

$C_{32}H_{46}F_2N_2O_2Si$; IR (KBr disk): $\nu = 2964$, 2871 (CH), 1648(C=N), 1565, 1477, 1452, 1426 (C=CH) cm^{-1}; ^1H NMR ($CDCl_3$, δ, ppm): 8.16 (2H, N=CH), 7.57 and 7.05 (4H, phenyl), 4.04 (4H, CH_2), 1.57 and 1.29 (36H, C–CH_3); ^{13}C NMR ($CDCl_3$, δ, ppm): C_7 164.7, C_1 159.3. C_6 116.8, C_8 53.9; ^{29}Si NMR ($CDCl_3$, δ, ppm): −187.1 ($^1J_{Si\text{-}F} = 172.6$ Hz).

X-ray data collection and structure determination of **1a**, **1b**, *and* **3**

Single crystals were mounted on glass fibres and transferred to the diffractometer. Data were collected on an Enraf-Nonius CAD-4 diffractometer using graphite monochromated Cu-K_α radiation ($\lambda = 1.5418$ Å) with ω-2Θ scans. The structures were solved by direct methods and refined by full matrix least squares on F^2 with anisotropic thermal parameters. Idealized positions for the hydrogen atoms were included for **1b** as fixed contributions by using a riding model. In **1a** and **3**, hydrogen atom positions were located and refined isotropically. Atomic coordinates, bond lengths, angles, and thermal parameters have been deposited at the Cambridge Crystallographic Data Centre (CCDC).

Table 8. Crystal data and structure refinement for compounds **1a**, **1b**, and **3**

	1a	**1b**	**3**
Formula	$C_{18}H_{20}N_2O_2$	$C_{32}H_{48}N_2O_2$	$C_{20}H_{21}F_2N_3O_2Si$
Formula weight	296.36	492,72	401.49
Temperature	293(2) K	293(2) K	172(2) K
Wavelength	1.54178 Å	1.54178 Å	1.54180 Å
Crystal system	monoclinic	monoclinic	monoclinic
Space group	$P2_1/n$ (#14)	C2/c (#15)	$P2_1/a$ (#14)
a	5.762(1) Å	18.921 (2) Å	12.3826(5) Å
b	20.450(1) Å	6.207(1) Å	10.8405(5) Å
c	6.804(1) Å	27.478(2) Å	13.8507(5) Å
β	104.45(1)°	106.55(1)°	98.800(5)°
V	776.4(2) $Å^3$	3093.4(6) $Å^3$	1837.34(13) $Å^3$
Z	2	4	4
Density (calculated)	1.268 g/cm^3	1.058 g/cm^3	1.451 g/cm^3
Absorption coefficient	0.666 mm^{-1}	0.499 mm^{-1}	1.503 mm^{-1}
F(000)	316	1080	840
Crystal size	0.2×0.5×0.6 mm	0.15×0.5×1.0 mm	0.2×0.5×0.6 mm
Θ range for data collection	4.32 to 74.89°	3.36 to 74.81°	3.23 to 74.98°
Index ranges	$-7\leq h\leq 6$ $-25\leq k\leq 25$ $0\leq l\leq 8$	$-23\leq h\leq 23$ $-2\leq k\leq 7$ $-34\leq l\leq 34$	$0\leq h\leq 15$ $-13\leq k\leq 13$ $-17\leq l\leq 17$
Reflections collected	3407	8734	7415
Independent reflections	1604 (*R*(int) = 0.0189)	3185 (*R*(int) = 0.0414)	3784 (*R*(int) = 0.0491)
Refinement method		Full-matrix least-squares on F^2	
Absorption correction	none	none	Semi-empirical from psi-scans
Max. and min. transmission	–	–	0.9984 and 0.8824
Data/restraints/parameters	1604/0/140	3185/4/201	3783/0/304
Goodness-of-fit on F^2	1.014	1.024	1.032
Final *R* indices ($I>2\sigma(I)$)	$R_1=0.0407$ $wR_2=0.1186$	$R_1=0.0557$ $wR_2=0.1479$	$R_1=0.0431$ $wR_2=0.1152$
R indices (all data)	$R_1=0.0449$ $wR_2=0.1241$	$R_1=0.0643$ $wR_2=0.1596$	$R_1=0.0482$ $wR_2=0.1200$
Largest diff. peak and hole	0.253 and $-0.244\,e\cdot Å^{-3}$	0.216 and $-0.266\,e\cdot Å^{-3}$	0.349 and $-0.453\,e\cdot Å^{-3}$

References

[1] (a) Corriu RJP and Young JC (1989) In: Patai S, Rappoport Z (eds) The Chemistry of Organic Silicon Compounds. Wiley, Chichester, 1241; (b) Holmes RR (1990) Chem Rev **90**: 17; (1996) **96**: 927; (c) Corriu RJP (1990) J Organomet Chem **400**: 81; (d) Chuit C, Corriu RJP, Reye C, Young JC (1993) Chem Rev **93**: 1371; (e) Corriu RJP Young JC (1991) In: Patai S, Rappoport Z (eds) The Silicon-Heteroatom Bond. Wiley, Chichester, pp 1 and 49; (f) Auner N, Probst R, Hahn

F, Herdtweck E (1993) J Organomet Chem **459**: 25; (g) Pfrommer B, Tacke R (1998) Europ J Inorg Chem 415; (h) Tacke R, Mühleisen M (1994) Angew Chem **13**: 1431
[2] Hobday MD, Smith TD (1972) Coord Chem Rev **9**: 311 and references cited therein
[3] Nishinaga A, Tajima K, Speiser B, Eichhorn E, Rieker A, Ohya-Nishiguchi H, Ishizu K (1991) Chem Lett 1403
[4] Pratt JM, Craig PJ (1973) Adv Organomet Chem **11**: 331
[5] Schrauzer GN, Sibert JW, Windgassen RJ (1968) J Amer Chem Soc **90**: 6681
[6] Schrauzer GN (1976) Angew Chem **88**: 465
[7] Cameron JH, Morgan PC, Turner SC (1990) J Chem Soc Chem Commun 1617
[8] Horowitz CP, Creager SE, Murray R (1990) Inorg Chem **29**: 1006
[9] Jacobsen EN, Zhang W, Güler ML (1991) J Amer Chem Soc **113**: 6703
[10] Jacobsen EN, Zhang W, Muci AR, Ecker JR, Deng L (1991) J Amer Chem Soc **113**: 7063
[11] (a) Ghose BN (1985) Acta Chimica Hungarica **118**: 191; (b) Siddiqi KS, Aqra FMAM, Shah SA, Zaidi SAA, (1993) Polyhedron **12**: 1967
[12] Compare for instance the wrong data of elemental analysis in [11b]. The calculated sum formula was given with $C_{14}H_{14}N_2Cl_2Si$ and the found data fitted this formula well. The correct formula should be $C_{16}H_{14}N_2O_2Cl_2Si$!
[13] Altmann R, Regen O, Schneider J (1984) Chemisch-technische Stoffwerte – eine Datensammlung. VEB Deutscher Verlag der Grundstoffindustrie, Leipzig
[14] Ruddick JNR, Sams JR (1975) J Inorg Nucl Chem **37**: 564
[15] (a)Breliere C, Carre F, Corriu RJP, Poirier M, Royo G, Zwecker J (1989) Organometallics **8**: 1831; (b) Carre F, Cerveau G, Chuit C, Corriu RJP, Reye C (1992) New J Chem **16**: 63; (c) Carre F, Chuit C, Corriu RJP, Mehdi A, Reye C (1995) Organometallics **14**: 2754
[16] Kalikhman I, Krivonos S, Stalke D, Kottke T, Kost D (1997) Organometallics **16**: 3255
[17] (a) Boyer-Elma K, Carré FH, Corriu RJP, Douglas WE (1995) Chem Commun 725; (b) Shim IW, Risen WM (1984) J Organomet Chem **260**: 171
[18] Pfeiffer P, Breit E, Lübbe E, Tsumaki T (1933) Liebigs Ann Chem **503**: 84, 127
[19] Larrow JF, Jacobsen EN (1994) J Org Chem **59**: 1939

Received July 8, 1998. Accepted (revised) August 20, 1998

Hexa(amino)disilanes with Saturated Cyclic Amino Ligands

Gerald Huber and **Hubert Schmidbaur***

Anorganisch-chemisches Institut der Technischen Universität München, D-85747 Garching, Germany

Summary. *Hexakis*-(N-aziridino)-, -(N-azetidino)-, -(N-pyrrolidino)-, and -(N-piperidino)-disilane ($Si_2(N(CH_2)_n)_6$, $n=2$, 3, 4, and 5) have been prepared from Si_2Cl_6 and an excess of the corresponding cyclic secondary amine in diethyl ether. For the synthesis of the piperidino compound, N-lithio-piperidine was required as a stronger nucleophile to overcome the increasing steric hindrance with this largest of the four amines. The volatile small ring compounds ($n=2$: **1**; $n=3$: **2**) are thermally labile and undergo decomposition upon storage at ambient temperature (mainly through oligomerization *via* opening of the strained small rings). Compounds **3** ($n=4$) and **4** ($n=5$) are stable if kept in an inert atmosphere.

Keywords. Disilane; Aminodisilane; Hexa(amino)disilane; Aziridino-disilane; Azetidino-disilane; Pyrrolidino-disilane; Piperidino-disilane.

Hexa(amino)disilane mit gesättigten cyclischen Aminliganden

Zusammenfassung. *Hexakis*-(N-aziridino)-, -(N-azetidino)-, -(N-pyrrolidino)- und -(N-piperidino)-disilan ($Si_2(N(CH_2)_n)_6$, $n=2$, 3, 4 und 5) wurden ausgehend von Si_2Cl_6 und einem Überschuß des entsprechenden sekundären cyclischen Amins in Diethylether dargestellt. Wegen des hohen sterischen Anspruchs der Piperidinliganden mußte zur Synthese der mit diesem größten Amin substituierten Verbindung zusätzlich N-Lithio-piperidin als stärkeres Nucleophil eingesetzt werden. Die beiden Verbindungen mit den kleinen Heterocyclen ($n=2$: **1**, $n=3$: **2**) sind flüchtig, thermisch labil und zersetzen sich allmählich bei Raumtemperatur (hauptsächlich durch Öffnung der kleinen gespannten Ringe und anschließende Oligomerisierung). Die Verbindungen **3** ($n=4$) und **4** ($n=5$) sind unter Inertgasatmosphäre stabil.

Introduction

Disilane (Si_2H_6) is known to be sensitive towards base-catalyzed degradation which leads to monosilane (SiH_4) and higher silanes. A similar reaction pattern has been observed for several substituted disilanes which also undergo rearrangement with cleavage of the Si–Si bond upon treatment with *e.g.* tertiary amines. A related process initiated by (amine) base is employed for the conversion of polychlorinated di- or polysilanes into the corresponding monosilanes [1–5].

---

* Corresponding author

It is therefore no surprise that very few amino substituted disilanes have been prepared in which the presence of an internal basic function is likely to promote decomposition. We have recently developed reliable synthetic pathways to a variety of amino-polysilanes which could be shown or are expected to be a new family of volatile precursors for the deposition of silicon nitride from the vapour phase [6, 7].

As part of this current program we became interested in the properties of disilanes fully substituted with small cyclic amino groups. These groups should enhance the volatility of the compounds and facilitate thermal elimination of fragments of the substituents.

Per(amino)disilanes are few in number in the literature [8–10]. A prominent example is *hexakis*(dimethylamino)disilane, first described by *Wiberg et al.* in 1965, but obtained in pure form and structurally characterized much later by *Verkade et al.* (in 1993). The dodeca-ethyl homologue was also included in this recent work [10].

We recently revisited the preparation, properties, and structure of several *tetrakis*(amino)monosilanes with cyclic amino substituents ($(CH_2)_nN$-, $n = 2, 3, 4$, and 5) [11, 12]. This work has provided useful reference data for the *disilane* investigations reported in this communication.

Results and Discussion

Hexakis(N-aziridino)dissilane (**1**) can be prepared from hexachlorodisilane and excess aziridine in diethyl ether in the temperature range from −60 to +25°C. A precipitate of aziridinium hydrochloride is formed, and from the mother liquor the product can be isolated in low yield (14%) through sublimation in a vacuum and recrystallization from pentane. Compound **1** is a colourless crystalline material, m.p. = 42°C, which is very sensitive to moisture and readily decomposes when stored at room temperature. Even freshly sublimed or crystallized samples show oligomerization products in their mass spectra, indicating oligomerization through aziridine ring opening.

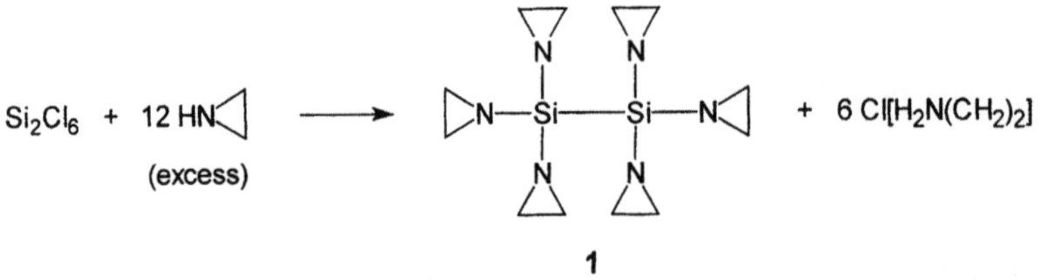

1

Scheme 1

Hexakis(N-*azetidino*)disilane (**2**) is obtained similarly with excess azetidine as the reagent for Si_2Cl_6. It is isolated as a colourless oil by vacuum distillation (b.p. = 120°C/0.01 mbar with decomposition, yield: 13%). Compound **2** is also sensitive to moisture, but thermally more stable. Its mass spectrum shows oligomerization products, although in lower intensity than observed for **1**.

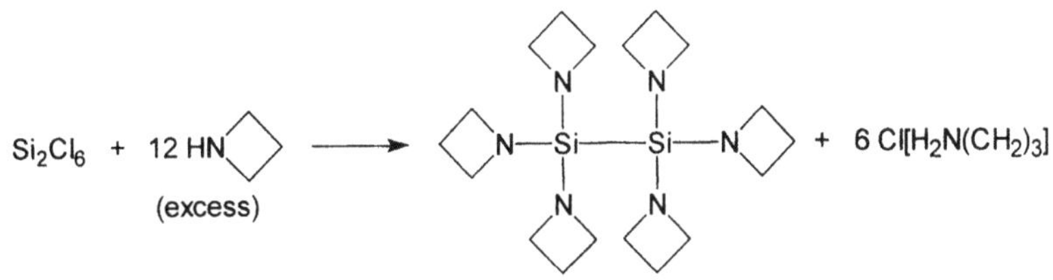

2

Scheme 2

Hexakis(N-*pyrrolidino*)disilane (**3**) is synthesized by an analogous procedure as described for **1** and **2**, but can be isolated as a crystalline solid (from pentane) in much higher yield (50%); it is less sensitive to temperature, air, and moisture. In the mass spectrum there are no peaks for ions with masses greater than the parent ion [$\mathbf{3}^+$]. Even after prolonged storage at room temperature no decomposition and oligomerization is observed.

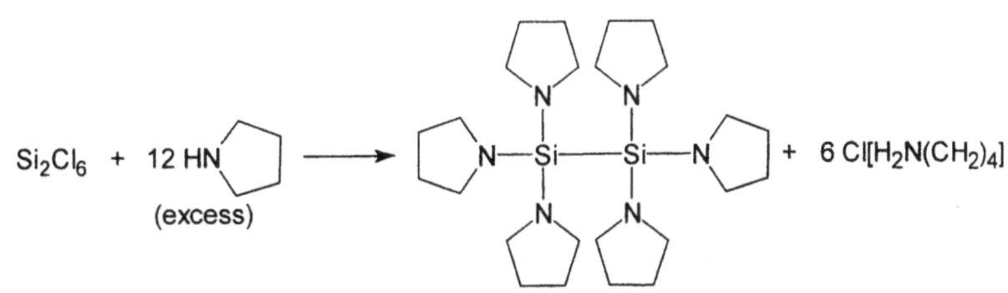

3

Scheme 3

Hexakis(N-*piperidino*)disilane (**4**) cannot be prepared directly from Si_2Cl_6 and excess piperidine; this reaction leads only to mixed chloro(N-piperidino)disilanes. Owing to steric hindrance, substitution of the chlorine atoms is becoming increasingly difficult as the number of piperidino groups is growing. The last two chlorine atoms therefore can only be replaced by the use of N-lithio-piperidine as a more powerful nucleophile. The overall yield of the fully substituted product is low (37%) even after long reaction times. As for compound **3**, purification of **4** is easily accomplished by crystallization from pentane. Both compounds have melting points beyond 200°C. Attempted vacuum sublimation/distillation is accompanied by decomposition. The mass spectrum shows the parent ion of the compound; no ions of higher mass were detected.

The NMR data of compounds **1–4** are in agreement with the proposed structures. The ^{29}Si resonances are difficult to detect owing to the adverse relaxation properties of the silicon nuclei in an asymmetrical environment of nitrogen quadrupolar nuclei. All nitrogen heterocycles undergo rapid inversion at the ring atoms, including the nitrogen atoms, rendering the CH_2 hydrogen atoms equivalent at all carbon atoms at room temperature (virtual C_{2v} symmetry for each ligand). This behaviour resembles that of the corresponding mononuclear tetra(amino)-silanes $Si(N(CH_2)_n)_4$, $n = 2–5$. The number of ^{13}C resonances is in agreement with

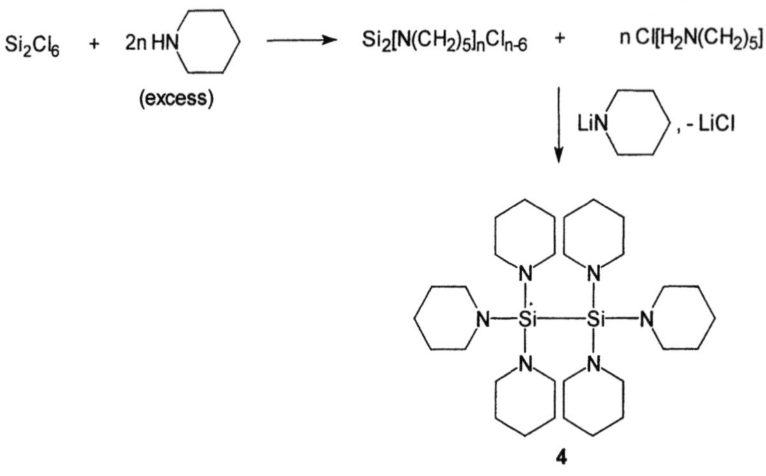

Scheme 4

at least local C_s symmetry at each ligand, or even virtual C_{2v} symmetry (on the NMR time scale). Attempts to grow single crystals of high quality were not successful.

Experimental

All experiments were carried out under an atmosphere of dry, pure nitrogen. Solvents and glassware were treated accordingly. Standard equipment was used throughout. Aziridine [13] and azetidine [14] were prepared following literature procedures. Due to the method of preparation, the sample of azetidine contained benzene which was quantified by ^{1}H NMR. *Tetrakis*(N-aziridino)silane was prepared as described earlier [12]. All other starting materials were commercially available. NMR: Jeol GX 400 and Jeol GX 270, solutions in C_6D_6 at 23°C unless otherwise stated; *TMS* as internal standard; MS: MAT 311 A (EI 70 eV).

Hexakis(N-aziridino)disilane (**1**)

A solution of aziridine (3.58 ml, 70 mmol) in 20 ml of diethyl ether was added within 15 min to a solution of hexa(chloro)disilane (0.86 ml, 5.0 mmol) in 50 ml of diethyl ether at −60°C. The reaction mixture was allowed to warm to room temperature. After stirring for 2 h at this temperature, the precipitate was separated and the solvent was removed from the filtrate under reduced pressure. The remaining colourless oil was purified by sublimation (100°C/0.01 mbar) and recrystallized from pentane.

Colourless crystals; yield: 0.21 g (14%); m.p.: 42°C; ^{1}H NMR: $\delta = 1.94$ (s, 24H, CH_2) ppm; $^{13}C\{^1H\}$ NMR: $\delta = 20.9$ (CH_2) ppm; $^{29}Si\{^1H\}$ NMR: $\delta = -12.1$ ppm; ^{29}Si NMR: $\delta = -12.1$ (m, $^3J_{Si,H} = 5$ Hz) ppm; MS (EI, 70 eV): $m/z = 576$ (4.8%), 502 (20.1%), 429 (15.0%), 355 (100%), 308 (M^+, 1.5%), 266 (M^+-$N(CH_2)_2$, 2.6%).

Hexakis(N-azetidino)disilane (**2**)

A solution of hexa(chloro)disilane (0.17 ml, 1.0 mmol) in 10 ml of diethyl ether was added dropwise to a solution of azetidine (21.9 mmol, 3 ml of a solution of azetidine containing 42% benzene) in 50 ml of diethyl ether at −60°C. The reaction mixture was allowed to warm to room temperature and stirring was continued at this temperature for 12 h. After filtration the solvent was removed under reduced pressure, and the residue was purified by distillation (120°C/0.01 mbar) to give a colourless oil.

Yield: 0.05 g (13%); ^{1}H NMR: $\delta = 1.54$ (quin, 12H, $^{3}J_{H,H} = 7$ Hz, CCH_2), 3.58 (t, 24H, $^{3}J_{H,H} = 7$ Hz, NCH_2) ppm; $^{13}C\{^{1}H\}$ NMR; $\delta = 16.3$ (CCH_2), 50.2 (NCH_2) ppm; $^{29}Si\{^{1}H\}$ NMR: not detected; MS (EI, 70 eV): $m/z = 430$ (2.2%), 410 (24.3%), 394 (M^{+}+2, 92.6%), 393 (M^{+}+1, 3.1%), 392 (M^{+}, 3.1%), 336 (M^{+}-$N(CH_2)_3$, 31.3%), 280 (336-$N(CH_2)_3$, 3.2%), 224 (280-$N(CH_2)_3$, 6.8%), 196 ($Si(N(CH_2)_3)_3$, 30.5%), 140 (196-$N(CH_2)_3$, 8.2%).

Hexakis(N-pyrrolidino)disilane (**3**)

Hexa(chloro)disilane (1.72 ml, 10 mmol) was added dropwise to a solution of pyrrolidine (16.52 ml, 200 mmol) in 250 ml of diethyl ether at 0°C. After stirring at room temperature for 72 h, the precipitate was separated, the solvent and excess pyrrolidine were removed from the filtrate *in vacuo*, and the residue was extracted with pentane. **3** crystallized from this solution at −30°C.

Colourless crystals; yield: 2.37 g (50%); m.p.: >200°C; ^{1}H NMR: $\delta = 1.68$ (m, 24 H, CCH_2), 3.13 (m, 24 H, NCH_2) ppm; $^{13}C\{^{1}H\}$ NMR: $\delta = 27.1$ (CCH_2), 46.6 (NCH_2) ppm; $^{29}Si\{^{1}H\}$ NMR: $\delta = -34.5$ ppm; MS (EI, 70 eV): $m/z = 476$ (M^{+}), 406 (M^{+}-$N(CH_2)_4$), 336 (406-$N(CH_2)_4$), 266 (336-$N(CH_2)_4$), 238 ($Si(N(CH_2)_4)_3$), 168 (238-$N(CH_2)_4$); $C_{24}H_{48}N_6Si_2$ (476.86); calcd: C 60.5, H 10.2, N 17.6; found: C 58.8, H 10.0, N 17.0.

Hexakis(N-piperidino)disilane (**4**)

Piperidine (8.26 ml, 83 mmol) was added dropwise to a solution of hexa(chloro)disilane (1.72 ml, 10 mmol) in 200 ml of diethyl ether at room temperature. After the reaction mixture had been stirred for 12 h, another 10 ml of piperidine (100 mmol) was added, and stirring continued for another 2 h at room temperature. The precipitate was then separated, and the filtrate was treated with a suspension of N-lithio piperidine in 50 ml of hexane/diethyl ether (30 mmol, from 5.00 ml (51 mmol) of piperidine and 17.65 ml (30 mmol) of a 1.7 *M* solution of *n*-butyllithium in hexane: *n*-butyllithium was added dropwise to a solution of the amine in diethyl ether (27 ml) at room temperature; the reaction mixture was stirred for 0.5 h and then used immediately). The reaction mixture was stirred for 72 h at room temperature, the solvent was removed *in vacuo*, and the residue was extracted with pentane. **4** crystallized from this solution at −30°C.

Colourless crystals; 2.09 g (37%); m.p.: >200°C; ^{1}H NMR: $\delta = 1.47$ (pseudo-quint, 24H, $^{3}J_{H,H} = 5$ Hz , NCH_2CH_2), 1.61 (pseudo-quint, 12 H, $^{3}J_{H,H} = 5$ Hz, $NCH_2CH_2CH_2$), 2.90 (pseudo-t, 24 H, $^{3}J_{H,H} = 5$ Hz NCH_2) ppm; $^{13}C\{^{1}H\}$ NMR: $\delta = 26.4$ ($NCH_2CH_2CH_2$), 29.2 (NCH_2CH_2), 50.2 (NCH_2)ppm; $^{29}Si\{^{1}H\}$ NMR: $\delta = -29.7$ ppm; MS (EI, 70 eV): $m/z = 561$ (M^{+}), 281 ($Si(N(CH_2)_5)_3$); $C_{30}H_{60}N_6Si_2$ (561.02); calcd.: C 64.2, H 10.8, N 15.0; found: C 61.6, H 10.9, N 14.2

Acknowledgements

This work was supported by the *Deutsche Forschungsgemeinschaft* and the *Fonds der Chemischen Industrie*.

References

[1] Ward LGL, MacDiarmid AG (1961) J Inorg Nucl Chem **21**: 287
[2] Abedine M, MacDiarmid AG (1963) Inorg Chem **2**: 608
[3] Schmidbaur H, Schuh H (1990) Naturforsch Z **45**B: 1679
[4] Trandell RF, Urry GJ (1978) J Inorg Nucl Chem **40**: 1305

[5] Schuh H, Schlosser T, Bissinger P, Schmidbaur H (1993) Z Anorg Allg Chem **619**: 1347
[6] Söldner M, Schier A, Schmidbaur H (1998) Inorg Chem **37**: 510
[7] Söldner M, Schier A, Schmidbaur H (1997) Inorg Chem **36**: 1758
[8] Höfler F, Peter W (1975) Naturforsch Z **30B**: 282
[9] Wiberg E, Stecher O, Neumaier A (1965) Inorg Nucl Chem Lett **1**: 33
[10] Wan Y, Verkade JG (1993) Inorg Chem **32**: 341
[11] Huber G, Schier A, Schmidbaur H (1997) Chem Ber/Recueil **130**: 1167
[12] Huber G, Jockisch A, Schmidbaur H (1998) Eur J Inorg Chem **1**: 107
[13] Allen CFH, Spangler FW, Webster ER (1963) Org Synth Coll **4**: 433
[14] Szmuszkovicz J, Kane MP, Laurian LG, Chidester CG, Scahill TA (1981) J Org Chem **46**: 3562

Received July 7, 1998. Accepted July 27, 1998

Synthesis and Structure of $P_7[Si(SiMe_3)_3]_3$ (Tri(hypersilyl)heptaphosphanortricyclane)

Harald Siegl, Werner Krumlacher, and **Karl Hassler***

Institute of Inorganic Chemistry, University of Technology, A-8010 Graz, Austria

Summary. By reaction of sodium-potassium phosphide (prepared from white or red phosphorus and sodium-potassium alloy) and chloro*tris*(trimethylsilyl)silane, $P_7[Si(SiMe_3)_3]_3$ has been prepared in high yields. An X-ray analysis proved that the molecule is composed of a P_7-nortricyclane cage substituted with three $Si(SiMe_3)_3$ groups and that two enantiomers are present in crystals grown from benzene. The ^{31}P NMR spectrum shows the complex splitting pattern of an $A[MX]_3$ spin system. ^{31}P chemical shifts and ^{31}P, ^{31}P one-bond and two-bond coupling constants have been calculated by numerical simulation of the spectrum.

Keywords. Tri(hypersilyl)heptaphosphanortricyclane; Synthesis; Crystal structure; $^{31}P\{^1H\}$ NMR data.

Synthese und Struktur von $P_7[Si(SiMe_3)_3]_3$ (Tri(hypersilyl)heptaphosphanortricyclan)

Zusammenfassung. Durch die Reaktion von Natrium-Kalium-Phosphid (darstellt aus rotem oder weißem Phosphor und Natrium-Kalium-Legierung) mit Chlor*tris*(trimethylsilyl)silan konnte $P_7[Si(SiMe_3)_3]_3$ in hoher Ausbeute dargestellt werden. Die Charakterisierung des mit drei $Si(SiMe_3)_3$-Gruppen substituierten P_7-Nortricylan-Käfigs erfolgte mit Hilfe der Röntgenstrukturanalyse. In aus Benzol gezüchteten Kristallen sind beide Enantiomere enthalten. Das ^{31}P-NMR-Spektrum zeigt das komplexe Aufspaltungsmuster eines $A[MX]_3$-Spinsystems. Die ^{31}P Verschiebungen und die ^{31}P, ^{31}P Kopplungskonstanten über eine oder zwei Bindungen wurden durch numerische Simulation des Spektrums berechnet.

Introduction

Due to its steric and electronic properties and facile availability, the *tris*(trimethylsilyl)silyl group (hypersilyl group) has been extensively used for the stabilization of exceptional structures comprising not only main group elements but also transition metals. Typical examples of main group compounds are $[(Me_3Si)_3Si]_2Sn \cdot LiCl(THF)_3)$ [1], $[(Me_3Si)_3Si]_2Sn$ [2], $(Me_3Si)_3SiBMe_2$ [3], $(Me_3Si)_3SiB{=}NCMe_3$ [4], and $(Me_3Si)_3SiSi(SiMe_3)_3$ [5]. *Tris*(trimethylsilyl)-acylsilanes have been found to rearrange to the corresponding silaethylenes [6], and formylsilanes ($(R_3Si)_3SiCHO$) are stabilized considerably [7]. Transition metal

* Corresponding author

compounds comprise $(Me_3Si)_3SiRe(CO)_5$ [8] or $(Me_3Si)_3SiAuPPh_2Me$ [9], just to name a few. It has been argued that *van der Waals* bonding contributions within the peripheral methyl groups of $(Me_3Si)_3SiSi(SiMe_3)_3$ compensate the reduced bond order of 0.83 of the central Si-Si-bond, an effect which is even more pronounced for hexa(*tert*-butyl)disilane with a Si-Si bond order of merely 0.26 [10, 11].

These unique properties of the hypersilyl group prompted us to investigate its potential use for the stabilization of structures contained in *Zintl* phases and *Zintl* ions composed of elements of groups 14 and 15. In the course of these experiments, we also developed a synthesis for the title compound, a molecule containing three weakly interacting hypersilyl groups which decrease the reactivity of the Si-P-bonds substantially. We will report on the reactions of $P_7[Si(SiMe_3)_3]_3$ with H_2O, BuLi, and KO^tBU as well as on the stabilization of Sb_7 and Bi_7 by the hypersilyl group in a forthcoming paper.

Results and Discussion

Molecular structure of tri(hypersilyl)heptaphosphanortricyclane

Colourless single crystals of $P_7[Si(SiMe_3)_3]_3 \cdot solv$ suitable for X-ray diffraction analysis were grown from benzene. Space group P-1 was suggested and later confirmed by a successful structure solution. As it is characterized by a centre of symmetry, the unit cell contains both enantiomers of $P_7[Si(SiMe_3)_3]_3$, along with two molecules of benzene. This is in contrast to $P_7(SiMe_3)_3$ which crystallizes enantiomorphically [12].

The molecular structure is shown in Fig. 1. The molecule possesses C_3-symmetry to a very good approximation, the small deviations being caused by packing effects within the crystal.

Apart from a large number of organosubstituted P_7-cages [15] as well as lithiated derivatives [16] that have been described in the literature, the molecular structures of three silyl substituted P_7-nortricyclane cages have been determined so far. Table 1 compares their relevant structural parameters with those of the title compound, their definition being given in Fig. 2. The P-P interatomic distances *a*, *b*, and *c* as well as the Si-P distance do not vary much with the bulkiness of the silyl group which increases in the order $SiMe_3 < SiPh_3 < Si(SiMe_3)_3 < Si^tBu_3$. The

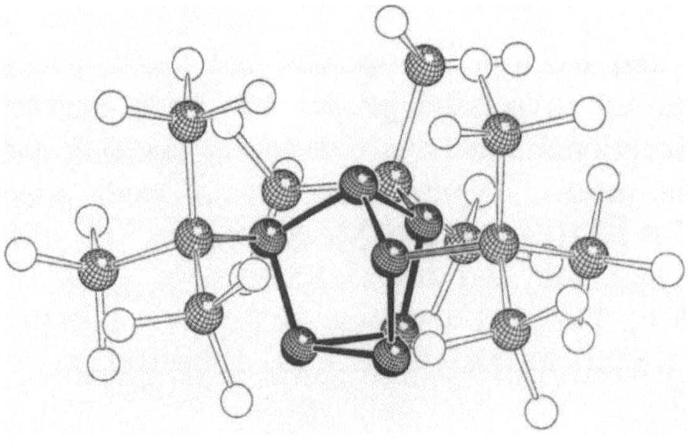

Fig. 1. Molecular structure of $P_7[Si(SiMe_3)_3]_3$; hydrogen atoms have been omitted

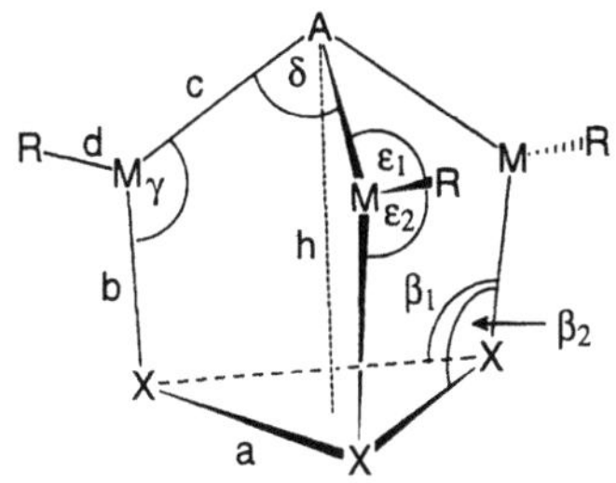

Fig. 2. Notation for the structural parameters and the $A[MX]_3$ spin system of P_7R_3

twisting of the cage with respect to the threefold axis of rotation, measured as the difference β_1-β_2 of the bond angles β_1 and β_2 (Table 1), increases moderately in the order $Si^tBu_3 > Si(SiMe_3)_3 > SiMe_3$, but surprisingly assumes its smallest value for $SiPh_3$. Notable changes occur within the exocyclic bond angles ε_1 and ε_2 (and to a lesser extent γ) of the equatorial phosphorus atoms. Increasing bulkiness forces the SiR_3-groups to rotate away from the apical P-atom (thereby increasing ε_1) and away from the neighbouring equatorial P-atom, thereby increasing not only ε_2 but also the nonbonded Si···P distance. The hypersilyl groups effectively surround and shield the P_7-cage as can be anticipated from Fig. 1. The smallest H···H nonbonded distances between two $Si(SiMe_3)_3$ groups range from 2.4 to 2.9 Å which is quite close to the twofold value of the *van der Waals* radius for hydrogen, given in the literature as 1.20–1.45 Å [17]. Therefore, the elongation observed for the P_7-cage of hypersilylheptaphosphane as expressed by the parameter h (Table 1) can be explained by *van der Waals* interactions within the hypersilyl groups, reducing slightly the distance between the equatorial P-atoms.

Table 1. Comparison of important intramolecular distances (Å) and bond angles (°) for silyl substituted heptaphosphanortricyclanes; for the meaning of the parameters, cf. Fig. 2

	$P_7(SiMe_3)_3$ [12]	$P_7(SiPh_3)_3$ [13]	$P_7[Si(SiMe_3)_3]_3$ [this work]	$P_7(Si^tBu_3)_3$ [14]
a	2.214(4)	2.217(3)	2.2176(11)	2.224(2)
b	2.192(4)	2.182(3)	2.1983(11)	2.185(2)
c	2.180(4)	2.180(6)	2.1836(9)	2.174(2)
d	2.288	2.286(6)	2.2884(10)	2.304(2)
h	3.154	3.144[a]	3.184	3.158
ε_1	107.6(4)	109.3(6)	111.27(4)	118.34(7)
ε_2	102.2(4)	101.6(7)	103.71(4)	109.90
γ	101.2(2)	102.2(2)	103.11(4)	102.59(7)
δ	98.3(2)	98.5(9)	97.52(4)	96.88
β_1	101.0(2)	101.9(5)	100.42(4)	99.55(7)
β_2	107.1(2)	106.5(2)	106.94(4)	107.35(7)
$\beta_1-\beta_2$	6.1	4.6	6.52	7.8

[a] Calculated by the authors

NMR-Spectroscopy

The P_7-cage was also characterized by ^{31}P and ^{29}Si NMR spectroscopy. The proton decoupled ^{31}P NMR spectrum shows the complex splitting pattern of an $A[MX]_3$ system, very similar to that of the analogous supersilyl compound [14]. It consists of three groups of signals corresponding to the spins A, M and X (as illustrated in Fig. 2) centred at $\delta = -100.6$, -7.3, and -161.6 ppm, respectively. From the symmetry of the spectrum it can be concluded that only the symmetric isomer (all hypersilyl groups possessing the same sense of rotation with respect to the threefold axis of rotation) is present. In order to determine the seven ^{31}P, ^{31}P-coupling constants characterizing the spin system, ($^1J_{AM}$, $^1J_{MX}$, $^1J_{XX'}$, $^2J_{AX}$, $^2J_{MM'}$, $^2J_{MX'}$, and $^2J_{MX''}$), a simulation of the spectrum using the parameters published for $P_7[Si^tBu_3]_3$ as a starting point and a subsequent iteration was performed. The calculated and the experimental spectra are presented in Fig. 3. The final chemical shifts and coupling constants are summarized in Table 2 and compared with the values reported for the analogous Si^tBu_3-substituted cage. The one-bond coupling constants between the equatorial and the apical as well as basal P-atoms ($^1J_{AM}$ and $^1J_{MX}$) are smaller than those of the supersilyl compound but larger than the values

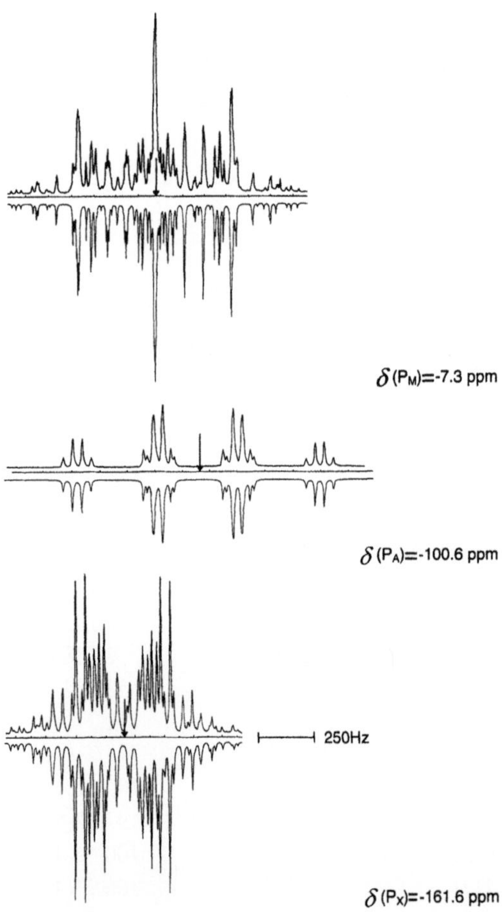

Fig. 3. A comparison of the measured $^{31}P\{H\}$ NMR spectrum (top) with the calculated one (bottom)

Table 2. ^{29}Si and ^{31}P NMR shifts (ppm relative to *TMS* for ^{29}Si and relative to H_3PO_4 for ^{31}P and coupling constants (Hz) of $P_7[Si(SiMe_3)_3]_3$ and $P_7(Si^tBu_3)_3^a$

	$P_7[Si(SiMe_3)_3]_3$	$P_7(Si^tBu_3)_3$ [14]
$\delta(P_A)$	−100.6	−113.0
$\delta(P_M)$	−7.3	−31.2
$\delta(P_X)$	−161.6	−175.6
$^1J_{AM}$	−349.2	−401.7
$^1J_{MX}$	−364.9	−431.1
$^1J_{XX'}$	−213.7	−205.0
$^2J_{AX}$	39.4	33.3
$^2J_{MM'}$	−13.7	−12.1
$^2J_{MX'}$	−14.6	−12.3
$^2J_{MX''}$	27.8	22.1
$\delta(^{29}Si)$	−92.5	
$\delta(^{29}SiMe_3)$	−7.8	
$^1J_{SiP}{}^b$	99	

[a] Recorded in benzene-d$_6$; [b] due to long-range coupling with other ^{31}P nuclei there was a significant peak broadening which could not be resolved any further and which reduced the accuracy of $^1J_{SiP}$ to ±5 Hz

reported for $P_7(SiMe_3)_3$, indicating that the exocyclic bond angles of the equatorial P-atoms (ε_1 and ε_2, Fig. 2) lie between these two extremes.

Experimental

General remarks

All operations with sodium, potassium, and phosphorus-silicon compounds were carried out under a nitrogen atmosphere. Solvents were distilled from potassium. The NMR experiments were performed in C_6D_6 at ambient temperature on a *Bruker* 300 MSL spectrometer. $^{29}Si\{^1H\}$ spectra were recorded at 59.6 MHz using the INEPT pulse sequence, $^{31}P\{^1H\}$ spectra at 121.4 MHz.

X-ray structure solution

The data collection was performed under a stream of N_2 at −70°C using a Siemens diffractometer with a CCD SMART detector and graphite monochromated molybdenum K_α radiation ($\lambda = 0.71073$ Å). The crystal was mounted on the tip of a glass fibre in an inert oil. A summary of the crystal data, data collection, and data refinement is given below. Detailed crystallographic data have been deposited at the Cambridge Crystallographic Data Centre, Cambridge, U.K., under the deposition number 102248.

Crystal data for $P_7[Si(SiMe_3)_3]_3$: $C_{27}H_{81}P_7Si_{12} + C_6H_6$, $M = 1037.9$ g/mol, triclinic (P-1), $a = 1400.72(1)$ pm, $b = 1414.31(2)$ pm, $c = 1825.98(3)$ pm, $\alpha = 84.216(1)°$, $\beta = 77.639(1)°$, $\gamma = 63.401(1)°$, $V = 3159.48 \times 10^6\ \text{pm}^3$, $Z = 2$, $\rho_{calc} = 1.091\ \text{g} \cdot \text{cm}^{-3}$; 27108 reflections (18895 independent, $1.14° < 2\Theta < 30.52°$; $h = -19$ to 20, $k = -19$ to 20, $l = -13$ to 26) were collected from a crystal of $0.4 \times 0.4 \times 0.3$ mm size; F(000) = 1116, $R_1[I > 2\sigma(I)] = 0.0385$ ($R = (\Sigma\|F_0| - |F_c\|)/\Sigma|F_0|$), $wR_2 = 0.1149$ ($wR_2 = (\Sigma(w(|F_0|^2 - |F_c|^2)^2)/\Sigma(w(F_0^4)))^{1/2}$), $GOF = 0.463$ ($GOF = (\Sigma w(|F_0|^2 - |F_c|^2)^2(n-m))^{1/2}$). The structure was solved by direct methods using XS [18] and refined by full-matrix least squares with SHELXL-93 [19] minimizing the residuals for F^2. No absorption correction was applied. Hydrogen atoms were included in the model at their calculated positions. The torsion angle of a methyl group was set to maximize the sum of the electron density at the three calculated hydrogen positions. Anisotropic displacement parameters were assigned to all non-hydrogen atoms and isotropic parameters used for the hydrogen atoms.

Synthesis of $P_7[Si(SiMe_3)_3]_3$

1.42 g (45.7 mmol) of red phosphorus [20] (or equivalently white phosphorus P_4 [21]) are placed in a 100 ml three-necked round bottom flask and dried carefully by repeatedly evacuating and flushing the flask with dry nitrogen. The phosphorus is then suspended in 60 ml of dry dimethoxyethane (*DME*), and 1.04 ml (24.8 mmol) of liquid sodium potassium alloy (prepared by melting together 0.1 g of Na and 0.81 g of K in *DME*) are added. The mixture is then refluxed for 24 h. Subsequently, the solvent is removed by evaporation *in vacuo*, and the phosphide is suspended in 50 ml of dry toluene. 6 g (21.3 mmol) of $(Me_3Si)_3SiCl$ (prepared from $(Me_3Si)_3SiH$ and CCl_4 [22]) and 5 ml of dry *DME* are added, and the reaction mixture is refluxed for 72 h. Without the addition of a few ml of *DME*, no reaction occurs even with prolonged refluxing. The salts are then separated by filtration and washed with dry toluene. The volume of the combined filtrates is then reduced to about 50 ml, and the clear yellow solution is cooled to -30°C. A colourless crystalline product separates in a total yield of 4.3 g (68.7%).

Acknowledgements

The authors thank the *Fonds zu Förderung der wissenschaftlichen Forschung* for financial support (project S 07905-CHE) and Prof. *U. Schubert* for his permission to have access to the X-ray diffractometer of the Institute of Inorganic Chemistry, Technical University of Vienna.

References

[1] Arif AM, Cowley AH, Elkins TN (1987) J Organomet Chem **325**: C11
[2] Klinkhammer KW, Schwarz W (1995) Angew Chem **107**: 1448
[3] Biffon W, Nöth H (1981) Z Naturforsch **B36**: 1509
[4] Haase M, Klingebiel U, Base R, Polk M (1986) Chem Ber **119**: 1117
[5] Gilman H, Harrell RL (1967) J Organomet Chem **9**: 67
[6] Brook AG, Nyberg S, Abdesaken F, Gutekunst B, Kalbery RKMR, Poon YC, Chang Y-M, Wong-Ng W (1982) J Am Chem Soc **104**: 5668
[7] Elsner FH, Woo H-G, Tilley TD (1988) J Am Chem Soc **110**: 314
[8] Couldwell MC, Simpson J, Robinson WT (1967) J Organomet Chem **107**: 323
[9] Meyer J, Willnecker J, Schubert U (1989) Chem Ber **122**: 223
[10] Wiberg N, Schuster H, Simon A, Peters K (1996) Angew Chem **98**: 100
[11] Bock H, Meuret J, Ruppert K (1993) J Organomet Chem **445**: 19
[12] Hönle W, Schnering HG v (1978) Z anorg allg Chem **440**: 171
[13] Mujica C, Weber D, Schnering HG v (1986) Z Naturforsch **B41**: 991
[14] Kovacs I, Baum G, Fritz G, Fenske D, Wiberg N, Schuster H, Karaghiosoff K (1993) Z anorg allg Chem **619**: 453
[15] Baudler M (1987) Angew Chem **99**: 429 and references cited therein
[16] Fritz G, Rothmann H, Matern E (1992) Z Anorg Allg Chem **610**: 33; Baudler M, Exner O (1983) Chem Ber 1268
[17] Huheey JE, Keiter EA, Keiter RL (1995) Anorganische Chemie. De Gruyter, Berlin
[18] SHELXTL 5.0, Siemens Crystallographic Research System

[19] Sheldrick GM (1993) SHELXL-93, Program for Crystal Structure Determination. Universität Göttingen, Germany
[20] Schmidbaur H, Bauer A (1995) Phosphorus, Sulfur and Silicon **102**: 217
[21] Krumlacher W, Siegl H, Hassler K, Organosilicon Chemistry IV. Wiley VCH (in press)
[22] Bürger H, Kilian W, Burczyk K (1970) J Organomet Chem **21**: 291

Received June 12, 1998. Accepted (revised) July 20, 1998

Solid-State Molecular Dynamics of $(Ph_3Si)Si(SiMe_3)_3$ Studied by Variable Temperature ^{13}C and ^{29}Si MAS NMR Spectroscopy

Xavier Helluy[1], **Jörg Kümmerlen**[2], **Christoph Marschner**[3], and **Angelika Sebald**[2,*]

[1] Rhône-Poulenc Rorer, Department of Pharmaceutical Sciences, Preformulation/Physical Analysis, Research Center of Vitry-Alfortville, F-94403 Vitry sur Seine Cedex, France

[2] Bayerisches Geoinstitut, Universität Bayreuth, D-95440 Bayreuth, Germany

[3] Institut für Anorganische Chemie, Technische Universität Graz, A-8010 Graz, Austria

Summary. One- and two-dimensional ^{13}C and ^{29}Si solid-state NMR experiments at variable temperature demonstrate that crystalline $(Ph_3Si)Si(SiMe_3)_3$ lacks all molecular symmetry and that internal $2\pi/3$ reorientation of all three crystallographically inequivalent $SiMe_3$ groups in the molecule occurs. Quantitative analysis of one-dimensional MAS NMR spectra and of two-dimensional NMR exchange experiments by spectral lineshape simulations yields exchange rate constants for internal $SiMe_3$ reorientation as a function of temperature. The activation energies determined from these kinetic data are $E_a = 21 \pm 4$, 22 ± 4, and $30 \pm 5\,kJ \cdot mol^{-1}$ for internal reorientation of the three inequivalent $SiMe_3$ groups in solid $(Ph_3Si)Si(SiMe_3)_3$.

Keywords. Molecular solid-state dynamics; Solid-state NMR; ^{13}C and ^{29}Si MAS NMR; Exchange rate constants; Spectral lineshape simulations.

Untersuchung der molekularen Dynamik von $(Ph_3Si)Si(SiMe_3)_3$ im Festkörper mittels ^{13}C- und ^{29}Si-MAS-NMR-Spektroskopie bei variabler Temperatur

Zusammenfassung. Ein- und zweidimensionale ^{13}C- und ^{29}Si-NMR-Experimente am Festkörper bei verschiedenen Temperaturen zeigen, daß kristallines $(Ph_3Si)Si(SiMe_3)_3$ keine molekulare Symmetrie besitzt und daß interne $2\pi/3$-Reorientierung der drei kristallographisch unterschiedlichen $SiMe_3$-Gruppen auftritt. Quantitative Analyse von eindimensionalen MAS-NMR-Spektren und zweidimensionalen Austauschexperimenten mittels Linienformsimulation ergibt die Geschwindigkeitskonstanten des obengenannten Austauschs als Funktion der Temperatur. Daraus erhält man für die interne Reorientierung der drei nichtäquivalentem $SiMe_3$-Gruppen in festem $(Ph_3Si)Si(SiMe_3)_3$ die folgenden Aktivierungsenergien: 21 ± 4, 22 ± 4 und $30 \pm 5\,kJ \cdot mol^{-1}$.

* Corresponding author

Introduction

$Si(SiMe_3)_4$ and $C(SiMe_3)_4$ are highly disordered crystalline phases at ambient temperatures, similar to the disordered crystalline phases which exist for several nonpolar, spherically shaped organic molecules such as *e.g.* adamantane [1]. Both compounds undergo structural phase transitions to less disordered low-temperature phases ($Si(SiMe_3)_4$ at $T<238$ K [2], $C(SiMe_3)_4$ at $T<213$ K [3]); the molecular dynamic properties of these two low-temperature phases have been investigated by variable temperature ^{13}C and ^{29}Si MAS NMR. The molecular point group symmetry in the low-temperature phases of $Si(SiMe_3)_4$ and $C(SiMe_3)_4$, as inferred from MAS NMR experiments, is C_3 [4–6] and is confirmed by X-ray diffraction analysis [7]. Despite identical molecular symmetry in these two low-temperature phases, the two compounds differ strongly with regard to the dynamic properties of their low-temperature phases. Quantitative analysis of one- and two-dimensional variable temperature ^{13}C and ^{29}Si MAS NMR experiments permits to distinguish and precisely determine exchange rate constants for internal $SiMe_3$ reorientation and for whole-molecule reorientation around preferred molecular axes as a function of temperature. The dynamic solid-state properties of the low-temperature phases of $Si(SiMe_3)_4$ and $C(SiMe_3)_4$ may be briefly summarized as follows. In the low-temperature phase of $Si(SiMe_3)_4$ we observe internal $SiMe_3$ reorientation and additional whole-molecule reorientation; the exchange rate constants for internal $SiMe_3$ reorientation always considerably exceed those for whole-molecule reorientation. The situation is different for solid $C(SiMe_3)_4$: over the entire temperature range of existence of the low-temperature phase, only whole-molecule reorientation is observed, whereas exchange rate constants for internal $SiMe_3$ reorientation are negligibly small [6]. We have interpreted this difference in the solid-state molecular dynamic properties of $Si(SiMe_3)_4$ and $C(SiMe_3)_4$ in relation to higher intramolecular steric crowding in $C(SiMe_3)_4$. This interpretation is qualitatively supported by empirical force-field calculations based on the gas-phase structures of $Si(SiMe_3)_4$ [8] and $C(SiMe_3)_4$ [9].

Here we report on the solid-state dynamic properties of a compound chemically closely related to $Si(SiMe_3)_4$. In $(Ph_3Si)Si(SiMe_3)_3$ (**1**) the molecular symmetry is lowered by chemical substitution as compared to the fully $SiMe_3$ substituted parent molecule. We have investigated the dynamic solid-state properties of **1** by variable temperature one- and two-dimensional ^{13}C and ^{29}Si CP/MAS NMR experiments. The experimental NMR results on **1** are compared to the dynamic properties of the low-temperature phase of $Si(SiMe_3)_4$.

Results and Discussion

Apart from internal reorientation of the methyl groups themselves (which will not be considered in the following) and possible π-flips of the phenyl groups in the $SiPh_3$ moiety, *a priori* there are two modes of molecular dynamics to be considered for solid $(Ph_3Si)Si(SiMe_3)_3$ (**1**):

i) internal reorientation of the individual R_3Si groups ($R=$ Ph, Me) around the $R_3Si\text{-}Si_{central}$ bond directions, leading to mutual exchange of the groups R within a SiR_3 unit (referred to as internal reorientation), and *ii*) reorientation of the $Si(SiMe_3)_3$ moiety around the Ph_3Si-$Si_{central}$ bond direction axis, leading to mutual exchange of the three $SiMe_3$ groups as a whole (referred to as $Si(SiMe_3)_3$ reorientation).

These two reorientational modes, if present, may occur simultaneously. Our experimental solid-state NMR tools to investigate and possibly distinguish these two reorientational modes consist of one- and two-dimensional variable temperature ^{13}C and ^{29}Si NMR experiments in conjunction with spectral lineshape simulations.

Many solid organosilicon compounds undergo structural phase transitions upon cooling or heating. Differential scanning calorimetry (DSC) of **1** gives no indication for the occurrence of such structural phase transitions in solid **1** in the temperature range of $T = 180$–$290\,K$. Accordingly, observation of temperature dependent spectral lineshapes in ^{13}C and ^{29}Si MAS NMR experiments on **1** cannot be ascribed to the effects of structural phase transitions in this temperature range.

Under MAS conditions, that is at MAS rates sufficiently high to average out all ^{29}Si chemical shielding anisotropies, ^{29}Si MAS NMR spectra of **1** can report on the molecular symmetry of **1** as well as – selectively – on $Si(SiMe_3)_3$ reorientational processes, whereas ^{13}C MAS NMR spectra of **1** will be affected by both internal R_3Si and $Si(SiMe_3)_3$ reorientation [6]. With this aspect in mind, we first consider the ^{29}Si MAS NMR spectra of **1** (Fig. 1). At $T = 150\,K$ we observe three barely resolved ^{29}Si resonances ($\delta_{iso} = -9.3, -9.9, -10.2\,ppm$), corresponding to three nonequivalent $SiMe_3$ groups, one resonance for the $SiPh_3$ group ($\delta_{iso} = -11.7\,ppm$), and one resonance for the central Si atom ($\delta_{iso} = -133.4\,ppm$). ^{29}Si CP NMR spectra of **1**, obtained under static conditions, display a ^{29}Si chemical shielding tensor with a large asymmetry parameter ($-\delta_{iso} = \sigma_{iso} = 133.4\,ppm$, $\sigma_{11} = 124.0\,ppm$, $\sigma_{22} = 132.6\,ppm$, $\sigma_{33} = 143.4\,ppm$, $\eta = 0.85$) for the resonance of the central Si atom (see Fig. 1). Accordingly, solid **1** lacks all molecular symmetry and, if present, we should be able to detect $Si(SiMe_3)_3$ reorientation from one- and two-dimensional variable temperature ^{29}Si MAS NMR experiments. Figure 2a shows the $SiMe_3/SiPh_3$ region of the ^{29}Si MAS NMR spectra of **1** in the

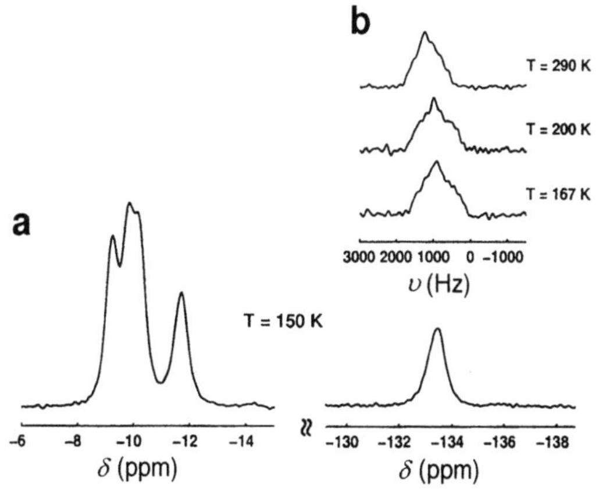

Fig. 1. ^{29}Si CP NMR experiments on **1** ($\omega_0/2\pi = 59.6\,MHz$): a) ^{29}Si CP MAS NMR ($\omega_{rot}/2\pi = 4.0\,kHz$, $T = 150\,K$); b) ^{29}Si CP NMR, static conditions, only the chemical shielding powder pattern of the ^{29}Si resonance of the central Si atom at three different temperatures is shown

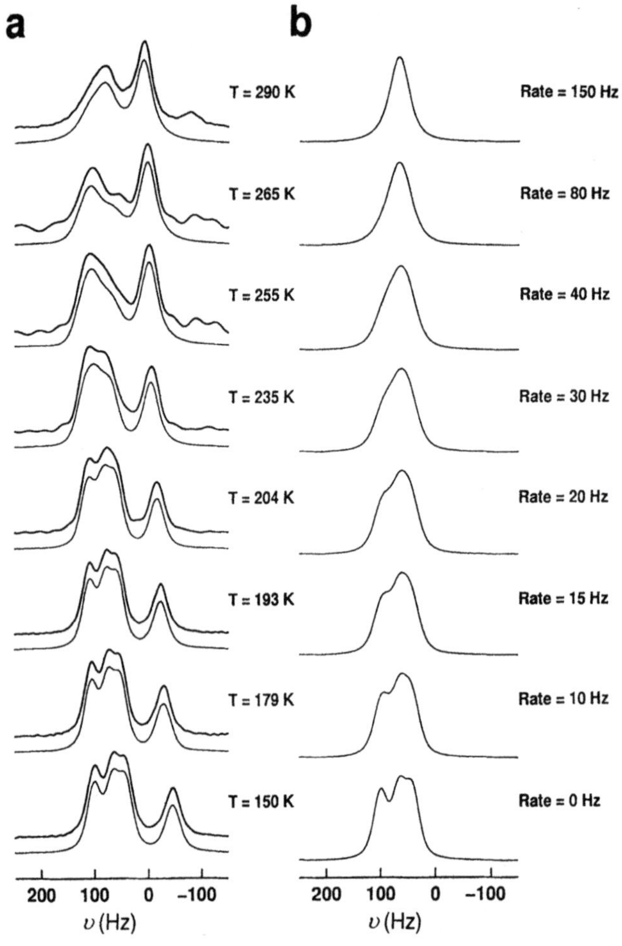

Fig. 2. Variable temperature ^{29}Si CP MAS NMR spectra of **1**; only the $SiPh_3/SiMe_3$ region is shown, temperatures are indicated ($\omega_0/2\pi = 59.6\,MHz$, $\omega_{rot}/2\pi = 4.0\,kHz$): a) upper traces represent experimental spectra, lower traces are the corresponding best-fit calculated spectra, assuming minor intrinsic temperature-dependence of $\delta_{iso}(^{29}Si)$ of the four ^{29}Si resonances of the $SiPh_3$ and $SiMe_3$ groups in **1**; b) calculated hypothetical ^{29}Si MAS NMR spectra of **1**, assuming mutual exchange of the three $SiMe_3$ groups by $Si(SiMe_3)_3$ reorientation

temperature range of $T = 150$–$290\,K$. It is tempting to ascribe the minor changes of isotropic chemical shifts in the $SiMe_3$ region, observed as a function of temperature, to $Si(SiMe_3)_3$ reorientation around the Ph_3Si-$Si_{central}$ bond direction and leading to coalescence of the three ^{29}Si resonances of the three $SiMe_3$ groups at higher temperatures. This explanation can be ruled out by calculating expected exchange broadened ^{29}Si NMR spectral lineshapes for mutual exchange of the three $SiMe_3$ groups in **1** and comparing them to the experimental ^{29}Si MAS NMR spectra. Temperature dependent spectral lineshapes caused by $Si(SiMe_3)_3$ reorientation around the Ph_3Si-$Si_{central}$ bond direction would have to display an asymmetric shape in the spectral $SiMe_3$ region with a shoulder at higher frequencies at temperatures approaching coalescence (Fig. 2b), whereas the experimentally observed ^{29}Si spectral lineshapes display an inversely asymmetric shape in this temperature region with an unresolved shoulder towards lower frequencies. Bearing in mind that the entire $SiMe_3$ region in the ^{29}Si MAS NMR spectra of **1** ($\omega_0/2\pi = 59.6$ MHz) only spans a frequency range of approximately 100 Hz and that we consider a temperature range of roughly 150 degrees, it is reasonable to ascribe the observed changes in ^{29}Si spectral lineshapes to minor intrinsic temperature shifts of $\delta_{iso}(^{29}Si)$. With this assumption, the temperature dependent ^{29}Si NMR spectral lineshapes can be well reproduced (Fig. 2a); the simulated spectral lineshapes correspond to intrinsic temperature shifts of $\delta_{iso}(^{29}Si)$ for the $SiMe_3$ and $SiPh_3$ groups in solid **1** of 0.003, 0.006, 0.005 ppm/K ($SiMe_3$) and 0.007 ppm/K ($SiPh_3$). Furthermore, within experimental error the ^{29}Si chemical shielding tensor of the resonance of the central Si atom does not change as a function of temperature (see Fig. 1), and also ^{29}Si 2D EXSY experiments on **1** (data not shown) under MAS ($T = 204$ and 224 K, mixing times of $\tau_{mix} = 4$ and 3 s) and under static conditions ($T = 290$ K, $\tau_{mix} = 5$ s and 50 µs) give no indication for the presence of a molecular dynamic process leading to mutual exchange of the three crystallographically nonequivalent $SiMe_3$ sites in **1**. In short, on the frequency scales accessible by one- and two-dimensional variable temperature ^{29}Si MAS NMR experiments (10^{-2}–$10^{3}\,s^{-1}$) on solid **1**, the molecule in the crystal lattice does not undergo $Si(SiMe_3)_3$ reorientational jumps around the $Ph_3Si\text{-}Si_{central}$ bond direction. The absence of this kind of dynamic disorder in solid **1** is in contrast to the dynamic properties of the (chemically) symmetrically substituted derivative $Si(SiMe_3)_4$. For the latter, whole-molecule reorientational jumps do occur at temperatures of $T \approx 170$–220 K with exchange rate constants of the order of $k \approx 10^{-1}$–$10^{2}\,s^{-1}$ [6].

Having established the complete lack of molecular symmetry and the absence of $Si(SiMe_3)_3$ reorientation for solid **1** by means of ^{29}Si MAS NMR, we may next inspect the solid-state dynamic properties of **1** as seen by one- and two-dimensional variable temperature ^{13}C MAS NMR. The lack of molecular symmetry for **1** leads us to expect up to eighteen resonances in the phenyl region and nine resonances in the methyl region of the ^{13}C MAS NMR spectrum of **1**. Not so surprisingly, even at $T = 150$ K (Fig. 3) these multiple carbon sites are only

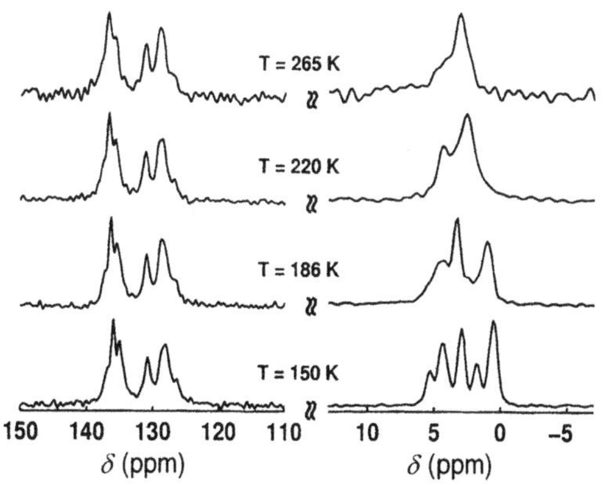

Fig. 3. ^{13}C CP MAS NMR spectra of **1** ($\omega_0/2\pi = 75.5$ MHz, $\omega_{rot}/2\pi = 4.0$ kHz) at different temperatures; left column displays ^{13}C-phenyl spectral region, right column ^{13}C-methyl spectral region

partially resolved in the ^{13}C MAS NMR spectrum ($\omega_0/2\pi = 75.5$ MHz). As far as the phenyl-^{13}C region is concerned, hardly any changes are observed in the ^{13}C MAS NMR spectra of **1** as a function of temperature. On the timescale of one-dimensional ^{13}C MAS NMR experiments in the temperature range of $T = 150$–290 K, we have no indication for the occurrence of either reorientation of the $SiPh_3$ group around the Ph_3Si-$Si_{central}$ bond direction or for π-flips of the individual phenyl groups in the $SiPh_3$ moiety. A different situation is found for the methyl-^{13}C region for which we do observe a strong temperature dependence in the ^{13}C MAS NMR spectra of **1** (Fig. 3). The only remaining possible dynamic process to account for these temperature dependent ^{13}C MAS NMR effects is internal $2\pi/3$ reorientation of the three crystallographically inequivalent $SiMe_3$ moieties in solid **1**. Given the incomplete spectral resolution in the methyl-^{13}C region of these ^{13}C MAS NMR spectra, it would be a rather hopeless task to immediately attempt quantitative analysis of these spectral lineshapes in terms of exchange rate constants of internal $SiMe_3$ reorientation as a function of temperature by means of spectral lineshape simulations. There are too many unknown parameters, such as the unknown assignment of the nine incompletely resolved ^{13}C resonances into groups of three, each group representing one $SiMe_3$ unit. Furthermore, with the three $SiMe_3$ groups in solid **1** being crystallographically inequivalent, these three $SiMe_3$ moieties do not necessarily have to undergo internal $SiMe_3$ reorientation with identical exchange rate constants k, k', and k''. Unequal exchange rate constants of internal $2\pi/3$ reorientation for crystallographically inequivalent ^{t}Bu groups in a Si^tBu_3 moiety [10] or $SiMe_3$ groups in a $C(SiMe_3)_3$ unit [11] have previously been observed.

Here we can take advantage of 2D ^{13}C MAS NMR EXSY experiments. As an example, the contour plot of a two-dimensional ^{13}C MAS NMR exchange experiment of **1** at $T = 169$ K and with a mixing time $\tau_{mix} = 250$ ms is displayed in Fig. 4. 2D ^{13}C EXSY experiments on **1**, obtained with different mixing times τ_{mix} and at various different temperatures, permit unambiguous assignment of the

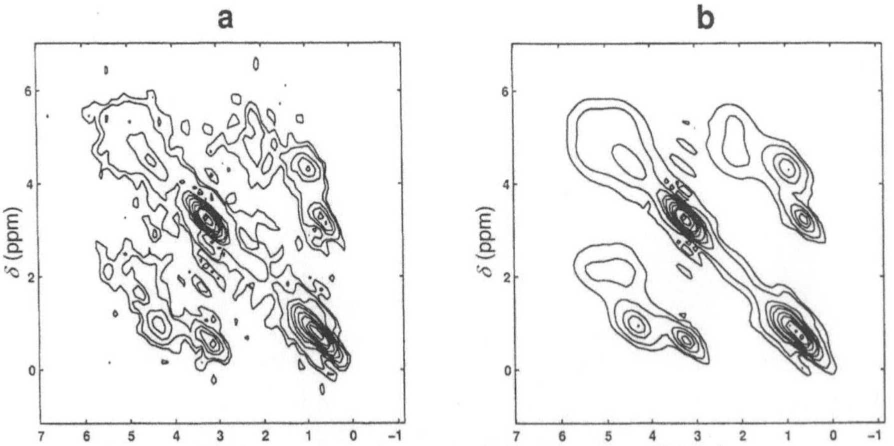

Fig. 4. Contour plots of 2D ^{13}C MAS EXSY spectra of **1** ($T = 169$ K, $\tau_{mix} = 250$ ms); only the ^{13}C-methyl region is shown; a) experimental spectrum, b) best-fit simulated spectrum

methyl-^{13}C spectral region into three groups of three methyl-^{13}C resonances, each representing one $SiMe_3$ group. We already know that the only molecular dynamic process we have to take into account for **1** is internal $2\pi/3$ $SiMe_3$ reorientation. Accordingly, off-diagonal peaks in the methyl-^{13}C spectral region in 2D ^{13}C EXSY spectra of **1** can only occur between methyl-^{13}C resonances belonging to methyl groups within the same $SiMe_3$ group. Careful inspection of 2D ^{13}C EXSY spectra of **1** (Fig. 4) reveals these "3×3 assignment connectivities" and thus provides valuable information input for subsequent spectral lineshape simulations of the one-dimensional variable temperature ^{13}C MAS NMR spectra of **1**. Beyond this assignment aspect from 2D ^{13}C EXSY spectra of **1**, differing relative off-diagonal intensities for ^{13}C-methyl resonances belonging to different $SiMe_3$ groups directly and clearly indicate unequal exchange rate constants k, k', and k'' for internal $2\pi/3$ $SiMe_3$ reorientation of the three crystallographically inequivalent $SiMe_3$ groups in solid **1**.

Most commonly, 2D ^{13}C EXSY experiments are carried out at temperatures low enough to be able to operate in the so-called slow-exchange regime. In this regime, exchange rate constants of the dynamic process under consideration are small enough to allow to take into account only exchange occuring during the mixing time of the two-dimensional NMR experiment and to neglect all motion occuring during the evolution and detection period. 2D ^{13}C EXSY NMR data obtained in this slow-exchange regime yield exchange rate constants in a straightforward manner directly from the relative integrated intensities of off-diagonal peaks as a function of the mixing time τ_{mix}; there is no need for a full simulation of the two-dimensional NMR experiment in order to extract kinetic parameters. For ^{13}C NMR experiments on solid **1**, this slow-exchange regime is only reached at inconveniently low temperatures ($T < 160$ K). Hence, 2D ^{13}C EXSY experiments on **1** have to be obtained in the so-called intermediate-exchange regime where exchange occuring during the evolution and detection period can no longer be neglected for meaningful analysis. In order to extract exchange rate constants from 2D ^{13}C EXSY experiments carried out in the intermediate-exchange regime, a full simulation of the two-dimensional NMR experiment, including the effects of exchange occuring during evolution and detection period, is necessary. Methods to fully analyze and simulate such experimental 2D EXSY data have been described in detail in the literature [12]; the contour plot of the best-fit simulated 2D ^{13}C EXSY spectrum of **1** is shown in Fig. 4b in comparison to the contour plot of the corresponding experimental spectrum. Qualitatively, it is obvious from inspection of the 2D ^{13}C EXSY experiment that internal $SiMe_3$ reorientation for one of the three $SiMe_3$ groups in **1** occurs at a lower exchange rate than for the other two $SiMe_3$ groups in the molecule. Of course, full simulation of such intermediate-regime 2D ^{13}C EXSY experiments yields exchange rate constants.

Based on the 2D ^{13}C EXSY information (signal assignment as well as knowing that internal $SiMe_3$ reorientation of one of the three $SiMe_3$ groups occurs at a considerably slower rate than for the other two $SiMe_3$ groups), we can now proceed and simulate the one-dimensional variable temperature ^{13}C MAS NMR exchange broadened lineshapes in terms of exchange rate constants of internal $SiMe_3$ reorientation. Formally, the exchange broadened lineshapes can be described as a superposition of three (independent) three-site exchange processes. The results

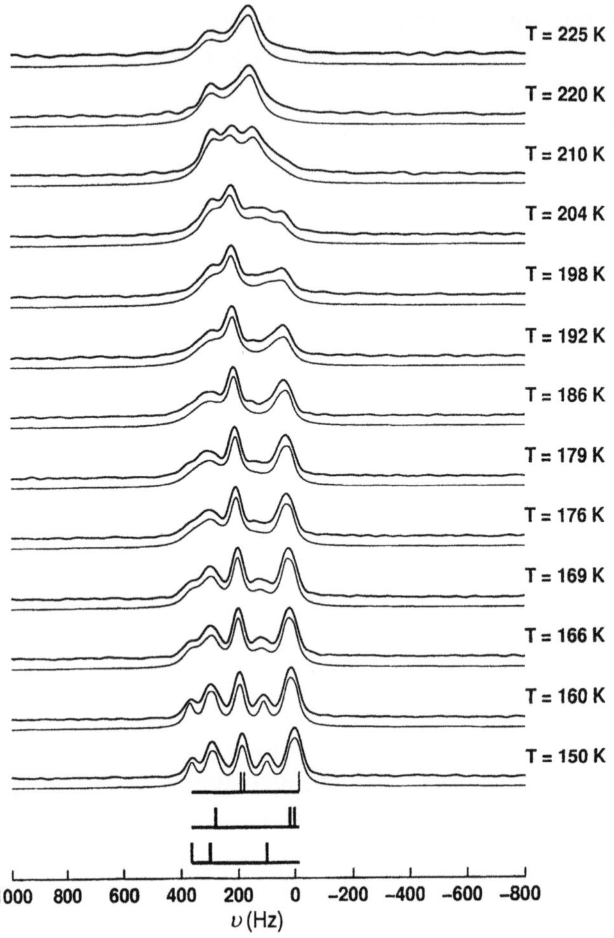

Fig. 5. Variable temperature ^{13}C MAS NMR spectra of **1**; only the ^{13}C-methyl region is shown and temperatures are indicated. Upper traces represent experimental spectra, lower traces are the corresponding best-fit calculated spectra; the bar code at the bottom indicates assignment of the ^{13}C-methyl resonances into groups of three, each group representing one $SiMe_3$ group; the smaller linewidth of the bars for one $SiMe_3$ group indicates slower exchange than for the other two $SiMe_3$ groups (see also Fig. 6)

of this iterative fitting procedure are shown in Fig. 5 where the ^{13}C-methyl region of experimental ^{13}C MAS NMR spectra of **1** in the temperature range of $T = 150-225\,K$ is compared to the best-fit calculated spectral lineshapes. The semilogarithmic plot of exchange rate constants *vs.* T^{-1} in Fig. 6 summarizes the exchange rate constants for internal $SiMe_3$ reorientation in solid **1** as determined by simulation of one- and two-dimensional variable temperature ^{13}C MAS NMR experiments. Over the temperature range considered, the exchange rate constants follow *Arrhenius*' law, and from the slope of the plots depicted in Fig. 6 we obtain activation energies $E_a = 21 \pm 4$, 22 ± 4, and $30 \pm 5\,kJ\,mol^{-1}$, respectively, for the internal reorientation of the three $SiMe_3$ groups in **1**. For $Si(SiMe_3)_4$, we find $E_a = 53 \pm 6\,kJ\,mol^{-1}$ for internal $SiMe_3$ reorientation [6]. For $Si(SiMe_3)_4$, the crystal structure has been determined [7], but not for **1**. Accordingly, any attempts

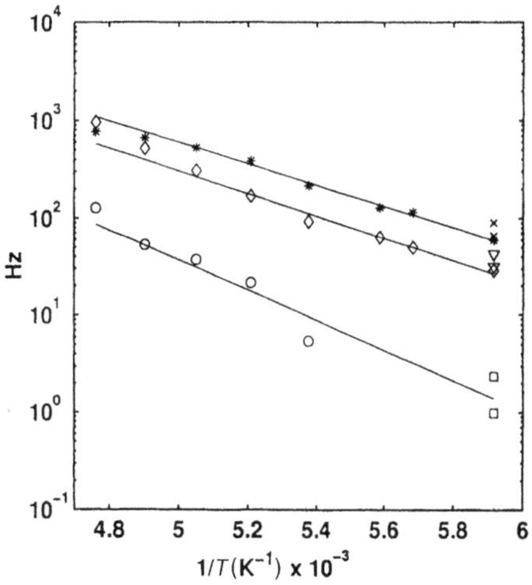

Fig. 6. Semilogarithmic plots of exchange rate constants k, k', and k'' for internal $SiMe_3$ reorientation in **1** *vs.* $1/T$; the symbols refer to exchange rate constants determined from ^{13}C MAS (∗,◇, ○) and ^{13}C 2D EXSY (×, ▽, □) NMR experiments; note the difference in exchange rate constants for the three inequivalent $SiMe_3$ groups in solid **1**

to rationalize these differences in E_a for internal $SiMe_3$ reorientation necessarily have to be speculative. A major contribution towards energy barriers for internal $SiMe_3$ reorientation of predominantly intramolecular nature seems a reasonable assumption, and thus implies a lesser degree of intramolecular steric hindrance of the $SiMe_3$ groups in **1** as compared to $Si(SiMe_3)_4$.

Experimental

Compound **1**, $(Ph_3Si)Si(SiMe_3)_3$, was synthesized following a published procedure [13].

NMR experiments and data analysis

All ^{13}C and ^{29}Si NMR experiments were carried out on a Bruker MSL 300 NMR spectrometer at *Larmor* frequencies of 75.5 (^{13}C) and 59.6 (^{29}Si) MHz, respectively. Standard 4 mm and 7 mm ZrO_2 rotors and double-bearing probes were used. For variable temperature MAS NMR experiments, purified N_2 gas served as drive and bearing gas. MAS frequencies were in the range of 2.0–4.5 kHz, ^1H $\pi/2$ pulse durations were 3–5 μs, recycle delays were 3–5 s, and contact times for *Hartmann-Hahn* cross-polarization were 1.5 ms (^{13}C) and 2.0 ms (^{29}Si). 2D exchange spectroscopy (EXSY) [14] employed phase cycling according to the TPPI method [15]. Spectral lineshape simulations of the exchange broadened one-dimensional ^{13}C and ^{29}Si MAS NMR spectra of **1** were carried out as described elsewhere [6, 10, 14, 16]. Some ^{13}C 2D EXSY spectra of **1** have been obtained in the so-called intermediate motional regime. Simulation of such 2D spectra hence requires a suitably extended exchange matrix formalism as described in the literature [12] where also motion during the evolution and detection periods of the experiment is taken into account and where T_1 relaxation during the mixing time is still negligible. Iterative fitting procedures employed the MATLAB [17] simplex and quasi-*Newton* routines for nonlinear least-squares minimization.

Isotropic chemical shifts $\delta_{iso}(^{13}C)$ and $\delta_{iso}(^{29}Si)$ are given with respect to external $SiMe_4$, chemical shielding tensor eigenvalues are reported according to *Haeberlen*'s convention [18]; $\delta_{iso} = -\sigma_{iso}$; $|\sigma_{33} - \sigma_{iso}| \geq |\sigma_{11} - \sigma_{iso}| \geq |\sigma_{22} - \sigma_{iso}|$; $\eta = (\sigma_{22} - \sigma_{11})\,(\sigma_{33} - \sigma_{iso})^{-1}$.

Acknowledgements

Support of this work by the *Deutsche Forschungsgemeinschaft* and the *Fonds der Chemischen Industrie* is gratefully acknowledged. We thank *A. Facher*, Bayreuth, for carrying out DSC measurements on **1**.

References

[1] Parsonage NG, Staveley LAK (1978) Disorder in Crystals. Clarendon Press, Oxford
[2] Murrill E, Breed LW (1971) Inorg Chem **10**: 641
[3] Dereppe JM, Magill JH (1972) J Phys Chem **76**: 4037
[4] Aliev AE, Harris KDM, Apperley DC (1993) J Chem Soc Chem Commun 251
[5] Aliev AE, Harris KDM, Apperley DC, Harris RK (1994) J Solid State Chem **110**: 314
[6] Helluy X, Kümmerlen J, Sebald A (1998) Organometallics (in press)
[7] Dinnebier RE, Dollase WA, Helluy X, Kümmerlen J, Sebald A, Pagola S, Stephens PW, van Smaalen S (1998) Acta Cryst B (submitted)
[8] Bartell LS, Clippard FB, Boates TL (1970) Inorg Chem **9**: 2436
[9] Beagley B, Pritchard RG, Titiloye JO (1988) J Mol Struct **176**: 81
[10] Kümmerlen J, Sebald A (1997) Organometallics **16**: 2971
[11] Helluy X, Kümmerlen J, Sebald A (1997) Organometallics **16**: 5218
[12] Schmidt-Rohr K, Spiess HW (1994) Multidimensional Solid-State NMR and Polymers. Academic Press, London
[13] Marschner C (1998) Eur J Inorg Chem 221
[14] Ernst RR, Bodenhausen G, Wokaun A (1987) Principles of Nuclear Magnetic Resonance in One and Two Dimensions. Oxford University Press, Oxford
[15] Marion D, Wüthrich K (1983) Biochem Biophys Res Commun **113**: 467
[16] Mehring M (1983) Principles of High Resolution NMR in Solids. Springer, Berlin
[17] MATLAB, Version 5.0 (1992) The Math Works Inc., Natick, MA
[18] Haeberlen U (1976) High Resolution NMR in Solids, Selective Averaging. Adv Magn Reson Supplement 1. Academic Press, New York

Received June 13, 1998. Accepted July 31, 1998

Reactions of a Cyclotrisilane with Chalcogen Transfer Reagents [1]

Manfred Weidenbruch[1,*], **Andreas Grybat**[1], **Wolfgang Saak**[1], **Eva-Maria Peters**[2], and **Karl Peters**[2]

[1] Fachbereich Chemie, Universität Oldenburg, D-26111 Oldenburg, Germany
[2] Max-Planck-Institut für Festkörperforschung, D-70506 Stuttgart, Germany

Summary. Treatment of hexa-*tert*-butylcyclotrisilane (**1**) with Et_3PTe afforded the telluratrisiletane **5** in high yield. Irradiation of **1** in the presence of phenyl isothiocyanate furnished the 1,4-dithia-2,5-disilacyclohexane-3,6-diimine ring compound **6**. The structures of **5** and **6** were determined by X-ray crystallography.

Keywords. Silicon; Chalcogens; Silaheterocycles.

Reaktionen eines Cyclotrisilans mit Chalkogen-Transfer-Reagentien [1]

Zusammenfassung. Einwirkung von Et_3PTe auf Hexa-*tert*-butylcyclotrisilan lieferte das Telluratrisiletan **5** in hoher Ausbeute. Belichten von **1** in Gegenwart von Phenylisothiocyanat ergab das 1,4-Dithia-2,5-disilacyclohexan-3,6-diimin-Ringsystem **6**. Der Aufbau von **5** und **6** wurde durch Röntgenstrukturanalysen gesichert.

Introduction

Cyclotrisilanes are the smallest and at the same time the most recently discovered representatives of the cyclooligosilanes [2]. The strain within the three-membered ring system is reflected by two typical reaction modes [3]. Under protection from light or in the absence of a catalyst, cleavage of one Si–Si bond occurs to give, depending on the type of the reaction partner, either open-chain trisilanes or ring expansion products. On the other hand, exposure to light results in the synchronous cleavage of two Si–Si bonds to furnish a silylene and a disilene, species which have been characterized by numerous trapping reactions.

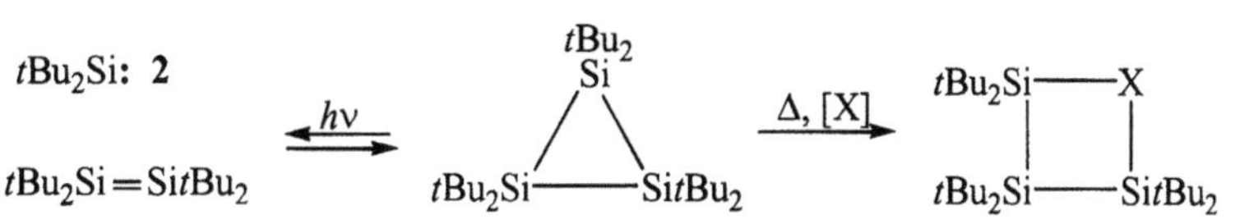

3 **1** **4**, X = S, Se

* Corresponding author

The behaviour of hexa-*tert*-butylcyclotrisilane (**1**) has been investigated widely; for example, it reacts with sulfur or selenium to furnish the ring expanded thia- or selenatrisiletanes **4**. In contrast, so far no reaction has been observed with tellurium [4]. Di-*tert*-butylsilylene (**2**) and tetra-*tert*-butyldisilene (**3**) are formed under photolytic conditions and undergo secondary addition reactions with various multiple bond systems to yield a wide variety of three- or higher membered ring systems [3]. We now report on the formation of a previously unknown telluratrisiletane as well as the photochemical reaction of **1** in the presence of phenylisothiocyanate.

Results and Discussion

Although **1** does not react with tellurium alone, addition of a small amount of triethylphosphane to a mixture of the components gives rise to a reaction, recognizable on account of the change in colour from yellow to colourless, which is complete within 90 minutes.

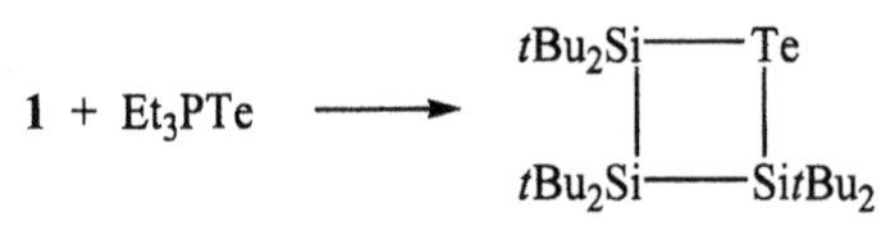

5

The NMR spectroscopic characterization of the colourless, crystalline telluratrisiletane **5** turned out not to be easy since the ^{1}H NMR spectrum surprisingly contained only one signal for the methyl protons. However, the ^{13}C NMR spectrum revealed two sets of signals for the primary and the quaternary carbon atoms. The ^{29}Si-NMR spectrum furnished more information in the form of two strongly deshielded signals at 31.19 and 68.84 ppm with the expected $^{1}J_{Si,Te}$, $^{2}J_{Si,Te}$, and $^{1}J_{Si,Si}$ coupling constants. Final verification of the structure of **5** as well as some noteworthy details were provided by an X-ray crystallographic analysis (Fig. 1).

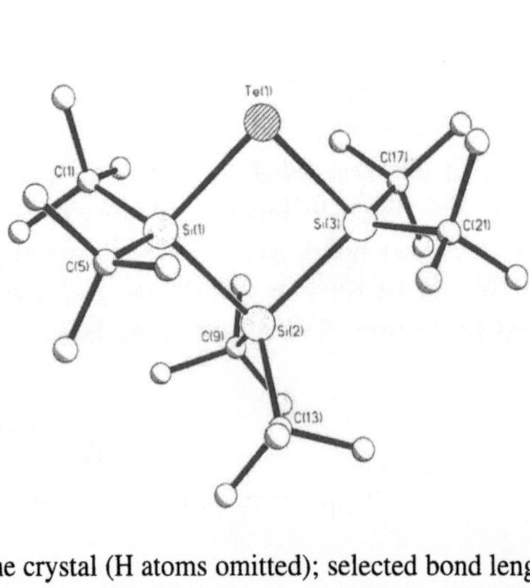

Fig. 1. Molecule of **5** in the crystal (H atoms omitted); selected bond lengths (pm) and angles (deg): Te–Si(1) 252.4(1), Te–Si(3) 252.76(9), Si(1)–Si(2) 251.25(12), Si(2)–Si(3) 251.35(12), Si(1)–Te–Si(3) 88.21(3), Si(2)–Si(1)–Te 91.54(3), Si(1)–Si(2)–Si(3) 88.79(4), Si(2)–Si(3)–Te 91.45(3)

As demonstrated by the 360° sum of the endocyclic angles, the four-membered ring of **5** is completely planar. The Si–Si bond lengths are considerably stretched and correspond to those in the cyclotrisilane **1**. This is not self-evident, since the ring strain should be reduced upon transformation of the three-membered ring in **1** to the four-membered ring in **5**. However, the concomitantly increasing interactions between the sterically demanding exocyclic substituents will effect increases in the Si–Si bond lengths as has been observed, for example, in hepta-*tert*-butylcyclotetrasilane [6].

Structurally characterized heterocyclic compounds containing silicon and tellurium atoms in the ring are still rather rare. The first cyclic compound of this type was obtained by *Jutzi et al.* from the reaction of decamethylsilococene with a phosphane telluride which surprisingly furnished a five-membered ring product containing two silicon and three tellurium atoms in the ring [7]. The corresponding telluradisilirane was obtained from the reaction of tetramesityldisilene with tellurium [8]. Finally, photolysis of **1** in the presence of a tellurophene afforded a four-membered ring, the 1,3-ditellura-2,4-disiletane system [9]. Thus, compound **5** supplements the series of structurally characterized smaller rings based on silicon and tellurium atoms.

In previous investigations we found that CS_2 and CO_2 could function as chalcogen transfer reagents similar to the phosphane chalcogenides. Thus, irradiation of **1** in the presence of CO_2 furnished hexa-*tert*-butylcyclotrisiloxane $(^{t}Bu_2SiO)_3$ as the final product. Analogous reactions with CS_2 led to various heterocyclic products containing silicon and sulfur in the rings [3, 10]. Isothiocyanates should be even better sulfur transfer reagents since they would afford the relatively stable isocyanides after abstraction of sulfur.

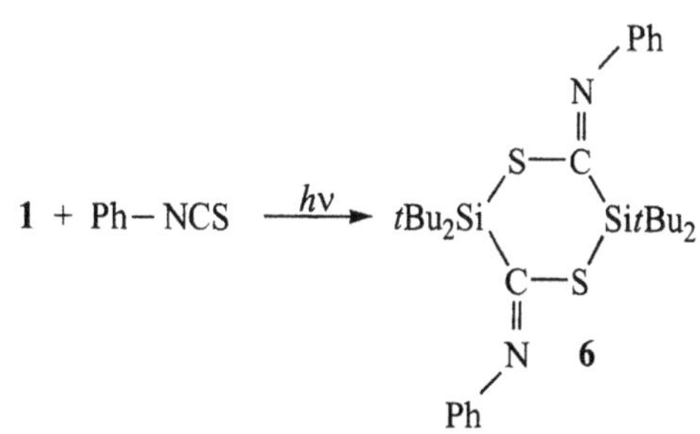

6

In fact, however, irradiation of **1** in the presence of phenyl isothiocyanate did not result in abstraction of sulfur. Instead, a compound was obtained for which analytical and spectroscopic data are indicative of a 1:1 adduct of silylene **2** with the isothiocyanate. Clarification of the structure of the product again required X-ray crystallography (Fig. 2) which identified the compound as the novel ring system 1,4-dithia-2,5-disilacyclohexane-3,6-diimine **6**. The heterocyclic six-membered ring of **6** adopts the chair conformation. The hardly stretched endo- and exocyclic bond lengths show that the compound is practically free of strain.

We can as yet only speculate on the mechanism of formation of **6**. One plausible pathway appears to consist of the addition of silylene **2** to the C=S double bond of the isothiocyanate with formation of a three-membered ring and subsequent σ-dimerization to furnish the isolated six-membered ring product. A comparable reaction sequence has been observed for the photolysis of **1** in the presence of alkyl

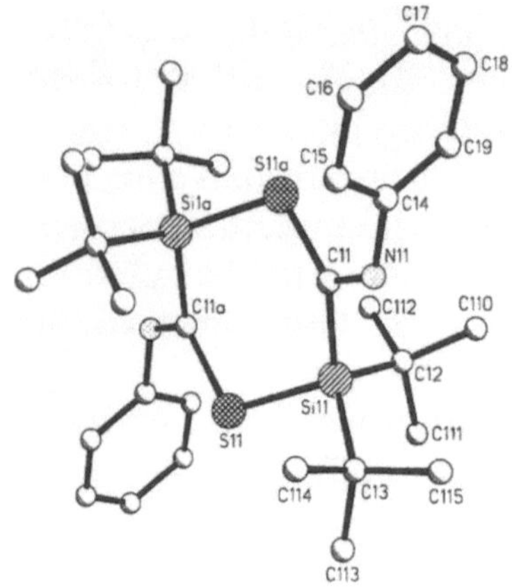

Fig. 2. Molecule of **6** in the crystal (H atoms omitted); selected bond lengths (pm) and angles (deg): S(1)–Si(1) 215.4(1), Si(1)–C(1) 189.8(3), S(1)–C(1A) 179.6(4), C(1)–N(1) 126.8(4), S(1)–Si(1)–C(1) 112.0(1), Si(1)–C(1)–S(1A) 124.7(2), Si(1)–S(1)–C(1A) 108.5(1), Si(1)–C(1)–N(1) 114.3(3), N(1)–C(1)–S(1A) 120.0(3)

or aryl cyanides. These reactions also proceed *via* a [2+1] cycloaddition of **2** to the C≡N bond with subsequent σ-dimerization to furnish 1,4-diaza-2,5-disilacyclohexa-3,6-dienes as final products [11].

Experimental

General procedures

All reactions were carried out in oven-dried glassware under an atmosphere of dry argon. Photolyses were carried out using a high-pressure mercury immersion lamp (Heraeus TQ 150). The ^{1}H and ^{13}C NMR spectra were obtained on a Bruker AM 300 spectrometer using C_6D_6 as solvent. The ^{29}Si NMR spectrum was recorded on a Bruker AMX 300 spectrometer. Mass spectra were recorded on a Vatrian-MAT 212 instrument. Elemental analyses were performed by *Analytische Laboratorien*, D-51779 Lindlar, Germany.

2,2,3,3,4,4-Hexa-tert-butyl-1-telluratrisiletane (**5**)

At room temperature, tellurium (0.21 g, 1.70 mmol) and a catalytic amount of triethylphosphane (3 drops) were added to a solution of **1** (0.71 g, 1.70 mmol) in 50 ml of petroleum ether (40–60°C). After 90 min the yellow colour of **1** had disappeared. To complete the reaction, the resulting mixture was stirred for 12 h at room temperature. Excess tellurium was filtered off, and the solvent was removed. Crystallization of the residue at −50°C from a minimum amount of *n*-hexane afforded 0.80 g (87%) of colourless crystals of **5**.

M.p.: 150–151°C; ^{1}H NMR: $\delta = 1.38$ (s) ppm; ^{13}C NMR: $\delta = 23.78$ (C_q), 29.05 (C_q), 32.91 (C_p), 34.13 (C_p) ppm (C_q and C_p refer to quaternary and primary carbon atoms); ^{29}Si NMR: $\delta = 31.19$ (s, Si(1), Si(3), $^{1}J_{Si,Te} = 260.0$ Hz, $^{1}J_{Si,Si} = 44.8$ Hz), 68.84 (s, Si(2), $^{2}J_{Si,Te} = 45.4$ Hz) ppm; MS (CI, isobutane): *m/z* (%) = 557 [MH$^+$] (100); $C_{24}H_{54}Si_3Te$ (554.54); calcd.: C 51.98, H 9.82%; found: C 51.79, H 10.02.

N,N-Diphenyl-2,2,5,5-tetra-tert-butyl-1,4-dithia-2,5-disilacyclohexane-3,6-diimine (**6**)

A solution of **1** (0.65 g, 1.50 mmol) and phenyl isothiocyanate (0.34 g, 2.5 mmol) in 80 ml of petroleum ether (40–60°C) was irradiated for 5 h at room temperature. The solvent was removed, and the red residue was redissolved in a minimum amount of *DME*. Cooling to −50°C furnished 0.12 g (21%) of colourless crystals of **6**.

M.p.: 109–110°C; ^{1}H NMR: $\delta = 1.21$ (s, 36 H), 6.95–7.21 (m, 10 H) ppm; ^{13}C NMR: $\delta = 23.67$ (C_q), 28.81 (C_p), 119.59, 124.86, 129.21, 151.43, 168.79 (CN) ppm; MS (CI, isobutane): *m/z* (%) = 556 [MH^+] (20); $C_{30}H_{46}N_2S_2Si_2$ (555.01); calcd.: C 64.92, H 8.30, N 5.05%; found: C 64.03, H 8.43, N 4.85.

X-ray structure analyses of **5** *and* **6**

Crystal and numerical data of the structure determinations are given in Table 1. In each case, the crystal was mounted in a thin-walled glass capillary. Data collection was performed at 296(2) K on a Siemens STOE AED 2 (**5**) or a Siemens P4 diffractometer (**6**) using graphite-monochromated MoK_α radiation.

The structures were solved by direct phase determination using the SHELX program systems and refined by full-matrix least-squares techniques against F^2 (**5**) or F (**6**) with the SHELXL [12] or SHELXTL PLUS program systems. Hydrogen atoms were placed in the calculated positions; all other atoms were refined anisotropically.

Table 1. Crystallographic data for **5** and **6**

	5	6
Empirical formula	$C_{24}H_{54}Si_3Te$	$C_{30}H_{46}N_2S_2Si_2$
Molar mass ($g \cdot mol^{-1}$)	555.54	505.01
a (pm)	928.1(1)	1150.2(1)
b (pm)	1984.6(1)	1169.8(1)
c (pm)	1634.8(1)	1246.5(1)
α (deg)	90	88.673(8)
β (deg)	90.06(1)	72.09(1)
γ (deg)	90	83.884(9)
$V \times 10^6$ (pm^3)	3011.1(4)	1623.4(3)
Z	4	2
D_{calcd} ($g \cdot cm^{-3}$)	1.223	1.135
Crystal system	monoclinic	triclinic
Space group	$P2_1/n$	$P\bar{1}$
Crystal dimensions (mm^3)	0.53×0.46×0.27	0.35×0.45×0.25
Data collection mode	$\omega - 2\theta$	ω
$2\theta_{max}$ (deg)	50	55
No. of reflections	5414	8580
No. of unique reflections	5296	7462
No. of observed reflections	4467	5295
Linear abs. coefficient (mm^{-1})	1.116	0.26
Data to parameter ratio	20.88	16.29
R (R_w) [$I > 2\sigma(I)$]	0.037	0.066 (0.066)
$wR2$ [all data]	0.098	0.069
Residual electron density (eÅ^{-3})	1.00; −0.32	0.50; −0.39

Additional material about the structure determination may be obtained from Fachinformationszentrum Karlsruhe, Gesellschaft für wissenschaftlich-technische Information mbH, D-76344 Eggenstein-Leopoldshafen, Germany, referring to the deposition numbers CSD-408 635 (**5**) and CSD-408 627 (**6**), the names of the authors, and the citation of the present paper.

Acknowledgements

Financial support of our work from the *Deutsche Forschungsgemeinschaft* and the *Fonds der Chemischen Industrie* is gratefully acknowledged. We thank Prof. *Marsmann*, Paderborn, for the ^{29}Si NMR spectrum.

References

[1] Silicon Compounds with Strong Intramolecular Steric Interactions, Part 67. Part 66: Weidenbruch M, Will P, Marsmann H, Peters K, von Schnering HG (1998) Z Anorg Allg Chem **624**: 15
[2] Masamune S, Hanzawa Y, Murakami S, Bally T, Blount JF (1982) J Am Chem Soc **104**: 1150
[3] Review: Weidenbruch M (1995) Chem Rev **95**: 1479
[4] Weidenbruch M, Schäfer A (1984) J Organomet Chem **269**: 231
[5] Schäfer A, Weidenbruch M, Peters K, von Schnering HG (1984) Angew Chem **96**: 311; Angew Chem Int Ed Engl **23**: 302
[6] Weidenbruch M, Kroke E, Pohl S, Saak W, Marsmann H (1995) J Organomet. Chem **499**: 229
[7] Jutzi P, Möhrke A, Müller A, Bögge A (1989) Angew Chem **101**: 1527; Angew Chem Int Ed Engl **28**: 1518
[8] Tan RPK, Gillette GR, Powell DR, West R (1991) Organometallics **10**: 546
[9] Weidenbruch M, Kirmaier L, Kroke E, Saak W (1997) Z Anorg Allg Chem **623**: 1277
[10] Brand-Roth B (1990) Thesis, University of Oldenburg, Germany
[11] Weidenbruch M, Schäfer A, Peters K, von Schnering HG (1986) J Organomet Chem **314**: 25
[12] Sheldrick GM, SHELXL 93 (1993) University of Göttingen, Germany

Received May 4, 1998. Accepted June 15, 1998

Zur Charakterisierung von Si-Polymeren durch flüssigchromatographische Verfahren in Kopplung mit MALDI-TOF-MS

Ralph-Peter Krüger[1,*], **Helmut Much**[1], **Günter Schulz**[1]
und **Eckhard Rikowski**[2]

[1] Bundesanstalt für Materialforschung und -prüfung (BAM), D-12205 Berlin, Germany
[2] GKSS-Forschungszentrum, Institut für Chemie, D-21502 Geesthacht, Germany

Zusammenfassung. Die Kopplung der drei flüssigchromatographischen Modi mit MALDI-TOF-Massenspektrometrie ist zur Bestimmung der molekularen, funktionalen und chemischen Verteilungen von Si-Polymeren geeignet. Silsesquioxane und Siloxane dienen als Beispiele für diese Art der Strukturerkennung. Die substanzspezifische Konstruktion einer SEC-Eichkurve mit Hilfe der MALDI-MS wird gezeigt.

Characterization of Si Polymers by Liquid Chromatographic Methods in Coupling with MALDI-TOF-MS

Summary. The coupling of the three liquid chromatographic modes with MALDI-TOF mass spectrometry is suitable for the determination of the molecular, functional, and chemical heterogeneities of Si polymers. Silsesquioxanes and siloxanes serve as examples for this method of structure identification. The substance specific construction of the SEC calibration curve by means of MALDI-MS is shown.

Keywords. Si polymer characterization; LC/MALDI-TOF-MS.

Einleitung

Si-Polymere gewinnen bei vielfältigen Anwendungen stark an Bedeutung. Für die umfassende Charakterisierung derartiger Produkte sind moderne Methoden, die Monomere, Oligomere und Polymere zuverlässig erfassen, notwendig. Dazu werden Methodenkopplungen der MALDI-TOF-MS (Matrix-Assisted Laser Desorption/Ionization Time-Of-Flight Mass Spectrometry) mit verschiedenen Modi der Flüssigchromatographie angewendet. Ziel der Untersuchungen ist die Charakterisierung der Produkte hinsichtlich der Molmassemittelwerte und -verteilung, der Funktionalitätsverteilung, der chemischen Heterogenität und der Polymerarchitektur (Ketten, Verzweigungen, Cyclen, Dendriten u.s.w.), um Beiträge zur Aufklärung des Polymerbildungsprozesses, zur kinetischen

* Corresponding author

Charakterisierung und Modellierung, zur Korrelation der Struktur mit den Werkstoffeigenschaften sowie zum Abbauverhalten zu erbringen.

Klassische Methoden zur Bestimmung der Molmasse von Oligomeren und Polymeren liefern nur Mittelwerte; Nebenbestandteile werden dabei unzureichend charakterisiert. Molmasseverteilungen werden mit flüssigchromatographischen Arbeitsweisen bestimmt, die Identifizierung der getrennten Spezies geschieht erfolgreich mit spektroskopischen Methoden, unter denen die MALDI-MS eine besondere Rolle einnimmt. Zur Anwendung kommen die drei Modi der Flüssigchromatographie (SEC: Size Exclusion Chromatography, LAC: Liquid Adsorption Chromatography, LACCC: LAC at Critical Conditions). Für die Charakterisierung der chemischen Heterogenität von Polymeren (z.B. verschiedene Endgruppen) ist die LACCC die Methode der Wahl. In früheren Untersuchungen wurden die kritischen Bedingungen jeweils durch Messung der Abhängigkeit des Elutionsvolumens von der Molmasse in unterschiedlichen gemischten Phasen ermittelt: Übereinstimmung der Elutionsvolumina unabhängig von der Molmasse charakterisieren den kritische Punkt. Zur Bestimmung wurden mindestens 4 Modellpolymere mit unterschiedlicher Molmasse untersucht. Durch die Darstellung des Retentionsfaktors $k' = t - t_0/t_0$ (t: Retentionszeit der Probe, t_0: Retentionszeit des ausgeschlossenen Volumens (exclusion limit)) in Abhängigkeit vom Volumenbruch der gemischten mobilen Phase ist es aber möglich, auch schon aus 2 Proben mit unterschiedlicher Molmasse den kritischen Punkt aus den Schnittpunkten der gemessenen Funktionen zu erhalten [1]. Die Identifizierung der getrennten Spezies erfolgt über MALDI-MS, die als hochempfindliches und schnelles Analysenverfahren neben absoluten Molmassewerten Informationen über Oligomerenverteilung, Molmassemittelwerte, Monomereinheiten, Kettendefekte und Endgruppen liefert [2–9]. Die MALDI-MS eröffnet aber auch neue Möglichkeiten der substanzspezifischen Eichung der SEC, wenn keine strukturell gleichen Eichstandards zur Verfügung stehen. Aus dem chromatographischen Lauf werden hierzu einzelne engverteilte Fraktionen entnommen und über MALDI eine Zuordnung der Molmasse zum Elutionsvolumen vorgenommen. Zur MALDI-TOF-MS an Si-Polymeren sind erst wenige Arbeiten veröffentlicht; stellvertretend sei auf Untersuchungen an Polydimethylsiloxanen, Polymethylphenylsilanen und Carbosilandendrimeren hingewiesen [10–15].

Ergebnisse und Diskussion

Im folgenden sollen Beispiele für die Charakterisierung von Si-Polymeren durch Anwendung der drei flüssigchromatographischen Modi in Kopplung mit MALDI-TOF-MS oder dieser spektroskopischen Methode alleine wiedergegeben werden.

MALDI-MS ohne Kopplung

Abbildung 1 zeigt das MALDI-Massenspektrum eines gemischt substituierten T_8-Silsesquioxangemisches. Ein Dendrimer auf T_8-Basis mit 24 Endgruppen wurde durch Hydrosilylierung von $Allyl_8T_8$ mit Trichlorsilan und nachfolgender Reaktion mit Allylmagnesiumbromid hergestellt. Auf diese Weise sollte ein G1-Dendrimer mit 24 Allylendgruppen synthetisiert werden. Im Spektrum ist auf

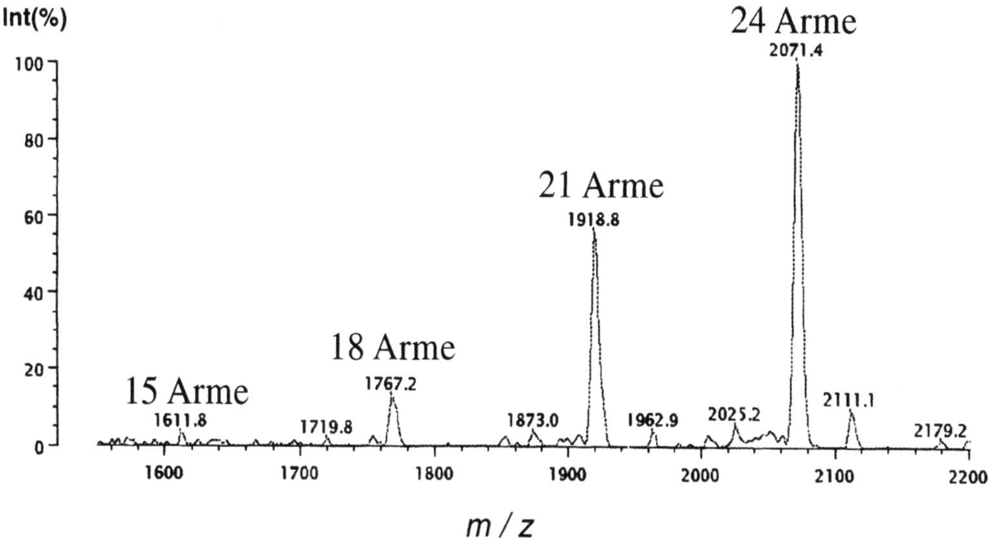

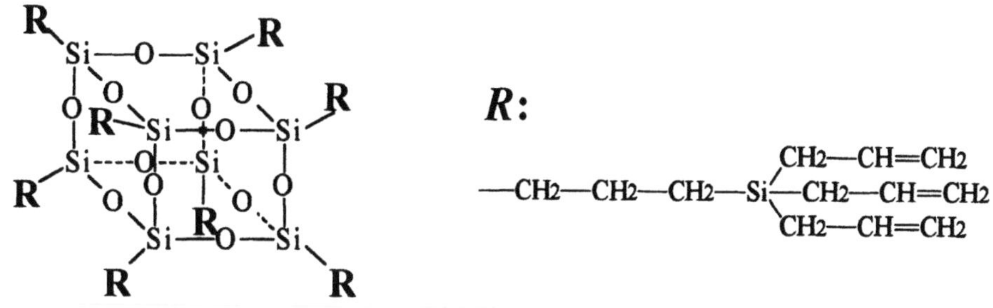

Abb. 1. MALDI-TOF-Massenspektrum eines Silsesquioxans angegebener Struktur (Lösungsmittel: *THF*, Matrix: 3-β-Indolacrylsäure/Silbertrifluoracetat, Ag-Addukte)

Grund der exakten Massebestimmung der am Würfel wirklich vorhandene Substitutionsgrad angezeigt.

Ein Beweis für die Selektivität der MALDI-MS ist im folgenden dargestellt. Silsesquioxane oder deren Gemische können u.a. durch Modifizierung eines vorhandenen Silsesquioxans hergestellt werden. Als Beispiel wird hier die nucleophile Substitution des Cl im $(Cl{-}C_3H_6)_8(SiO_{1.5})_8$ durch NCS-Gruppen gezeigt [16]. Durch mehrfaches Ausführen dieser Substitutionsreaktion ist die Darstellung von $(NCS{-}C_3H_6)_8(SiO_{1.5})_8$ aus $(Cl{-}C_3H_6)_8(SiO_{1.5})_8$ möglich. Bei der in dieser Weise durchgeführten Synthese eines Octasilsesquioxans konnte das Vorhandensein vieler Substitutionszustände nachgewiesen werden. Es erfolgte hier nicht nur die vollständige Umsetzung

$$(Cl{-}C_3H_6)_8(SiO_{1.5})_8 + 8\,NCS^- \rightarrow (NSC{-}C_3H_6)_8(SiO_{1.5})_8 + 8\,Cl^-,$$

sondern es entstand parallel ein Produktgemisch $(NCS{-}C_3H_6)_x(Cl{-}C_3H_6)_y(SiO_{1.5})_8$ bzw. $A_xB_yT_8$ mit $x+y=8$ (Abb. 2).

Die Produkte liegen als Na-Addukte vor (Lösungsmittel: *THF*, Matrix: 2,4,6-Trihydroxyacetophenon).

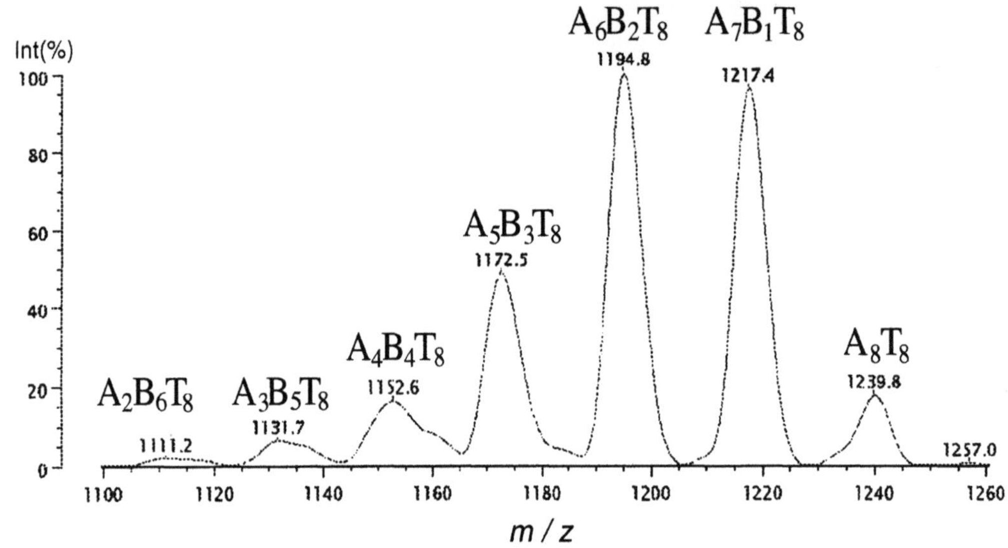

Abb. 2. MALDI-Nachweis verschiedener Substitutionszustände eines Silsesquioxans

Kopplung LACCC/MALDI-TOF-MS

Abbildung 3 zeigt das Chromatogramm eines Polydimethylsiloxans 1000 unter kritischen Bedingungen (mobile Phase 27.0% (v/v) Toluol und 73.0% (v/v) Isooctan). Aus den während des chromatographischen Laufs entnommenen kleinen Eluatmengen im jeweiligen Peakmaximum konnte mit der MALDI-TOF-MS die Identifizierung als cyclische ($RM = 0$ Da) und lineare, OH-terminierte ($RM = 18$ Da) Spezies vorgenommen werden (*RM*: Restmasse). Die Monomereinheit (*ME*) von 74 Da entspricht der angegebenen Grundstruktur. Im Peak 2 ist eine Unterverteilung zu erkennen, es handelt sich um noch vorhandene kleine Gehalte an Cyclen.

Kopplung SEC/MALDI

Octaalkylsilsesquioxane ($Ethyl_8(SiO_{1.5})_8$ bis $Butyl_8(SiO_{1.5})_8$) werden durch HCl-katalysierte Hydrolyse von Alkyltrichlor- bzw. Alkylmethoxysilanen in Methanol [17–19] oder durch vollständige Hydrosilylierung von Octahydridosilsesquioxanen [20, 21] mit den entsprechenden Alkenen dargestellt ($Pentyl_8(SiO_{1.5})_8$ bis $Decyl_8(SiO_{1.5})_8$):

$$8\,R\mathrm{Si}X_3 + 12\,\mathrm{H_2O} \rightarrow (R\mathrm{SiO}_{1.5})_8 + 24\,\mathrm{H}X \quad (R = \text{Ethyl, Propyl, } n\text{-Butyl}; X = \text{Cl, OMe})$$

$$\mathrm{H_8(SiO_{1.5})_8} + 8\mathrm{H_2C{=}CH}R \xrightarrow{\mathrm{H_2PtCl_6}} (R\mathrm{CH_2CH_2})_8(\mathrm{SiO_{1.5}})_8 \quad (\text{Alkene: 1-Penten bis 1-Decen})$$

Für das Octasilsesquioxan $R_8(SiO_{1.5})_8$ mit verschieden langen Alkylarmen wurde mittels MALDI-MS die genau für diese Substanzklasse geltende SEC-Eichkurve erstellt. Die Chromatogramme im SEC-Modus sind in Abb. 4 oben dargestellt. Beachtenswert ist, daß überraschenderweise neben den erwarteten Substanzen ein

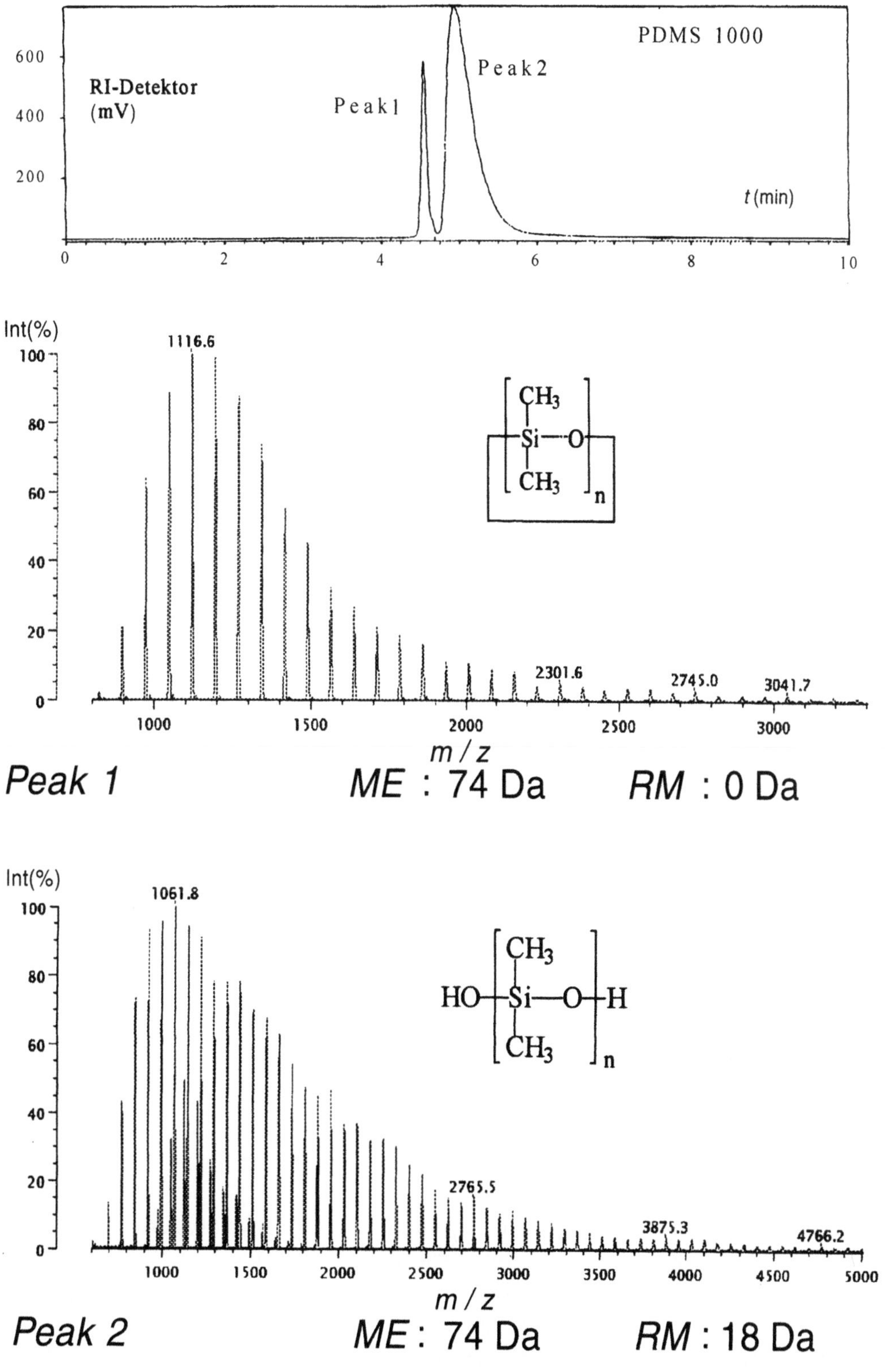

Abb. 3. Chromatogramm eines Polydimethylsiloxans unter kritischen Bedingungen und MALDI-Massenspektren der LACCC-Fraktionen (Lösungsmittel: *THF*, Matrix: 2,4,6-Trihydroxyacetophenon, Dotierung mit LiCl)

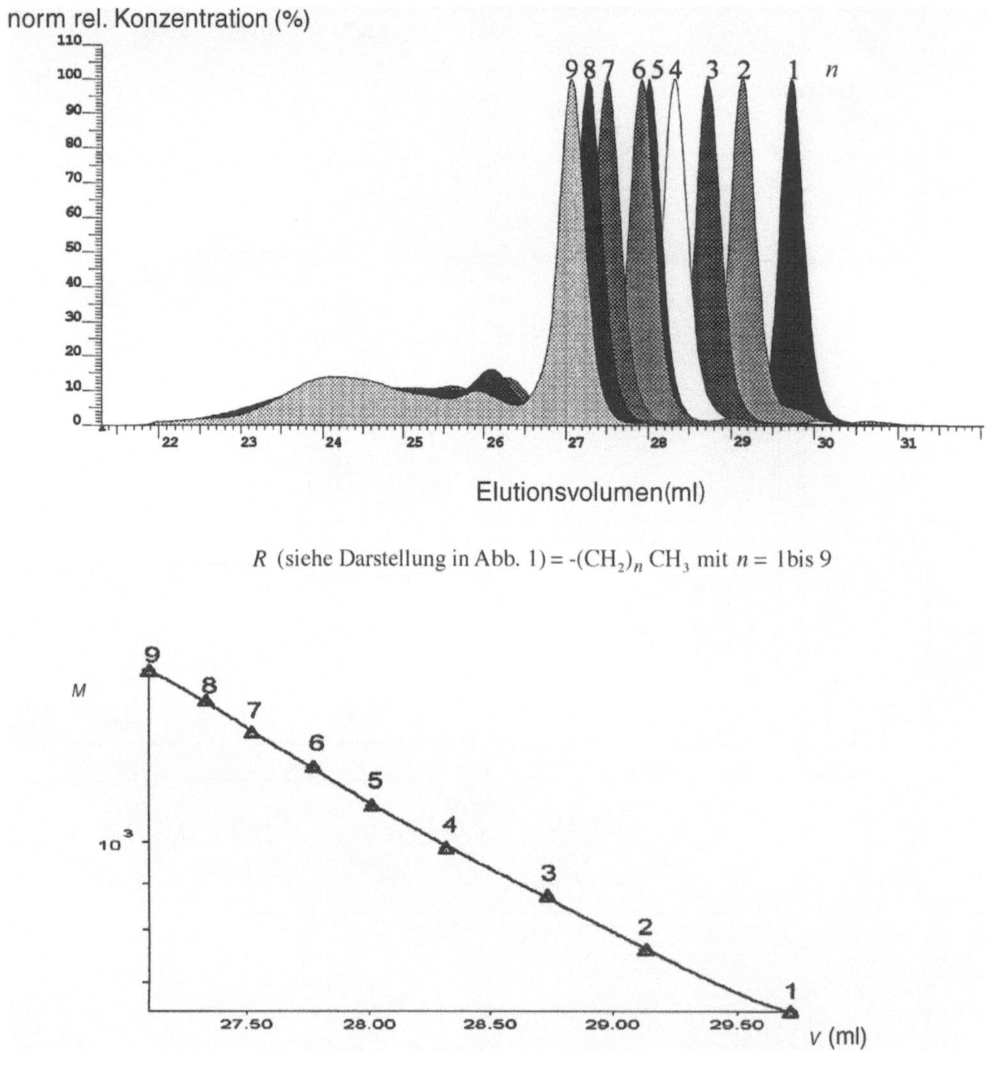

Abb. 4. Chromatogramme im SEC-Modus von neun verschieden substituierten Octasilsesquioxanen (oben) und die mittels MALDI-TOF-MS erstellte substanzspezifische SEC-Eichkurve ($n = 1$ bis 9)

großer Anteil höhermolekularer Verbindungen enthalten ist. Die gefundene SEC-Eichkurve ist unten eingetragen.

Gemische höherer Silsesquioxane konnten durch basenkatalysierte Käfigumlagerung von Octasilsesquioxanen synthetisiert werden [22].

Die MALDI-Untersuchung eines mittels Gerüstumlagerung hergestellten Gemisches aus T_8- und T_{10}-Silsesquioxan (Octylarme) ergab, daß neben den erwarteten auch der T_{12}-Polyeder enthalten war (Abb. 5 oben). Diese Mischung wurde in der SEC getrennt (Abb. 5 unten); die Fraktionen wurden mit Hilfe der MALDI-MS untersucht.

Erwartungsgemäß zeigten sich in Fraktion 5 (Hauptfraktion) die niedermolekularsten Silsesquioxane (T_8 und T_{10}). Fraktion 4 enthielt die drei schon mittels

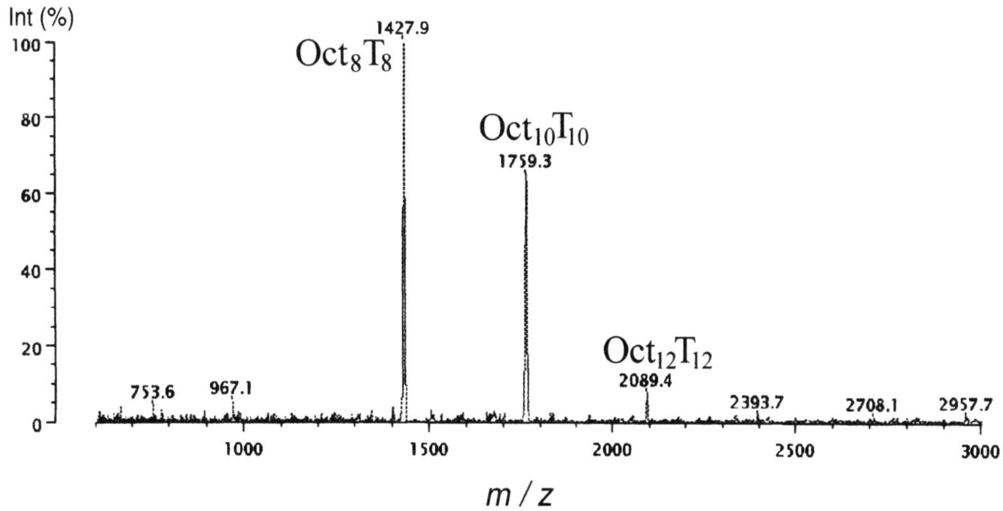

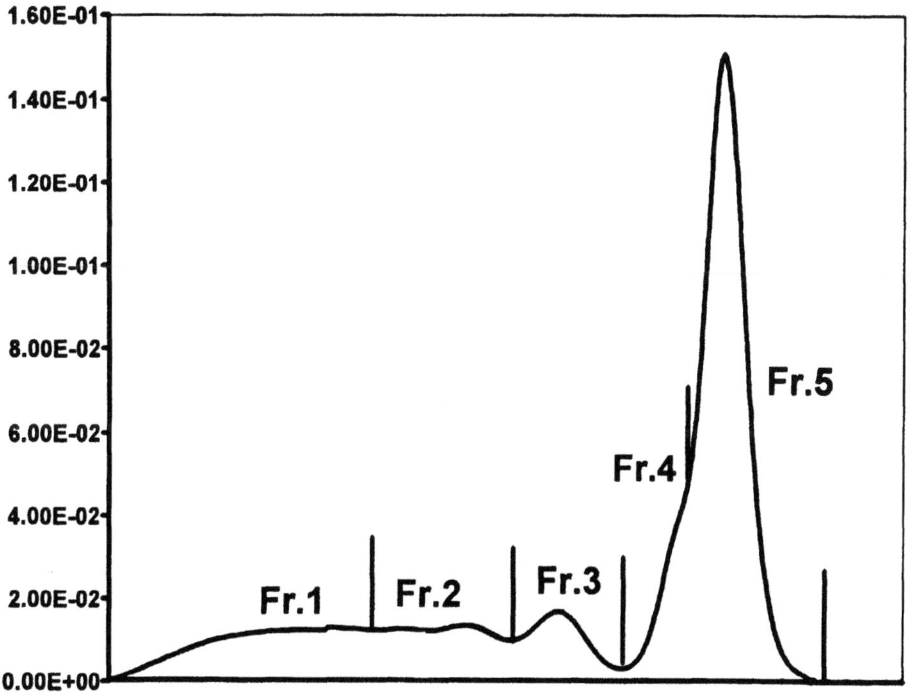

Abb. 5. MALDI-TOF-Massenspektrum eines Oct_8T_8- und $Oct_{10}T_{10}$-Silsesquioxangemisches (oben, Na-Addukte) und dessen SEC-Fraktionierung

MALDI-MS ohne SEC-Trennung nachgewiesenen Polyedergrößen. In den höhermolekularen Fraktionen konnten überraschend mit MALDI-MS deutlich größere Silsesquioxane nachgewiesen werden (bis $\geq T_{36}$, siehe Abb. 6).

Diese Methodenkopplung erlaubt es, auch SEC-Eichkurven von hyperverzweigten Si-Polymeren zu erstellen.

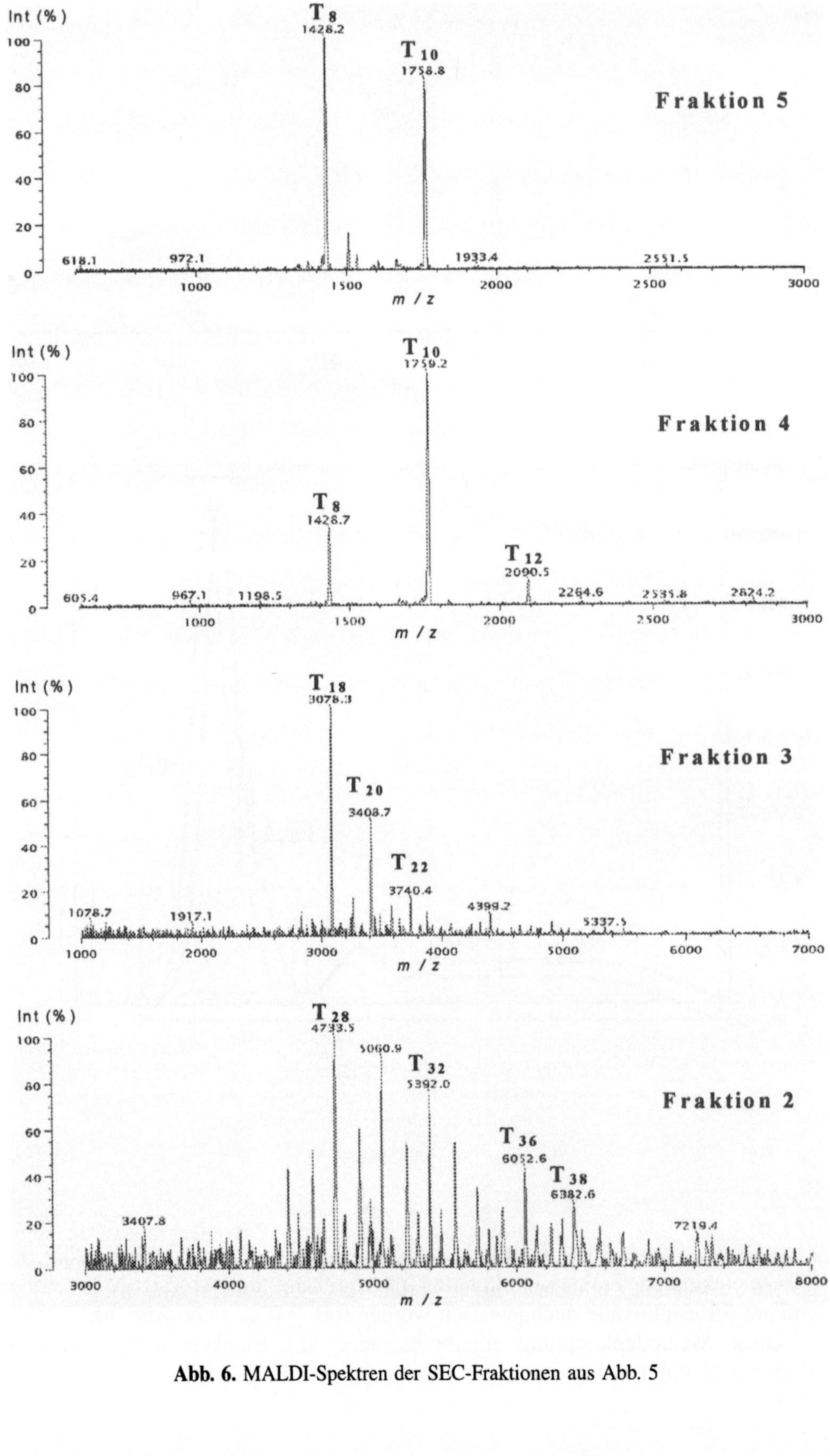

Abb. 6. MALDI-Spektren der SEC-Fraktionen aus Abb. 5

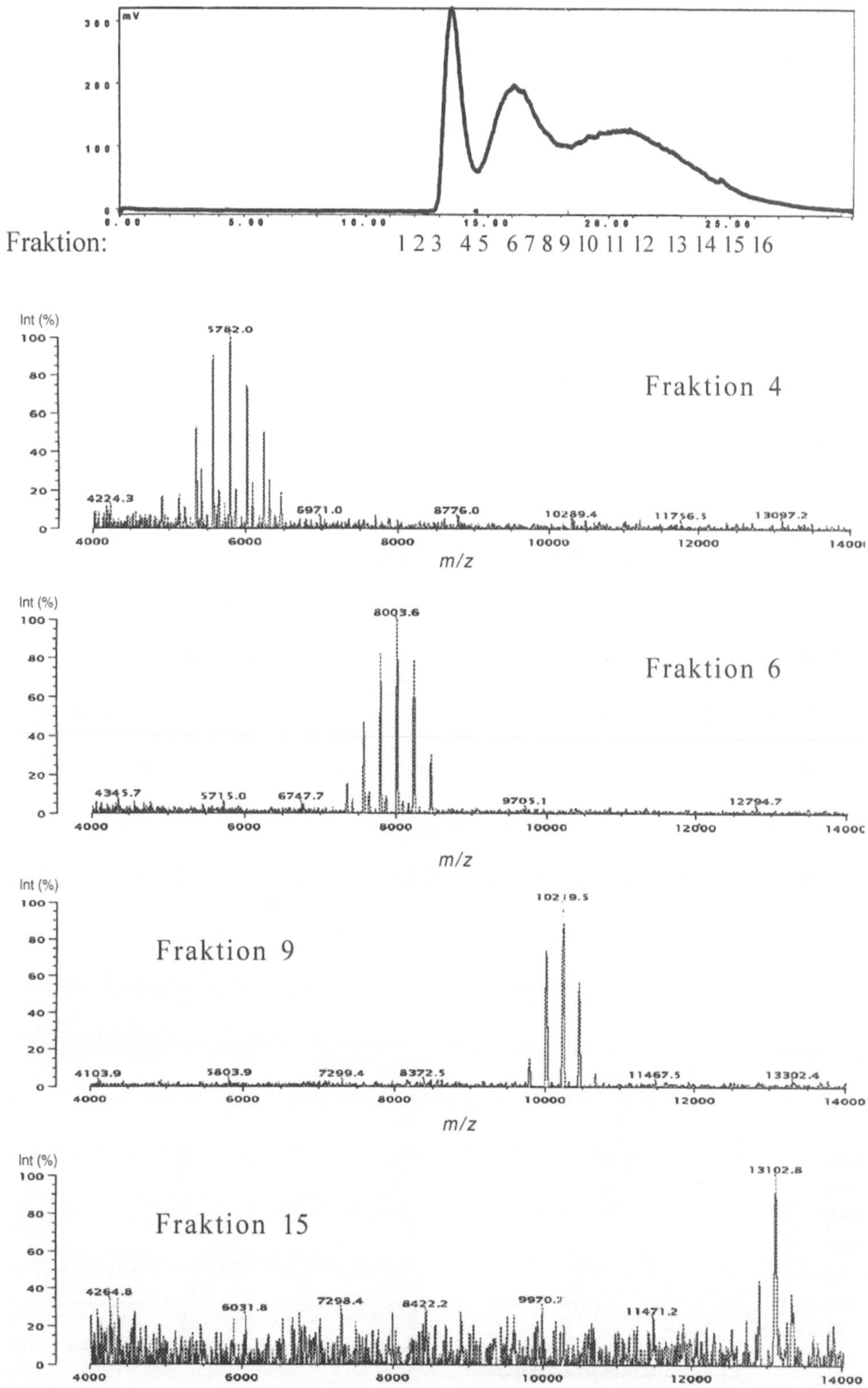

Abb. 7. Chromatogramm eines PDMS-Standard-Gemisches nahe dem kritischen Punkt und MALDI-Spektren ausgewählter LAC-Fraktionen

Kopplung LAC/MALDI

Unsere früheren Untersuchungen an siliciumorganischen Polymeren haben gezeigt, daß im Adsorptionsmodus in Gemischen nahe den kritischen Bedingungen bzw. im LAC-Modus eine Auftrennung in isokratischer Elution bis hin zu hohen Molmassen möglich ist. Abbildung 7 zeigt oben das LAC-Chromatogramm einer Mischung aus drei verschiedenen Polydimethylsiloxan-Standards (mobile Phase: 45.0% (v/v) *THF* und 55.0% (v/v) Methanol). An der Zeitachse ist die Zuordnung der Fraktionen angegeben, die off-line mit der sehr empfindlichen MALDI-TOF-Massenspektrometrie gemessen wurden. Beispielhaft sind nachfolgend vier MALDI-Spektren von an unterschiedlichen Stellen des Chromatogramms entnommenen Fraktionen zu sehen.

Die engere Verteilung mit steigender Molmasse im LAC-Modus ist deutlich erkennbar; die Selektivität nimmt im Gegensatz zum SEC-Mechanismus merklich zu (siehe Abb. 8). Dies resultiert aus den verschiedenen Anstiegen der SEC- und LAC-Eichkurven im höhermolekularen Bereich und erklärt, daß im LAC-Modus die höheren Molmassen mit verbesserter Auflösung bestimmt werden können ($\Delta V_1 < \Delta V_2$).

Die Leistungsfähigkeit der LAC sei noch an einem weiteren Silsesquioxanbeispiel gezeigt. Abbildung 9 (links) zeigt die Chromatogramme (SEC-Modus)

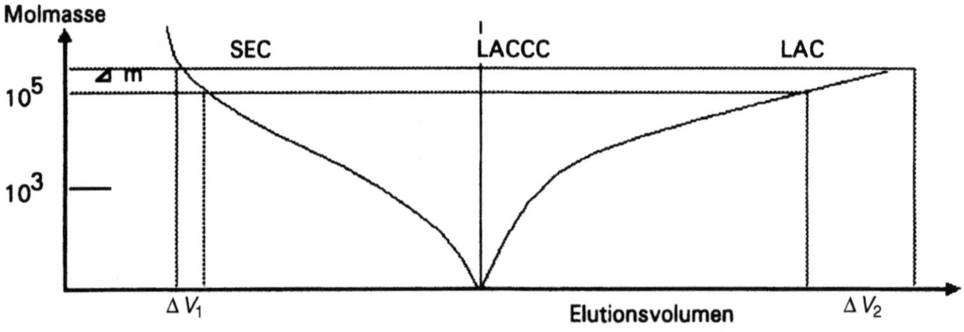

Abb. 8. Schematische Darstellung des Elutionsverhaltens von Polymeren in den drei LC-Modi

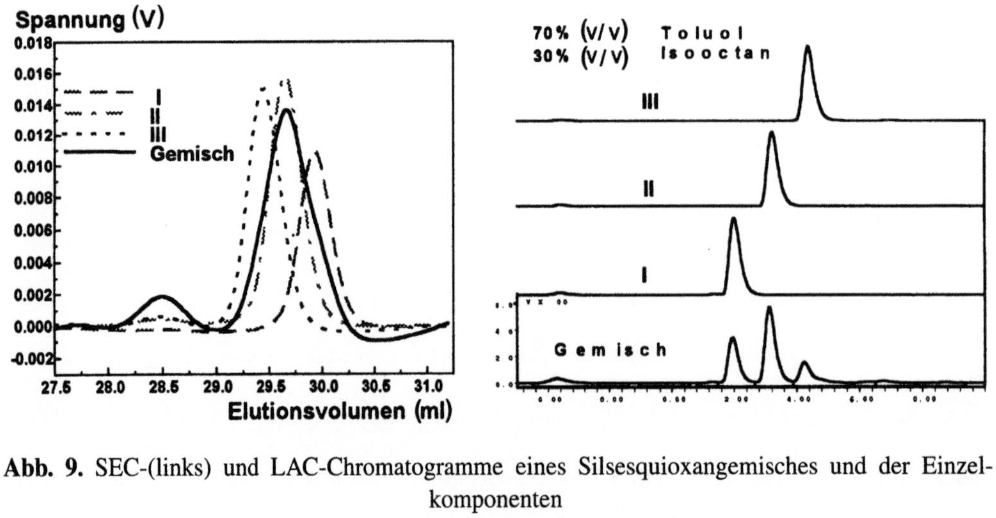

Abb. 9. SEC-(links) und LAC-Chromatogramme eines Silsesquioxangemisches und der Einzelkomponenten

des T_8-, T_{10}- und T_{12}-Polyeders (**I** bis **III**, Chlorpropylarme) sowie des Gemisches, in welchem keine Auftrennung der einzelnen Spezies mehr erfolgt. Die LAC ist im Gegensatz dazu in der Lage, im gleichen Gemisch die Einzelkomponenten anzuzeigen (Abb. 9, rechts). Der Vergleich der Einzelchromatogramme in beiden Abbildungen läßt erwartungsgemäß die unterschiedliche Elutionsreihenfolge erkennen.

Experimentelles

Die MALDI-TOF-MS-Messungen wurden an einem Kratos Kompact MALDI III-Instrument (Shimadzu) unter Verwendung eines Stickstofflasers ($\lambda = 337$ nm) durchgeführt (Polarität positiv, Beschleunigungsspannung 20 kV, Reflektron-Modus). 0.5 µl der Probenlösung (1 mg/ml) und 0.5 µl der Matrixlösung (25 mg/ml) wurden auf einem Deckgläschen gemischt und dann auf den Probenträger aufgebracht. In einigen angegebenen Fällen wurde dotiert. Vor der Messung wurde das Lösungsmittel oder Lösungsmittelgemisch abgedampft.

Die SEC-Messungen erfolgten an einem Flüssigchromatographen 1090 (Hewlett Packard) mit einem RI-Detektor Optilab DSP (Wyatt Technology). Als stationäre Phase diente ein Styrol-Divinylbenzol-Gel, als mobile Phase Toluol (Fluß 1 ml/min, injizierte Menge 100 µl, drei 5 µm-SDV-Säulen (PSS Mainz, je 300×8 mm, 10^5, 1000 und 100 Å)).

Die LACCC-Messungen wurden mit einem Marathon Probengeber (Jasco), einer Pumpe 880-PU (Jasco) und einem RI-Detektor Typ 7511 (ERC) durchgeführt (Macherey-Nagel-Säule Nucleosil 100 Å, 5 µm, 200×4 mm, Fluß: 0.5 ml/min.

Die LAC-Messungen erfolgten mit der genannten Gerätekombination (Detektor: ELSD Sedex 45 (ERC), Säulen: Nucleosil 300 Å, C-18 und Nucleosil 1000 Å, C-18).

Dank

Wir danken der *Deutschen Forschungsgemeinschaft* für die Förderung der Arbeiten sowie Dr. *Frey* (Freiburg), Prof. *Marsmann* (Paderborn), Prof. *Schnabel* und Herrn *Just* (beide Berlin), die an den Arbeiten der hier gezeigten Beispiele beteiligt waren, für die Zusammenarbeit sowie CI *Laging* und CI *Wittke* (beide Berlin) für experimentelle Mitarbeit.

Literatur

[1] Krüger R-P, Much H, Schulz G (1995) InCom Düsseldorf 20.–24. 3. 1995
[2] Hillenkamp F, Karas M, Beavis RC, Chait BT (1991) Anal Chem **63**: 1193 A
[3] Karas M, Bahr U, Deppe A, Stahl B, Hillenkamp F (1992) Makromol Chem, Macromol Symp **61**: 397
[4] Krüger R-P (1995) GIT Fachz Lab **39**: 189
[5] Krüger R-P, Much H, Schulz G (1996) Int J Polymer Analysis & Characterization **2**: 221
[6] Krüger R-P, Much H, Schulz G (1996) GIT Fachz Lab **40**: 398
[7] Trathnigg B, Maier B, Schulz G, Krüger R-P, Just U (1996) Macromol Symp **110**: 231
[8] Krüger R-P, Much H, Schulz G, Wachsen O (1996) Macromol Symp **110**: 155
[9] Wachsen O, Reichert KH, Krüger R-P, Much H, Schulz G (1997) Polymer Degradation and Stability **55**: 225
[10] Montaudo G, Montaudo MS, Puglisi C, Samperi F, Sepulchre M (1996) Macromol Chem Phys **197**: 2615
[11] Wu Z, Biemann K (1997) Int J Mass Spectrom Ion Proc **165/166**: 349
[12] Lorenz K, Mülhaupt R, Frey H, Rapp U, Mayer-Posner FJ (1995) Macromol **28**: 6657

[13] Sheiko SS, Eckert G, Ignat'eva G, Muzafarov AM, Spickermann J, Räder HJ, Möller M (1996) Macromol Rapid Commun **17**: 283
[14] Tang X, Dreifuss PA, Vertes A (1995) Rapid Commun Mass Spectrom **9**: 1141
[15] Williams JB, Gusev I, Hercules DM (1996) Macromol **29**: 8144
[16] Dittmar U, Hendan BJ, Flörke U, Marsmann HC (1995) J Organomet Chem **489**: 185
[17] Olsson K (1958) Ark Kemi **13**: 367
[18] Olsson K, Grönwall C (1961) Ark Kemi **17**: 529
[19] Olsson K, Axen C (1964) Ark Kemi **22**: 237
[20] Herren D, Bürgy H, Calzaferri G (1991) Helv Chem Acta **74**: 24
[21] Calzaferri G, Herren D, Imhof R (1991) Helv Chim Acta **74**: 1278
[22] Rikowski E, Marsmann HC (1997) Polyhedron **16**: 3357

Received June 12, 1998. Accepted (revised) September 24, 1998

Phase Behaviour of Poly(di-*n*-decylsilane)

Christian Mueller[1], **Holger Frey**[2], and **Claudia Schmidt**[1,*]

[1] Institut für Makromolekulare Chemie, Universität Freiburg, D-79104 Freiburg, Germany
[2] Freiburger Materialforschungszentrum, Universität Freiburg, D-79104 Freiburg, Germany

Summary. The phase behaviour of poly(di-*n*-decylsilane) (*PD10S*) was investigated by DSC and ^{29}Si as well as ^{2}H solid state NMR spectroscopy. For ^{2}H NMR experiments, the polymer was deuterated in the α-position of the side chains. Depending on the thermal history of the sample, DSC shows two transitions at 303 and 320 K with varying intensity ratio. Above 320 K, only the conformationally disordered mesophase exists. NMR spectra prove the existence of two crystalline states **I** and **II**. State **I** is either a pure phase with a complex backbone structure or a superposition of several crystalline modifications, whereas state **II** probably corresponds to a single phase with transoid backbone conformation. Either state can be obtained separately.

Keywords. Polysilanes; Poly(di-*n*-decylsilane); Phase behaviour; Mesophase; ^{29}Si solid state NMR; ^{2}H solid state NMR.

Phasenverhalten von Poly(di-*n*-decylsilan)

Zusammenfassung. Das Phasenverhalten von Poly(di-*n*-decylsilan) (*PD10S*) wurde mittels DSC und ^{29}Si- sowie ^{2}H-Festkörper-NMR-Spektroskopie untersucht. Für die ^{2}H-NMR-Experimente wurde das Polymer in der α-Position der Seitenketten deuteriert. Abhängig von der thermischen Vorgeschichte der Probe zeigt die DSC zwei Übergänge (bei 303 und 320 K) mit variablem Intensitätsverhältnis. Oberhalb von 320 K existiert nur die konformativ ungeordnete Mesophase. NMR-Spektren beweisen die Existenz von zwei kristallinen Zuständen **I** und **II**. Zustand **I** ist entweder eine reine Phase mit komplexer Struktur des Rückgrats oder eine Überlagerung mehrerer kristalliner Modifikationen, während Zustand **II** wahrscheinlich einer reinen Phase mit transoider Rückgratkonformation entspricht. Jeder Zustand kann separat erhalten werden.

Introduction

Polysilanes (polysilylenes) with organic side chains, (-SiR_2-)$_n$, are linear inorganic-organic hybrid polymers with unusual properties due to their one-dimensional Si chains. They exhibit, for instance, strong UV absorptions and photoconductivity [1, 2]. Furthermore, poly(di-*n*-alkylsilanes) with side chains in the range of 4 to 14 methylene units exhibit an order-disorder transition from the crystalline phase to a hexagonal, columnar mesophase (μ) with conformational disorder, resulting in peculiar changes in their electro-optical properties [1]. Whereas the lower

* Corresponding author

homologues, poly(di-*n*-propylsilane) (*PD3S*) to poly(di-*n*-hexylsilane) (*PD6S*), show only one crystalline modification at atmospheric pressure and therefore only one phase transition from crystalline to mesophase [2], polysilanes with longer alkyl side chains (from 8 to 10 methylene units) show two phase transitions with an intermediate phase below the mesophase [1, 3]. In poly(di-*n*-decylsilane) (*PD10S*) the two phase transitions occur at approximately 308 K and 328 K [1, 4]. At least two different crystalline phases with λ_{max} of 350 and 378 nm have been observed for *PD10S* [1]; the solid state composition was reported to vary with molecular weight and thermal history. Further heating into the mesophase leads to a UV absorption with λ_{max} of 318 nm.

Some disagreement still exists in the literature regarding the structure of the low temperature (**I**) and the intermediate phase (**II**) of *PD10S*. Based on the UV absorption maxima, *van der Laan et al.* [4] assume an all trans (TT) backbone structure and a TGTG′ structure for the low-temperature and the intermediate crystalline phase, respectively. *KariKari et al.* [5], however, report only on one phase, probably the low-temperature phase, for which they find X-ray patterns indicative of a TGTG′ or possibly a T_3GT_3G' conformation. Solid state NMR studies, which can contribute to the elucidation of the crystal structure, have, to our knowledge, not been reported.

We selected *PD10S* (besides other poly(di-n-alkylsilanes)) for an ongoing systematic study of the molecular dynamics of polysilanes by solid state NMR and of their polymorphism at elevated pressure for several reasons. First of all, *PD10S* exists in different crystalline phases at atmospheric pressure. Furthermore, of the known alkyl substituted polysilanes it exhibits the highest value of λ_{max}, and its charge carrier mobility is extraordinarily high [4]. For our deuteron experiments, a sample of *PD10S* deuterated in the α-position of the side chains was synthesized. First ^{2}H and ^{29}Si solid state NMR results on the phase behaviour of *PD10S* are presented in the following.

Results and Discussion

Figure 1 shows differential scanning calorimetry (DSC) traces of *PD10S* samples with different thermal and crystallization histories. After conventional precipitation and fractionation, *PD10S* shows a DSC trace (2nd heating run) as represented by curve *a* of Fig. 1 with a dominating first phase transition at 304.1 K ($\Delta H = 11$ kJ/mol) followed by a minor transition at 321.7 K with a much smaller transition enthalpy of $\Delta H = 2$ kJ/mol. When a sample is annealed at 313 K between the two transitions and slowly cooled to 243 K, the subsequent heating runs show that the low temperature transition vanishes. After annealing for 17 h, both transitions are still observable, but after annealing for 24 h only the transition at higher temperature remains ($T = 322.9$ K, $\Delta H = 10.6$ kJ/mol) as shown in curve *b*. A similar DSC trace with only the high temperature transition at $T = 323.9$ K ($\Delta H = 11.7$ kJ/mol) is observed for the first heating run when the sample is not precipitated but slowly crystallized from solution at elevated temperature (curve *c*). When either type of sample with only one transition, precipitated and annealed or slowly crystallized, is cooled rapidly (at 2 K/min) from the mesophase, the subsequent heating run shows two transitions of similar magnitude.

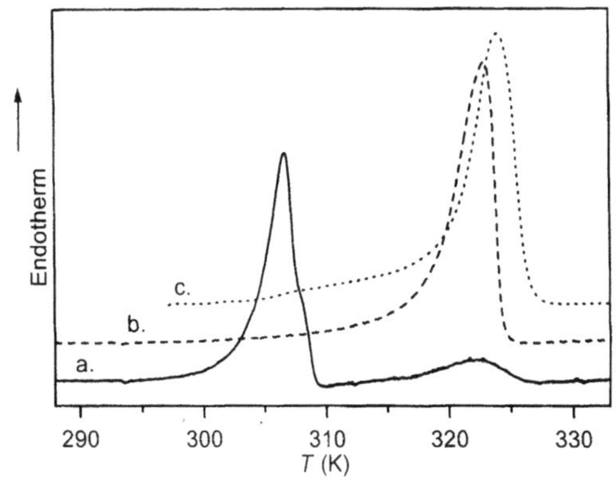

Fig. 1. DSC traces of *PD10S*; *a*: fractionated polymer sample, 2nd heating run; *b*: fractionated sample after annealing for approximately 24 hours; *c*: slowly crystallized sample as received (1st heating run)

Figure 2 depicts ^{29}Si CP-MAS solid state NMR spectra. The spectra *a–d* were obtained for a fractionated sample upon heating. The first spectrum at 260 K (Fig. 2a) corresponds to the low temperature crystalline state which is referred to as state **I** in the following. It shows three resonance lines at −19.7, −22.5, and −30.8 ppm which are unusually narrow as compared to the spectra of other polysilanes. The spectrum at 303 K (Fig. 2b) is a superposition of the three-line spectrum of state **I**, a broad resonance line at −20.7 ppm attributed to a second crystalline state (**II**), and the sharp resonance line of the mesophase at −24.6 ppm. It is commonly observed for polysilanes and other polymers with similar mesophases that the phase transitions are broad and that the resonances of the higher-temperature phases are observed well below the temperatures of the DSC peaks. Therefore, it is not surprising to find the mesophase peak already at 290 K upon heating (spectrum not shown). At 305 K (Fig. 2c), the low-temperature state **I** has disappeared, and only the peaks of the mesophase and of state **II** remain. A spectrum of the pure mesophase at 322 K is shown in Fig. 2d. Very slow crystallization from the mesophase during cooling at a rate of 0.5 K/h down to 260 K yields a crystalline *PD10S* consisting only of state **II** as can seen from Fig. 2e. In a series of fast crystallization experiments (not shown), spectra analogous to Fig. 2a were obtained. This shows, in agreement with the DSC results, that the low-temperature state can be recovered.

^{2}H NMR spectra of fractionated α-deuterated *PD10S* are shown in Fig. 3. The spectrum at a temperature of 255 K, well below the first phase transition, is a typical powder spectrum with a quadrupole splitting ($\Delta\nu$) of about 121 kHz, evidencing that the polymer sample is immobile except for small amplitude motions of the side chains. Above the transition to the mesophase at 322 K, motionally narrowed powder spectra are obtained with splittings of about 23 kHz. A splitting of this size is characteristic of the mesophase of polysilanes [6, 7]. The spectral shape reflects the axial symmetry of the molecular motions, and the narrow width proves the highly dynamic nature of both side and main chains. The spectrum recorded at 304 K, near the transition to the intermediate state, is a superposition of several fractions: the almost rigid one with a splitting of 119 kHz

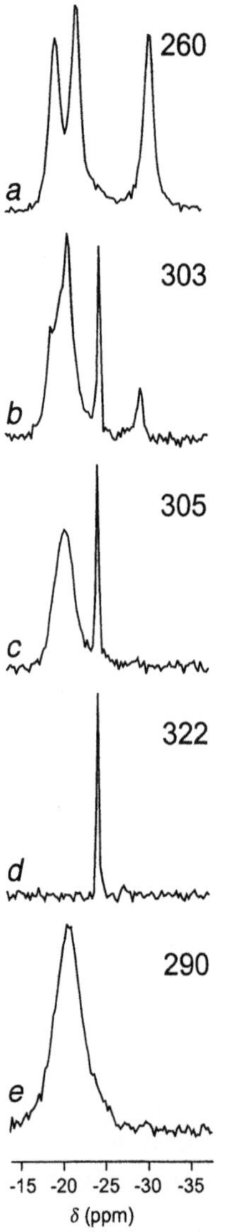

Fig. 2. ^{29}Si CP/MAS NMR spectra; *a–d*: variable temperature series of fractionated *PD10S* upon heating; *e*: after slow cooling from the mesophase (temperatures in K)

at this temperature (probably state **I**), a more mobile fraction with a reduced quadrupolar splitting of about 85 kHz (probably state **II**), the beginning mesophase ($\Delta\nu$ of 23 kHz) , and finally a small amount of an amorphous fraction giving rise to the narrow isotropic peak in the center. The spectra in the mesophase are similar for all polysilanes studied so far [6, 7], indicating that the dynamic processes are the same in all compounds. These spectra are motionally narrowed by a factor of about 1/5, making evident that the Si chain is also involved in the motions. If the polymer backbone were rigid and free rotation of the side chains were the only motion, the spectrum would be narrowed by a factor of only about 1/3 compared to the rigid powder pattern. Both side chains and main chain are conformationally disordered and mobile in the mesophase.

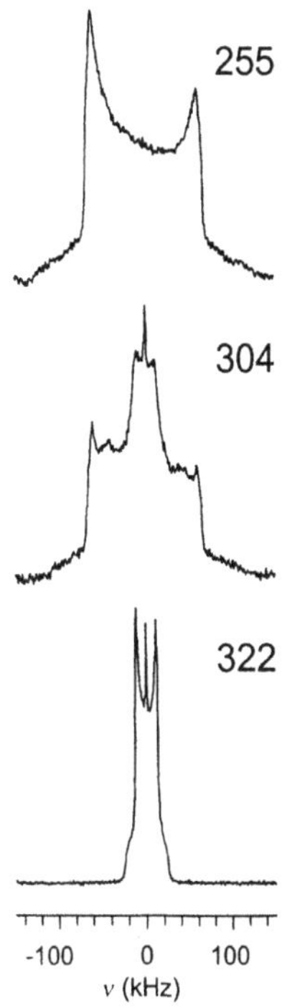

Fig. 3. ^{2}H NMR spectra of α-deuterated *PD10S* (temperatures are given in K)

DSC and NMR data together clearly show the existence of two different crystalline states below the mesophase. Either one can be obtained separately: the low-temperature state (**I**) by rapid cooling from the mesophase, and the intermediate-temperature state (**II**) by slow crystallization from the mesophase or by annealing at temperatures where the intermediate state and the mesophase coexist. Obviously, the formation of state **I** is favoured kinetically, whereas state **II** is the thermodynamically more stable form, at least at intermediate temperatures.

Regarding the structure of the two distinct states, ^{29}Si NMR spectroscopy provides important information. The three resonances of state **I** rule out a single phase with a TT backbone conformation. The spread in the chemical shifts of the three peaks is too large to be accounted for by mere intermolecular packing effects. The present data, however, do not allow to decide whether or not state **I** corresponds to a single phase. Assuming a single phase, at least three non-equivalent Si sites must exist in the elementary cell. A simple sequence such as TGTG′ or T_3GT_3G' can probably be ruled out because it does not account for the observed number of peaks with different chemical shifts. Alternatively, the three NMR peaks may arise from two or even three coexisting crystalline modifications. The spectra of state **II** are very similar to those of *PD6S* with respect to peak position and width. For this compound, only one modification with a TT backbone conformation is known [8–10]; therefore, we assign a single phase with transoid structure to state **II**. X-ray experiments, which are expected to allow a more precise determination of the phase behaviour, are presently underway.

Experimental

Dichlorodidecylsilane was polymerized using the well-known *Wurtz*-type coupling reaction with Na [1]. For the polymer with α-deuterated side chains, a deuterated monomer was prepared in a four-step synthesis, starting from decanoic acid anhydride which was reduced with $LiAlD_4$. The resulting α-deuterated decanol was converted to the corresponding *n*-decyl bromide using aqueous HBr; from this, the respective *Grignard* derivative was prepared. A *THF* solution of the *Grignard* compound was added to a solution of $SiCl_4$ in pentane to yield the α-deuterated dichlorodidecylsilane monomer. The crude polymers were precipitated twice from methanol/*THF* and fractionated with methanol from toluene solution to obtain polymers with a monomodal mass distribution. Alternatively, slowly crystallized samples of *PD10S* were prepared by dissolving 0.01–0.02 weight% polymer in cyclohexanol at 433 K. After cooling slowly within several hours, the turbid solution was filtered through 0.45 μm PTFE filters, and the precipitate was dried *in vacuo*. The poly(di-*n*-decylsilanes) were characterized by ^{1}H, ^{13}C, and ^{29}Si as well as two-dimensional ^{1}H–^{13}C COSY NMR spectroscopy in toluene-d_6 solution on a Bruker ARX 300 MHz spectrometer. Gel permeation chromatography (GPC), calibrated to polystyrene standards, yields a molecular weight M_w (weight average) of about 200000 g/mol with $M_w/M_n = 1.6$ for the fractionated non-deuterated sample.

Differential scanning calorimetry (DSC) was performed with a Perkin-Elmer DSC 7 at scan rates of 1 K/min. All samples were subjected to a first heating and cooling cycle before data acquisition unless indicated otherwise. Variable temperature ^{2}H and magic angle spinning (MAS) ^{29}Si solid state NMR measurements were carried out on a Bruker MSL 300 MHz spectrometer at resonance frequencies of 46.073 MHz and 59.625 MHz for ^{2}H and ^{29}Si, respectively. Deuteron spectra were obtained with a quadrupole echo pulse sequence with 90° pulses of 3.9 μs and a recycle delay of 3 s. 1024 scans were averaged. ^{29}Si MAS spectra were measured using the cross-polarization (CP) technique. A double-bearing MAS probe at spinning rates of about 2.5 to 3 kHz was used. The 90° pulse length was set to 4.5 μs with a CP contact time of 3 ms and a recycle delay of 4 s. 512 spectra were accumulated. ^{29}Si chemical shifts (referenced to *TMS*) were calibrated with Q_8M_8 (Bruker).

Acknowledgements

We thank *Beate Gloderer* for support with the synthesis and *Alfred Hasenhindl* for help with the NMR measurements. We gratefully acknowledge financial support by the *Deutsche Forschungsgemeinschaft* (Schwerpunktprogramm *Spezifische Phänomene in der Silicium-Chemie*.

References

[1] Miller R, Michl J (1989) Chem Rev **89**: 1359
[2] West R (1986) J Organometallic Chem **300**: 327
[3] Farmer BL, Miller RD, Rabolt JF, Fleming WW, Fickes GN (1988) Bull Am Phys Soc **33**: 1657
[4] van der Laan GP, de Haas MP, Hummel A, Frey H, Möller M (1996) J Phys Chem **100**: 5470
[5] KariKari EK, Greso AJ, Farmer BL, Miller RD, Rabolt JF (1993) Macromolecules **26**: 3937
[6] Mueller C, Schmidt C, Frey H (1996) Macromolecules **29**: 3320
[7] Mueller C, Schmidt C, Koopmann F, Frey H (submitted) Polymorphism and Molecular Mobility of Polysilanes and Poly(silylenemethylenes). In: Auner N, Weis J (eds) Organosilicon Chemistry – From Molecules to Materials, vol IV. Wiley-VCH, Weinheim
[8] Gobbi GC, Fleming WW, Sooriyakumaran R, Miller RD (1986) J Am Chem Soc **108**: 5624
[9] Schilling FC, Bovey FA, Lovinger AJ, Zeigler JM (1986) Macromolecules **19**: 2660
[10] Patnaik SS, Farmer BL (1992) Polymer **33**: 4443

Received June 12, 1998. Accepted (revised) September 8, 1998

Synthese, Funktionalisierung und Vernetzung von Poly(silylenphenylenen)

Wolfram Uhlig

Laboratorium für Anorganische Chemie der Eidgenössischen Technischen Hochschule Zürich, ETH-Zentrum, CH-8092 Zürich

Synthesis, Functionalization, and Cross-Linking Reactions of Poly(silylenephenylenes)

Summary. Treatment of 1,4-*bis*(bromomagnesium)benzene with *p*-anisyl-silyl-*bis*(triflates) gave poly(*p*-anisyl-silylenephenylenes) in high yields. The anisyl-silyl bond was cleaved selectively by triflic acid leading to triflate substituted poly(silylenephenylenes). Conversions of these polymers with nucleophiles afforded functionalized derivatives. Hydrosilylation reaction between silicon-vinyl and silicon-hydrogen derivatives results in polymer networks which may serve as interesting preceramic materials. The structures of the polymers were proven by NMR spectroscopy (^{29}Si, ^{13}C, ^{1}H).

Keywords. Organosilicon polymers; Poly(silylenephenylenes); Silyl triflates.

Einleitung

Polymere, in denen Organosiliciumeinheiten mit π-Elektronensystemen, beispielsweise Phenylen- [1–6], Ethenylen- [7], Ethinylen- [8–15], Diethinylen- [16–19], Thienylen- [20–22], Furylen- [23, 24] oder Butinylenbausteine [25, 26], in der Hauptkette regelmäßig alternieren, finden in den letzten Jahren wachsendes Interesse. Dieses ist vor allem durch ihre potentielle Nutzung als Funktionsmaterialien (Photoresists, Halbleiter, organische Vorstufen für siliciumhaltige Keramiken) begründet. In der Regel werden diese Polymere durch eine *Wurtz*-Reaktion aus *Bis*(chlorosilyl)-substituierten Verbindungen erhalten. Einen weiteren Zugang stellt die Kupplungsreaktion von Dilithioderivaten der π-Systembausteine mit organischen Dichlorosilanen oder Dichlorodisilanen dar. Verbindungen, die nach diesen Methoden hergestellt werden, enthalten jedoch immer einen geringen Prozentsatz Siloxyeinheiten, die auf der Hydrolyse und Kondensation unumgesetzter Chlorosilyleinheiten im resultierenden Polymer beruhen. Siloxyeinheiten in der Polymerkette bedeuten jedoch eine Unterbrechung der Elektronendelokalisation und senken daher die Photoaktivität und Leitfähigkeit der Polymere. Andererseits führt eine unvollständige Umsetzung zur Verringerung der Molmassen. Dieses wirkt sich wiederum negativ auf die thermischen Eigenschaften aus, da die keramische Ausbeute sinkt. Es hat daher nicht an der Suche nach alternativen Synthesemöglichkeiten gefehlt. Genannt seien besonders Ringöffnungspolymerisationen mittels katalytischer Mengen von Alkyllithiumverbindungen [8, 10], Polymerisationen von Diethinylsilanen mittels Pd-Katalysatoren [27, 28], von Diethinyldisilanen mit Rh-Katalysatoren [25, 26] oder von Phenylendisilanen mit Zirkoniumverbindungen [29]. Auch die Umsetzung von 1,4-*Bis*(bromomagnesium)benzol mit Diorganodichlorsilanen [30] bietet eine Alternative.

Bei Untersuchungen zur Chemie der als "Superelektrophile" bezeichneten oligomeren und polymeren Silyltriflate wurden effektive Zugänge zu vielseitig modifizierten Organosiliciumpolymeren gefunden [31–36]. Dabei gelang es, Copolymere mit regelmäßig alternierenden Struktureinheiten aufzubauen sowie fertige Polymerketten gezielt zu funktionalisieren und zu vernetzen. Silyltriflate besitzen gegenüber anderen elektrophilen Reagenzien R_3SiX eine Reihe wesentlicher Vorzüge. Gemeinsam mit den Iodiden verfügen sie über die höchste Reaktivität aller Silylierungsreagenzien [37, 38]. In der Anwendung sind sie diesen aufgrund ihrer Redoxstabilität jedoch überlegen. Außerdem werden bei der Umsetzung mit Nucleophilen keine dem Metall-Halogen-Austausch analogen Prozesse beobachtet. Schließlich besteht ein wesentlicher Vorteil darin, daß sie häufig leichter und selektiver hergestellt werden können als analoge Halogenderivate. Diese Erkenntnis beruht darauf, daß Silicium-Element-Bindungen von Trifluormethansulfonsäure ($TfOH = CF_3SO_3H$) gemäß Schema 1 in selektiver Weise gespalten werden. Die Spaltungsgeschwindigkeit der Bindung Si-*X* sinkt in der Reihenfolge *X* = NEt_2 > Allyl > *p*-Anisyl > *p*-Tolyl > Phenyl > Cl > Ethinyl > H > Alkyl signifikant [39–42]. Der Spaltungsmechanismus und die Spaltungsreihenfolge der Silicium-Phenyl-Bindungen sind dabei von besonderem Interesse, da aromatisch substituierte Silanderivate leicht hergestellt werden können.

In der vorliegenden Arbeit werden diese Syntheseprinzipien zur Darstellung neuer funktionalisierter und vernetzter Poly(silylenphenylene) genutzt. Das Interesse richtete sich dabei besonders auf die Synthese von Verbindungen, die den Anforderungen eines effektiven präkeramischen Polymers entsprechen. Derartige Polymere sollen entweder hochverzweigt sein (Netzwerkpolymere) oder

$$R_2SiXY + TfOH \longrightarrow R_2SiX(OTf) + HY$$

$$R_2SiX(NEt_2) + 2\,TfOH \longrightarrow R_2SiX(OTf) + Et_2NH_2OTf$$

X, *Y* = NEt_2, Allyl, *p*-Anisyl, *p*-Tolyl, Phenyl, Cl, Ethinyl, H; Alkyl

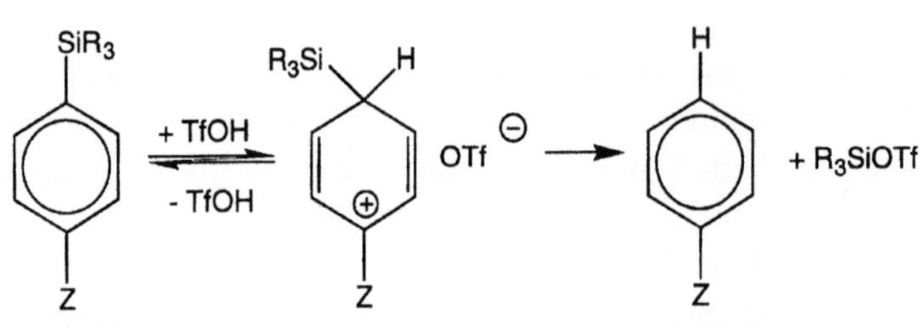

Z = OCH_3, CH_3, H

Schema 1

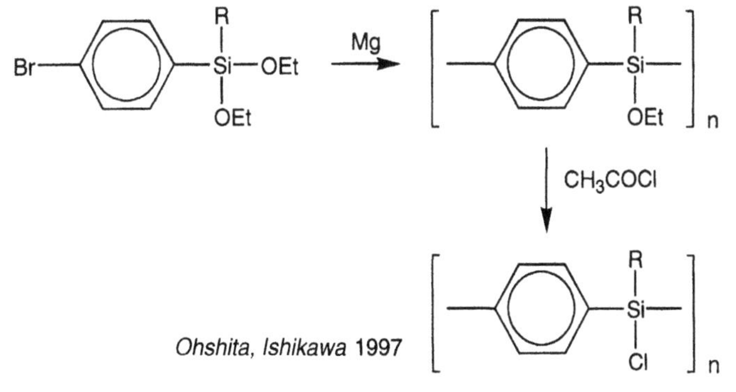

Schema 2

funktionelle Gruppen besitzen, die während der Pyrolyse zu Verzweigungen führen. Auch hohe Molekulargewichte und Löslichkeit sind erwünschte Eigenschaften [43].

Eine effektive Synthese von Polymeren mit einem alternierenden Arrangement von funktionalisierten Siliciumatomen und Phenyleneinheiten wurde bisher lediglich von *Ohshita* und *Ishikawa* [4] gemäß Schema 2 beschrieben.

Ergebnisse und Diskussion

Die Poly(silylenphenylene) **1–3** wurden in glatter Reaktion aus Silyl*bis*(triflaten) und 1,4-*Bis*(bromomagnesium)benzol gemäß Schema 3 erhalten. Der Einsatz von Silyltriflaten anstatt der entsprechenden Chlorsilane führt neben kürzeren Reaktionszeiten unter milden Bedingungen (60 min, 25°C) auch zu deutlich höheren Molmassen. Voraussetzung ist allerdings eine exakte Gehaltsbestimmung der verwendeten *Grignard*-Verbindung.

Die Protodesilylierung des Poly(silylenphenylens) **1** mit Trifluormethansulfonsäure stößt im Gegensatz zu anderen phenylierten Organosiliciumpolymeren auf die Schwierigkeit, daß vorrangig – aber nicht ausschließlich – die Polymerkette gespalten wird. Zwischen den verschiedenen aromatischen Struktureinheiten des Polymers besteht folglich keine Selektivität gegenüber dem elektrophilen Angriff der Säure. Schema 4 vergleicht das Reaktionsverhalten phenylierter Poly(silylene) [44], Poly(silylenemethylene) [34], Poly(silylenealkine) [12] und Poly(silylenephenylene). Die Umsetzung des Polymers unter Erhalt der Kette ist daher an die Verwendung einer besseren Abgangsgruppe am Silicium gebunden.

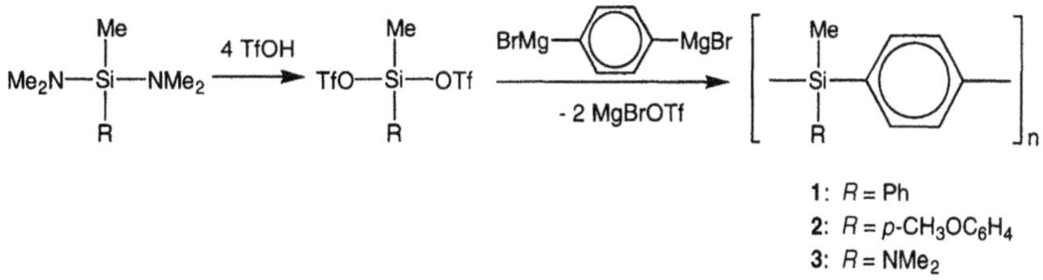

Schema 3

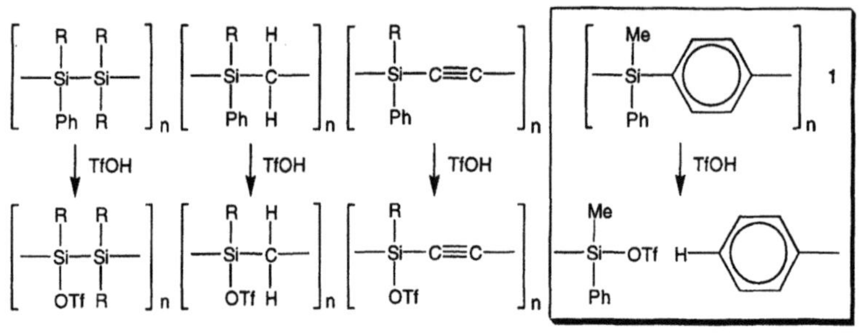

Schema 4

Die Polymere **2** und **3** erfüllen diese Voraussetzung, denn sowohl die Amino- als auch die *p*-Anisylgruppe stehen in der oben angegebenen Reaktivitätsreihe vor dem Phenylsubstituenten. Die *p*-Anisylverbindung **2** besitzt dabei gegenüber **3** bezüglich der Herstellung (Schema 3) den Vorteil der Hydrolysestabilität. Das vereinfacht die Aufarbeitung und Isolierung. Ein zweiter Vorteil besteht darin, daß die Spaltung einer Anisyl-Silicium-Bindung nur die äquimolare Menge an Trifluormethansulfonsäure erfordert. Die Spaltung einer Amino-Silicium-Bindung benötigt aufgrund der Ammoniumsalzbildung die doppelte Säuremenge. Daher wird **2** bei den nachfolgenden Umsetzungen als Ausgangsverbindung verwendet. Die Reaktion mit Trifluormethansulfonsäure in Toluol führt gemäß Schema 5 unter Abspaltung von Anisol zum triflatsubstituierten Polymer **4**. Dieses fällt nach Entfernung des Lösungsmittels als rotbraunes, stark hydrolyseempfindliches, viskoses Öl an. Daher wird in der Regel auf die Isolierung verzichtet und die Lösung von **4** mit Nucleophilen zu stabileren Folgeprodukten umgesetzt. In Schema 6 sind einige charakteristische Folgereaktionen mit Organometallverbindungen bzw. Lithiumalanat zusammengefaßt.

Die Umsetzung des Si-H-haltigen Polymers **5** mit der Vinylverbindung **6** im Sinne einer katalytischen Hydrosilylierung sollte eine gezielte intermolekulare Vernetzung möglich machen. Das Reaktionsprodukt ist jedoch in organischen Lösungsmitteln unlöslich und konnte nur ungenügend spektroskopisch charakterisiert werden. Um lösliche vernetzte Polymere zu erhalten, wurde versucht, die vernetzenden Gruppen im Molekül zu verdünnen. Ausgehend von **2** wurden Polymere synthetisiert, in denen nur ein Teil der Anisylgruppen durch Vinylgruppen bzw. H-Atome ersetzt war. Setzt man **2** mit Trifluormethansulfonsäure im molaren Verhältnis 2:1 um, so wird erwartungsgemäß nur die Hälfte der

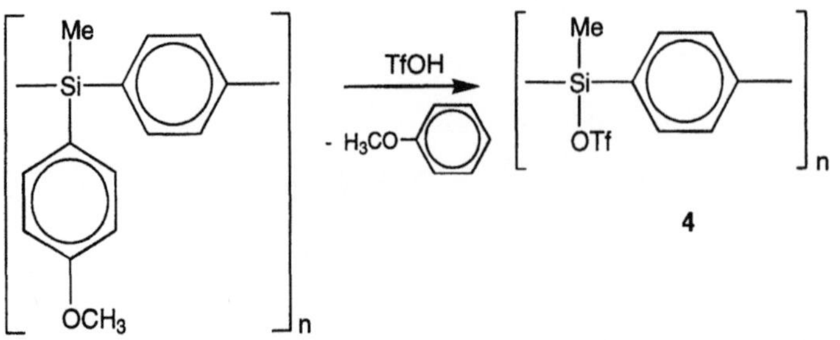

Schema 5

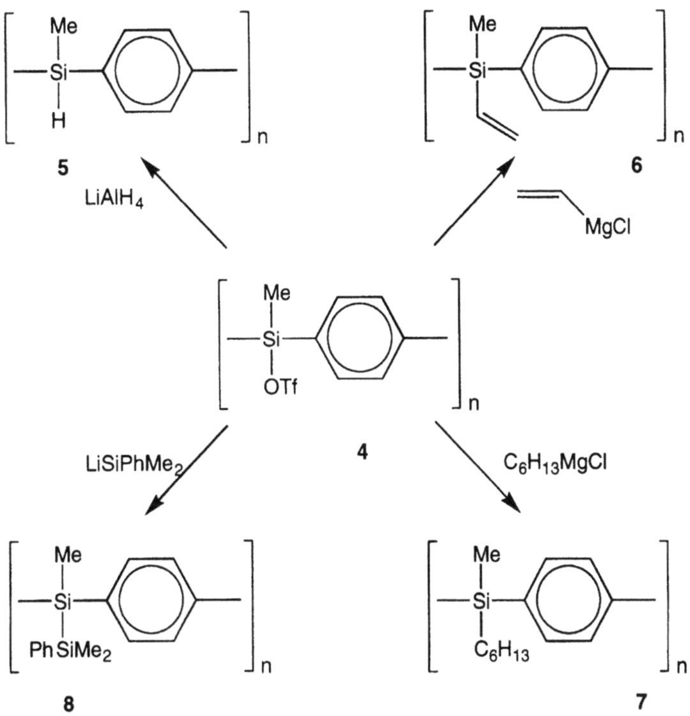

Schema 6

Siliciumatome funktionalisiert. Breite Signale im ^{29}Si-NMR-Spektrum belegen jedoch, daß die Triflatsubstituenten statistisch im Molekül verteilt werden und die Regularität der Polymerstruktur verloren geht. Partiell funktionalisierte Poly(silylenphenylene) mit weitgehend regulär alternierender Struktur sollten sich jedoch erhalten lassen, wenn die Polymerhauptkette aus phenylenverbrückten Di- oder Trisilaneinheiten aufgebaut wird. Untersuchungen zur Funktionalisierung an Oligosilanen mittels Trifluormethansulfonsäure hatten gezeigt, daß die Spaltung von Silicium-Element-Bindungen stufenweise erfolgt [45]. Somit kann aus jeder Di- oder Trisilaneinheit des Polymers genau ein Substituent abgespalten werden. Die Synthese des disilanverbrückten Polymers **10** ist in Schema 7 dargestellt.

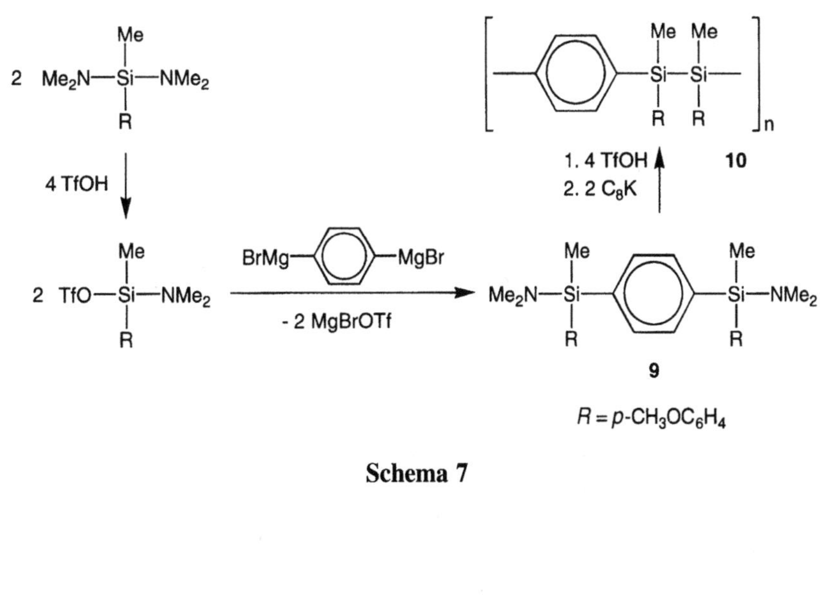

Schema 7

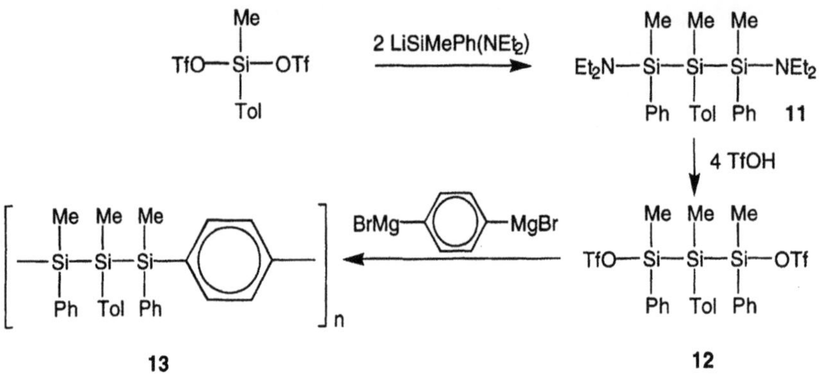

Schema 8

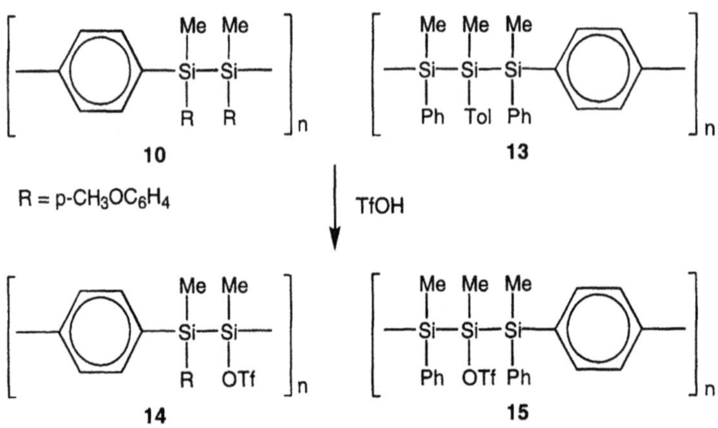

Schema 9

Zunächst wird das 1,4-*Bis*-(*p*-anisyl-dimethylaminomethylsilyl)benzol (**9**) aufgebaut. In einer anschließenden Umsetzung mit CF_3SO_3H werden die Aminogruppen gegen Triflatsubstituenten ausgetauscht. Die nachfolgende Reduktion mit C_8K führt zum Polymer **10**. Die bei Raumtemperatur durchgeführte Reduktion muß invers erfolgen (langsame Zugabe des Reduktionsmittels zum Silyltriflat); anderenfalls liegt Kaliumgraphit im Überschuß vor und reduziert auch die Abgangsgruppe. Darüberhinaus kann überschüssiges C_8K die Polymerhauptkette spalten. Zur Synthese des trisilanverbrückten Polymers **13** wird zunächst das Trisilan **11** hergestellt, welches an den terminalen Siliciumatomen Aminogruppen und am zentralen Si-Atom einen *p*-Tolylsubstituenten trägt. Durch Umsetzung mit Trifluormethansulfonsäure erhält man das 1,3-*Bis*(triflat)derivat **12**, welches gemäß Schema 8 mit 1,4-*Bis*(bromomagnesium)benzol zum Polymer **13** reagiert.

Setzt man das Polymer **10** mit einem Äquivalent Trifluormethansulfonsäure um, so wird in jeder Disilaneinheit eine Anisylgruppe gegen den Triflatsubstituenten ausgetauscht (Schema 9). Die Signale des ^{29}Si-NMR-Spektrums von **14** (Abb. 1) sind relativ schmal und belegen den regulären Kettenaufbau. Dabei ist zu berücksichtigen, daß durch Diastereomerenbildung eine Verbreiterung der Signale erwartet werden muß. Bei der Reaktion von **13** mit einem Äquivalent der Säure

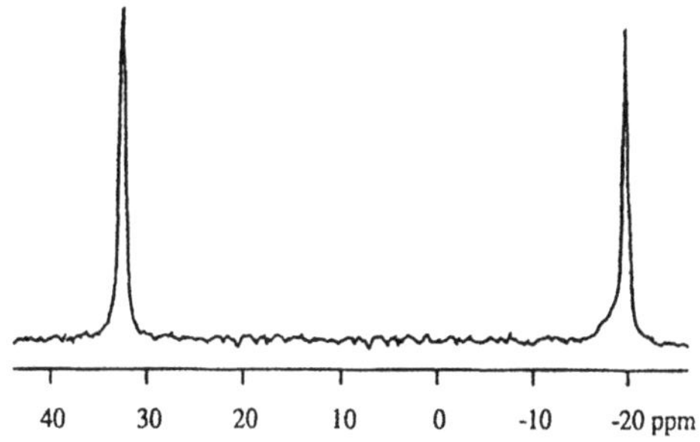

Abb. 1. ^{29}Si-NMR-Spektrum des Polysilylenphenylens **14** in $CDCl_3$

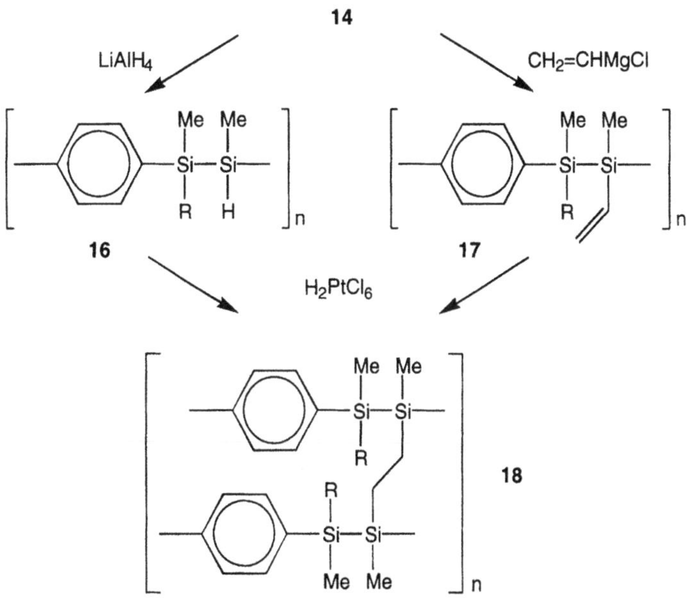

Schema 10

zum Polymer **15** erfolgt ausschließlich die Abspaltung der Tolylgruppe aus jeder Trisilaneinheit gemäß Schema 9.

14 und **15** können in Analogie zum Poly(silylenphenylen) **4** durch Umsetzung mit Nucleophilen weiter modifiziert werden. Im folgenden werden beispielhaft zwei Vernetzungsreaktionen vorgestellt. **14** reagiert mit Lithiumalanat zum Si-H-Derivat **16** bzw. mit Vinylmagnesiumchlorid zum Vinyl-Silyl-Derivat **17**. In Analogie zu Arbeiten von *Corriu* [46] wurde die Hydrosilylierung mit H_2PtCl_6 in Chlorbenzol durchgeführt, mittels ^{1}H-NMR-Spektroskopie verfolgt und mit dem Verschwinden des Si-H-Signals beendet (Schema 10). Im resultierenden vernetzten Poly(silylenphenylen) **18** werden 90% Si–CH_2–CH_2–Si-Gruppen (β-Hydrosilylierung) und 10% Si–CH(CH_3)–Si-Gruppen (α-Hydrosilylierung) gefunden. Das Formelbild von **18** berücksichtigt nur die dominierende β-Hydrosilylierung.

Schema 11 zeigt am Beispiel des Poly(silylenphenylens) **15** eine Vernetzungsreaktion mittels einer Alkinbrücke. **15** wird mit Ethinylmagnesiumchlorid umgesetzt. Das funktionalisierte Polymer **19** kann anschließend erneut metalliert werden und reagiert mit **15** zum vernetzten Derivat **20**. Die stufenweise

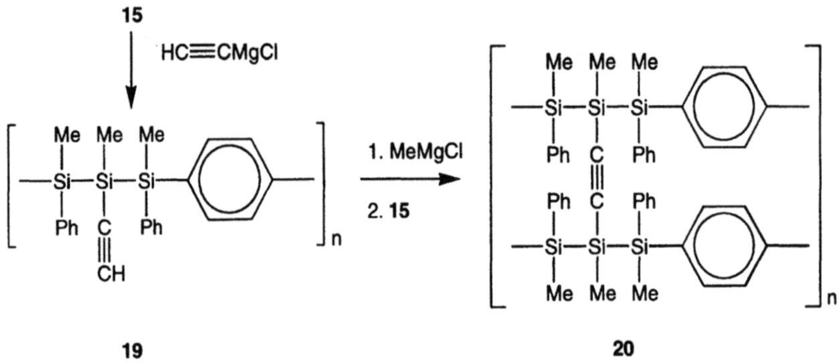

Schema 11

Metallierung des Alkins ist gegenüber der Umsetzung von **15** mit LiC≡CLi zu bevorzugen, da nur auf diesem Wege zwingenderweise eine intermolekulare Vernetzung erzielt wird, was wiederum zu einem hohen Molekulargewicht führt.

Alle linearen Polymere sind in Aromaten und Chloroform gut löslich. Bei den vernetzten Vertretern **18** und **20** nimmt die Löslichkeit erwartungsgemäß ab. In heißem Toluol oder Xylol sind sie jedoch gut, in Chloroform mäßig löslich. Die Molmassen M_w der hydrolysestabilen linearen Derivate wurden mittels GPC bestimmt und liegen im Bereich zwischen 10000 und 30000 g/mol (M_w:M_n = 1.8–2.5). Diese Werte sind jedoch an die exakte Einhaltung der Stöchiometrie (Gehaltsbestimmung der *Grignard*-Lösung) bei der Polykondensation gebunden. Alle Verbindungen wurden mittels ^{1}H-, ^{13}C- und ^{29}Si-NMR-Spektroskopie charakterisiert. Bei einem Polymerisationsgrad von $n = 70-120$ gelang die Detektion von terminalen Siliciumatomen mittels ^{29}Si-NMR in der Regel nicht. Mit Ausnahme der Triflatderivate, die als hydrolyseempfindliche Öle anfallen, wurden die Polymere als hellgelbe Pulver erhalten. Die Zusammensetzung der Verbindungen wurden elementaranalytisch geprüft.

Die thermischen Eigenschaften einiger Polymere wurden mittels thermogravimetrischer Analyse (TGA) untersucht. Abb. 2 zeigt die TGA-Profile von **6**, **7** und

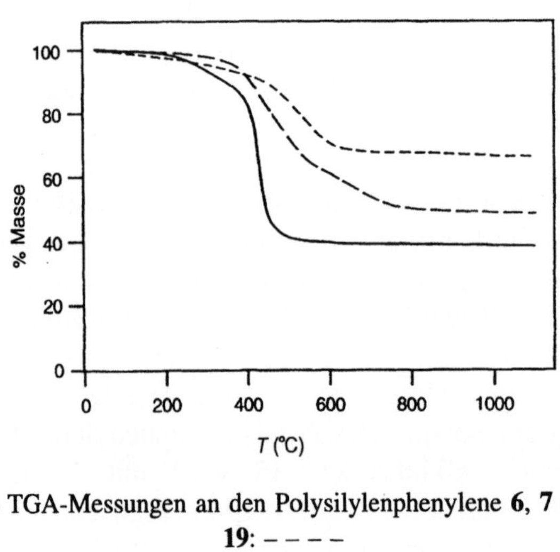

Abb. 2. Vergleichende TGA-Messungen an den Polysilylenphenylene **6**, **7** und **19**; **6**: - -, **7**: ——, **19**: – – – –

Tabelle 1. Physikalische Daten der Polysilylenphenylene **1–8**

	R	Ausb. (%)	Fp. (°C)	M_w	*PD*	$\delta(^{29}Si)$ (ppm)	$\delta(^{13}C)$ (ppm)	$\delta(^{1}H)$ (ppm)
1	Ph	84	210–225	21300	1.8	−13.0	−3.0 (MeSi), 127.3–137.5 (Ph)	0.81 (MeSi), 7.15–7.83 (Ph)
2	*p*-Anisyl	79	155–167	24800	1.9	−15.1	−3.6 (MeSi), 53.0 (OMe), 115.1, 121.1–139.4, 160.0 (Ph)	0.74 (MeSi), 3.51 (OMe), 6.92–8.05 (Ph)
3	Me_2N	78	viskoses Öl	–	–	−8.4	−1.6 (MeSi) 30.5 (NMe_2), 134.4, 136.9 (Ph)	0.65 (MeSi), 2.39 (NMe_2), 7.54 (Ph)
4	TfO	95	viskoses Öl	–	–	+32.9	6.3 (MeSiOTf), 117.9 (CF_3), 133.0, 135.8 (Ph)	0.88 (MeSi), 7.67 (Ph)
5	H	76	67–78	16400	2.4	− 19.0	−4.8 (MeSi), 134.9, 136.1 (Ph)	0.62 (MeSi), 4.81 (SiH), 7.49 (Ph)
6	Vinyl	84	167–174	17800	2.0	−14.8	−3.8 (MeSi), 135.0, 135.9 ($CH_2=CH$), 133.9, 136.8 (Ph)	0.64 (meSi), 5.74–6.40 (Vi), 7.59 (Ph)
7	Hexyl	91	71–81	24100	2.2	−8.1	−3.1 (MeSi), 12.5–28.4 (Hex), 134.2, 138.5 (Ph)	0.53 (MeSi), 1.02–1.35 (Hex), 7.39 (Ph)
8	$PhSiMe_2$	77	180–195	21900	2.3	−24.8 −21.9	−4.5 (MeSi), −2.6 (Me_2Si), 126.5–139.0 (Ph)	0.53 (Me_2Si), 0.78 (MeSi), 7.21–7.69 (Ph)

19 als typische Beispiele. Die linearen Polymere zeigen einen signifikanten Masseverlust zwischen 450 und 550°C. Der gesamte Masseverlust bei 1000°C beträgt zwischen 30 und 60%. Der höchste Masseverlust wird beim Polymer **7** (60%) beobachtet, welches keine funktionellen Seitengruppen trägt. Der Masseverlust der Polymere **6** (51%) und **19** (36%) mit ungesättigten Seitengruppen ist hingegen deutlich geringer. Wie *Ishikawa* und Mitarbeiter in einer früheren Arbeit am Beispiel des Poly(dimethylsilylenephenylens) zeigten [47], beginnt die Pyrolyse mit einer homolytischen Spaltung der Phenylen-Silicium-Bindung. Die Wechselwirkung der entstehenden Radikale mit Vinyl- oder Ethinylgruppen kann zu Vernetzungen und damit zur Reduzierung des Masseverlustes führen. So läßt sich beispielsweise zeigen, daß nach fünfzehnminütigem Erhitzen auf 500°C die C≡C-Streckschwingung ($\nu = 2036\,cm^{-1}$) im IR-Spektrum von **19** vollständig verschwunden ist und durch eine neue C=C-Streckschwingung bei $1608\,cm^{-1}$ ersetzt wurde. Dies steht in Übereinstimmung mit Arbeiten von *Corriu* [18] und *Barton* [48].

Tabelle 2. Physikalische Daten der Polysilylenphenylene **10** und **13–20**

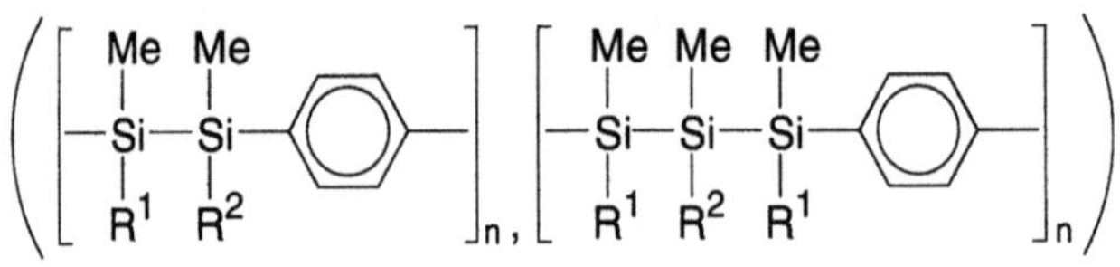

	R^1	R^2	Ausb. (%)	Fp. (°C)	M_w *PD*	$\delta(^{29}Si)$ ppm	$\delta(^{13}C)$ ppm	$\delta(^{1}H)$ ppm
10	*p*-Anisyl	*p*-Anisyl	67	143–157	15000 2.1	−25.5	−4.5 (MeSi), 52.4 (OMe), 114.8, 122.5–138.1, 161.6 (Ph)	0.65 (MeSi), 3.57 (OMe), 6.84–7.98 (Ph)
13	Ph	*p*-Tolyl	74	188–199	13700 1.8	−44.2 (SiTol); −26.0 (SiPh)	−5.1, −3.6 (MeSi), 22.8 (Me), 128.0–139.5 (Ph)	0.51–0.83 (MeSi), 2.45 (Me), 7.13–7.98 (Ph)
14	*p*-Anisyl	OTf	92	viskoses Öl	–	−19.8 (SiAnis); +33.6 (SiOTf)	−1.4, +5.9 (MeSi), 54.8 (MeO), 118.0 (CF_3), 116.8, 122.7–138.0, 159.5 (Ph)	0.76, 0.86 (MeSi), 3.56 (MeO), 7.07–7. (Ph)
15	Ph	OTf	95	viskoses Öl	–	−23.2 (SiPh); +28.9 (SiOTf)	−2.3, +4.8 (MeSi), 118.0 (CF_3), 126.7–139.8 (Ph)	0.62–0.86 (MeSi), 7.14–7. (Ph)
16	*p*-Anisyl	H	76	137–150	10200 2.3	−38.7 (SiH); −26.9 (SiAnis)	−5.0, −3.6 (MeSi), 51.1 (OMe), 116.2, 122.5–139.9, 161.2 (Ph)	0.60, 0.77 (MeSi), 3.48 (OMe), 4.7 (SiH), 7.06–7.9 (Ph)
17	*p*-Anisyl	Vinyl	81	154–168	10800 2.5	−24.7 (SiAnis); −21.9 (SiVi)	−3.4, −2.5 (MeSi), 53.0 (OMe), 114.9, 123, 0–140.5, 161.8 (Ph), 134.6, 135.8 (Vi)	0.57, 0.81 (MeSi), 3.56 (OMe), 5.65–6. (Vi), 6.92–8.05 (Ph)
18	*p*-Anisyl	CH_2CH_2	67	>250 (Zers.)	39000 3.6	−23.6 (SiAnis); −18.5 ($SiCH_2$)	−3.6, −1,9 (MeSi), 4.4 ($SiCH_2$), 53.7 (OMe), 114.1, 122.7–139.8, 159.5(Ph)	0.49, 0.74 (MeSi), 0.91 $SiCH_2$, 3.51 (OMe), 6.92–8.05 (Ph)
19	Ph	C≡CH	70	>250 (Zers.)	9900 2.5	−55.8 (SiC≡CH); −24.9 (SiPh)	−4.7, −2.8 (MeSi), 81.6 (SiC≡), 96.8 (≡CH), 127.3–137.5 (Ph)	0.66–0.87 (MeSi), 2.57 (≡CH), 7.15–7.83 (Ph)
20	Ph	C≡CSi	76	>250 (Zers.)	–	−58.5 (SiC≡CSi); −25.6 (SiPh)	−3.8, −1.9 (MeSi), 112.6 (Sic≡), 127.3–137.5 (Ph)	0.61–0.83 (MeSi), 7.15–7.83 (Ph)

Schlußfolgerungen

Die vorliegende Arbeit zeigt, daß funktionell substituierte Poly(silylenphenylene) mit regelmäßiger Struktur über Triflatzwischenstufen hergestellt werden können. Die bei Raumtemperatur schnell und quantitativ verlaufende Umsetzung von Poly(*p*-anisylsilylenphenylenen) mit CF_3SO_3H sowie die folgenden nucleophilen Substitutionen gestatten den Einbau verschiedener funktioneller Gruppen in die Polymerkette. Bei geeignete Wahl der Substituenten kann im Anschluß eine Vernetzung durch intermolekulare Hydrosilylierung erfolgen. Die Untersuchungen der thermischen Eigenschaften der Polymere durch TGA zeigen, daß ungesättigte Substituenten sowie Vernetzung der Polymere den Masseverlust bei hohen Temperaturen senken. Die hier vorgestellte Funktionalisierungsmethode ist für technische Anwendungen vermutlich zu teuer. Der entscheidende Vorteil besteht jedoch darin, daß eine relativ große Anzahl unterschiedlich strukturierter Poly(silylenphenylene) mit verhältnismäßig geringem Aufwand für materialwissenschaftliche Untersuchungen hergestellt werden kann.

Experimentelles

Alle Arbeiten wurden in getrockneten Lösungsmitteln unter Schutzgas ausgeführt. Die Trifluormethansulfonsäure sollte vor der Verwendung unter Argon destilliert werden. Einzelheiten zur Reinigung sind in Lit. [38] (Kapitel 7) dargelegt. CF_3SO_3H ist stark hygroskopisch. Spuren von Wasser in der Säure haben immer unerwünschte Siloxanbildung als Nebenreaktion zur Folge. Die Säure ist mit den meisten gebräuchlichen Lösungsmitteln nicht mischbar oder zersetzt diese. Sie mischt sich jedoch mit Diethylether. Diese Mischung ist im Rahmen der angegebenen Reaktionszeiten hinreichend stabil. Durch Verwendung dieser Mischung wird die Säurekonzentration am Reaktionsort gegenüber dem Einsatz der reinen Säure stark vermindert. Nebenreaktionen können dadurch unterdrückt werden.

Die NMR-Spektren wurden mit einem Gerät AC-250 der Firma Bruker gemessen. Mit Ausnahme der triflatsubstituierten Verbindungen wurden die ^{29}Si-NMR-Spektren unter Zusatz des Relaxationsreagenzes Cr(*acac*)$_3$ aufgenommen. Die chemischen Verschiebungen triflatsubstituierter Siliciumatome schwanken in Abhängigkeit von der Konzentration der Lösungen bis zu 3 ppm. Die Molmassen wurden in Toluol mittels GPC bestimmt (bezogen auf Polystyrenstandard). Silyltriflate sollten nicht längere Zeit in *THF* gehandhabt werden, da sie in diesem Lösungsmittel als Polymerisationsinitiatoren wirken können [49]. Umsetzungen mit Grignardlösungen in *THF* sind jedoch durchführbar, da die Silyltriflate sehr schnell mit der Organometallverbindung reagieren.

Die Ausgangsverbindungen $BrMg–C_6H_4–MgBr$ [50], HC≡CMgCl [51], $(Et_2N)MePhSiLi$ [52, 53], C_8K [54, 55] sowie die aminosubstituierten Silyltriflate [39] wurden nach Literaturvorschriften hergestellt.

Darstellung der Poly(silylenphenylene) **1–3** und **13**

Zu einer Lösung von 0.01 mol des entsprechenden Silyl(*bis*triflates) in 250 ml Toluol tropft man unter Rühren bei Raumtemperature eine frisch bereitete und titrierte Lösung von 1,4-*Bis*(bromomagnesium)benzol innerhalb von 10 min. Man läßt 1 h nachrühren und filtriert anschließend das ausgefallene Magnesiumsalz ab. Man entfernt das Lösungsmittel im Vakuum und erhält **1**, **2** und **13** als hellgelbe Pulver, die mit Chloroform/Ethanol umgefällt werden können. Nach erneuter Filtration werden Lösungsmittelreste im Hochvakuum entfernt. **3** fällt als hochviskoses Öl an. Die Einzeldaten sind den Tabellen 1 und 2 zu entnehmen.

Darstellung der triflatsubstituierten Poly(silylenphenylene) **4**, **14** und **15**

5 mmol des entsprechenden Poly(silylenphenylenes) **2**, **10** oder **13** werden in 200 ml Toluol gelöst. Bei 0°C tropft man eine Lösung von 0.75 g (5 mmol) CF_3SO_3H in 50 ml Diethylether innerhalb von 10 min zu. Die Reaktionsmischung wird 30 min bei Raumtemperature gerührt. Nach Entfernen des Lösungsmittels fallen **4**, **14** und **15** quantitativ als hochviskose, extrem hydrolyseempfindliche Öle an. Die spektroskopischen Daten dieser Verbindungen sind in den Tabellen 1 und 2 enthalten. Für Folgeumsetzungen mit Nucleophilen wird jedoch auf eine Isolierung verzichtet und die Reaktionslösung ohne weitere Reinigungsoperationen eingesetzt.

Darstellung des Poly(silylenphenylens) **10**

0.02 mol *Bis*(dimethylamino)-*p*-anisylmethylsilan werden in 250 ml Ether gelöst und bei 0°C unter Rühren mit 0.04 mol CF_3SO_3H, gelöst in 50 ml Ether, versetzt. Das ausgefallene Ammoniumsalz wird abfiltriert. Unter Rühren versetzt man die Lösung mit 0.01 mol *Bis*(bromomagnesium)benzol. Nach zwei Stunden wird vom Niederschlag abfiltriert und das Lösungsmittel im Vakuum entfernt. **9** resultiert als Öl und wird im Vakuum deștilliert (Kp.: 165–168°C/0.001 Torr; 54%).

0.01 mol **9** werden in 250 ml Toluol gelöst und mit 0.04 mol CF_3SO_3H, gelöst in 50 ml Ether, versetzt. Das ausgefallene Ammoniumsalz wird abfiltriert. Zu der gerührten Lösung gibt man innerhalb von 2 h portionsweise 0.02 mol C_8K, welches im Laufe der Zeit seine Farbe von goldbronze nach schwarz ändert. Das Reduktionsmittel darf nicht im Überschuß und nicht zu schnell zugegeben werden, da ansonsten auch die Polymerkette gespalten wird. Nach 12 h wird vom Graphit abfiltriert. Man entfernt das Lösungsmittel im Vakuum. Der resultierende Rückstand wird im Chloroform aufgenommen und mit Isopropanol gefällt. **10** resultiert nach Filtration und Trocknung im Hochvakuum als gelber Feststoff (Tabelle 2).

Darstellung der Trisilane **11** und **12**

Zu einer Lösung von 0.02 mol Tolylmethylsilyl*bis*(triflat) in 250 ml Ether tropft man bei −20°C innerhalb von 1 h eine frisch bereitete und gekühlte Lösung von 0.04 mol $(Et_2N)MePhSiLi$. Man erwärmt auf Raumtemperatur und entfernt anschließend das Lösungsmittel im Vakuum. Nach Zugabe von 200 ml Toluol fällt CF_3SO_3Li als weißer Feststoff aus und wird abfiltriert. Der Rückstand wird eingedampft; **11** resultiert als klare, hydrolyseempfindliche Flüssigkeit.

0.01 mol **11** werden in 200 ml Toluol gelöst und bei 0°C mit einer Lösung von 0.04 mol CF_3SO_3H in 25 ml Ether versetzt. Nach 15 min wird vom ausgefallenen Ammoniumsalz abfiltriert und das Lösungsmittel abdestilliert. Man kontrolliert die Vollständigkeit der Umsetzung mittels NMR-Spektroskopie. Das extrem hydrolyseempfindliche Öl wird ohne weitere Aufarbeitung für Folgereaktionen eingesetzt.

11: ^{1}H-NMR ($CDCl_3$): $\delta = 0.29$, 0.36 (MeSi), 1.13 (*Me*CH_2N), 2.45 (*Me*C_6H_5), 2.83 (NCH_2), 7.04–7.89 (ph) ppm; ^{13}C-NMR ($CDCl_3$): $\delta = -2.7$, 0.3 (MeSi), 15.9 (*Me*CH_2N), 24.1 (*Me*C_6H_5), 43.6 (NCH_2), 127.5–140.1 (Ph) ppm; ^{29}Si-NMR ($CDCl_3$): $\delta = -42.5$ (*p*-TolMeSi), −15.1 ($MeSiNEt_2$).

12: ^{1}H-NMR ($CDCl_3$): $\delta = 0.35$, 0.64 (MeSi), 2.52 (*Me*C_6H_5), 6.97–8.04 (Ph) ppm; ^{13}C-NMR ($CDCl_3$): $\delta = -3.0$, +5.1 (MeSi), 26.8 (*Me*C_6H_5), 118.5 (CF_3), 126.9–138.7 (Ph) ppm; ^{29}Si-NMR ($CDCl_3$): $\delta = -41.7$ (*p*-TolMeSi), +35.0 (MeSiOTf).

Darstellung der substituierten Poly(silylenphenylene) **6–8**, **17**, **19**

Eine Lösung von 5 mmol des entsprechenden triflatsubstituierten Poly(silylenphenylens) **4**, **14** oder **15** in 200 ml Toluol wird bei 0°C unter Rührenmit der äquimolaren Menge der entsprechenden

Organoelementverbindung umgesetzt. Man rührt 2 h bei Raumtemperatur nach, trennt das ausgefallene Salz ab und entfernt das Lösungsmittel im Vakuum. Die Polymere resultieren als gelbe Feststoffe, die aus Chloroform/Isopropanol umgefällt werden können. Nach Filtration werden sie im Hochvakuum von Lösungsmittelresten befreit. Die Einzeldaten sind in den Tabellen 1 und 2 enthalten.

Darstellung der Si-H-funktionellen Poly(silylenphenylene) **5** und **16**

Eine Lösung von 10 mmol des entsprechenden triflatsubstituierten Poly(silylen-phenylens) **4** oder **14** in 100 ml Toluol tropft man bei 0°C zu einer Suspension von 5 mmol Lithiumalanat in 100 ml Diethylether. Man rührt 3 h bei Raumtemperatur und hydrolysiert anschließend mit Wasser. Nach Abtrennen der organischen Phase wird die wäßrige Phase mit Chloroform extrahiert. Die vereinigten organischen Phasen werden über Magnesiumsulfat getrocknet. Nach Entfernen des Lösungsmittels im Vakuum verbleiben **5** (vgl. Lit. [44]) und **16** als gelbe Feststoffe, die aus Chloroform/Isopropanol umgefällt werden können. Die Einzeldaten sind in den Tabellen 1 und 2 enthalten.

Darstellung des vernetzten Poly(silylenphenylens) **18**

0.01 mol **16** und 0.01 mol **17** werden in 200 ml Chlorbenzol gelöst. Unter Rühren wird eine Lösung von 0.1 mmol H_2PtCl_6 in 10 ml Chlorbenzol langsam zugetropft. Nach einer stark exothermen Periode sinkt die Temperatur allmählich und wird für 24 h bei 40°C gehalten. Man bricht die Reaktion ab, wenn die Si-H- und die Si-Vinyl-Signale im ^{1}H-NMR-Spektrum kaum mehr zu beobachten sind. Das Chlorbenzol wird im Vakuum entfernt. Man löst den Rückstand in Toluol und filtriet, um den Katalysator zu entfernen. **18** wird als hellgelber Feststoff aus der Toluollösung durch Fällung mit Isopropanol isoliert (Tabelle 2).

Darstellung des vernetzten Poly(silylenphenylens) **20**

0.01 mol **19** werden in 250 ml Toluol gelöst. Bei Raumtemperatur tropft man eine etherische Lösung von Methylmagnesiumbromid (0.01 mol) innerhalb von 15 min zu und rührt anschließend 2 h nach. Zu der entstandenen Suspension tropft man bei 0°C eine Lösung von 0.01 mol **15** in 50 ml Toluol und rührt 12 h bei Raumtemperatur. Die Reaktionslösung wird mit Wasser hydrolysiert. Nach Abtrennen der organischen Phase extrahiert man die wäßrige Phase zweimal mit Chloroform. Die vereinigten organischen Phasen werden über Magnesiumsulfat getrocknet. Nach Entfernen des Lösungsmittels im Vakuum verbleibt ein hellbrauner Feststoff. Dieser wird erneut in Chloroform gelöst und mit Isopropanol gefällt. **20** wird als gelber Feststoff erhalten (Tabelle 2).

Dank

Diese Arbeit wurde vom *Schweizer Nationalfonds zur Förderung der Wissenschaften* unterstützt. Der Autor dankt weiterhin Herrn Prof. *R. Nesper* für fördernde Unterstützung sowie der *Wacker Chemie GmbH* (*Burghausen*) für Chemikalienspenden.

Literatur

[1] Ishikawa M, Ni H, Matsuzaki K, Nate K, Yokono H (1984) J Polym Sci Polym Lett Ed **22**: 669
[2] Nate K, Ishikawa M, Ni H, Watanabe H, Saheki Y (1987) Organometallics **6**: 1673
[3] Ishikawa M, Nate K (1988) In: Inorganic and Organometallic Polymers ACS Symp Ser, Washington, DC, **360**: 209

[4] Ohshita J, Yamashita A, Hiraoka T, Shinpo A, Kunai A, Ishikawa M (1997) Macromolecules **30**: 1540
[5] Uhlig W (1994) Helv Chim Acta **77**: 972
[6] Kira M, Tokura S (1997) Organometallics **16**: 1100
[7] Ohshita J, Kanya D, Ishikawa M, Yamanaka T (1989) J Organomet Chem **369**: C 18
[8] Ishikawa M, Hasegawa Y, Kunai A, Yamanaka T (1989) Organometallics **8**: 2741
[9] Iwahara T, Hayase S, West R (1990) Macromolecules **23**: 1298
[10] Ishikawa M, Hatano T, Hasegawa Y, Horio T, Kunai A, Miyai Y, Ishida T, Tsukihara T, Yamanaka T, Koike T, Shioya J (1992) Organometallics **11**: 1604
[11] Ijadi-Magshoody S, Pang Y, Barton TJ (1990) J Polym Sci Part A Polym Chem **28**: 955
[12] Tretner C, Zobel B, Hummeltenberg R, Uhlig W (1994) J Organomet Chem **468**: 63
[13] Habel W, Oelschläger A, Sartori P (1995) J Organomet Chem **494**: 157
[14] Habel W, Häusler W, Sartori P (1997) J prakt Chem **339**: 288
[15] Habel W, Moll A, Sartori P (1997) J prakt Chem **339**: 291
[16] Ishikawa M, Hasegawa Y, Kunai A, Yamanaka T (1990) J Organomet Chem **381**: C 57
[17] Ijadi-Magshoody S, Barton TJ (1990) Macromolecules **23**: 4485
[18] Bredford JL, Corriu R, Gerbier Ph, Querin C, Henner B, Jean A, Kuhlmann T, Garnier F, Yasser A (1992) Organometallics **11**: 2500
[19] Uhlig W (1994) Organometallics **13**: 2843
[20] Hu S, Weber WP (1989) Polym Bull **21**: 133
[21] Ohshita J, Kanaya D, Ishikawa M, Koike T, Yamanaka T (1991) Macromolecules **24**: 2106
[22] Chicart P, Corriu R, Moreau J, Garnier F, Yasser A (1991) Chem Mate **3**: 8
[23] Hong HH, Weber WP (1989) Polym Bull **22**: 363
[24] Ohshita J, Kanaya D, Ishikawa M, Yamanaka T (1990) Chem Express **5**: 489
[25] Ohshita J, Furumori K, Ishikawa M, Yamanaka T (1989) Organometallics **8**: 2084
[26] Ohshita J, Matsuguchi A, Furumori K, Hong R, Ishikawa M, Koike T, Yamanaka T, Shioya J (1992) Macromolecules **25**: 2134
[27] Corriu R, Douglas WE, Yang Z, Karakus Y, Gross GH, Bloor D (1993) J Organomet Chem **455**: 69
[28] Corriu R, Douglas WE, Yang Z (1990) J Polym Sci Polym Lett Ed **28**: 431
[29] Tilley TD (1993) Chem Mater **5**: 1487
[30] Ohshita J, Kanaya D, Ishikawa M (1993) Appl Organomet Chem **7**: 269
[31] Uhlig W (1996) Chem Ber **129**: 733
[32] Uhlig W (1997) Polym Adv Techn **8**: 731
[33] Uhlig W (1997) J Organometal Chem **545**: 281
[34] Uhlig W (1997) Z Naturforsch **52b**: 577
[35] Uhlig W (1998) J Polym Sci Part A Polym Chem **36**: 725
[36] Uhlig W, In: Gielen M (ed) Advanced Topics in Solid State Organometallic Chemistry (eingereicht)
[37] Hergott HH, Simchen G (1981) Liebigs Ann 1718
[38] Emde H, Domsch D, Feger H, Frick U, Götz A, Hergott HH, Hofmann K, Kober W, Krägeloh K, Oesterle T, Steppan W, West W, Simchen G (1982) Synthesis 1
[39] Uhlig W, Tretner C (1994) J Organometal Chem **467**: 31
[40] Bassindale AR, Stout T (1984) J Organometal Chem **271**: C1
[41] Uhlig W, Tzschach A (1989) J Organometal Chem **378**: C1
[42] Rüdinger C, Beruda H, Schmidbaur H (1992) Chem Ber **125**: 1401
[43] Birot M, Pillot JP, Dunoguès J (1995) Chem Rev **95**: 1443
[44] Uhlig W (1991) J Organometal Chem **402**: C45
[45] Uhlig W (1991) J Organometal Chem **421**: 189
[46] Boury R, Corriu RJP, Leclerq D, Mutin Ph, Planeix JM, Vioux A (1991) Organometallics **10**: 1457

[47] Ohshita J, Ishii M, Ueno Y, Yamashita A, Ishikawa M (1994) Macromolecules **27**: 5583
[48] Pang Y, Ijadi-Maghsoody S, Barton TJ (1993) Macromolecules **26**: 5671
[49] Hrkach JS, Matyjaszewski K (1995) J Polym Sci Part A Polym Chem **33**: 285
[50] Rieke RD, Bales SE (1973) J Chem Soc Chem Commun 879
[51] Holmes AB, Sporikou CN (1987) Organic Syntheses **65**: 61
[52] Tamao K, Kawachi A, Ito Y (1992) J Am Chem Soc **114**: 3989
[53] Tamao K, Kawachi A, Ito Y (1993) Organometallics **12**: 580
[54] Fürstner A, Weidmann H (1988) J Organomet Chem **354**: 15
[55] Uhlig W (1996) J Organomet Chem **516**: 147

Received May 15, 1998. Accepted (revised) June 18, 1998

$Sr_{12}Mg_{17.8}Li_{2.2}Si_{20}$, die erste *Zintl*-Phase mit einer Si_3-Kette

Reinhard Nesper* und **Steffen Wengert**

Laboratorium für Anorganische Chemie, Eidgenössische Technische Hochschule Zürich, Schweiz

$Sr_{12}Mg_{17.8}Li_{2.2}Si_{20}$, the First *Zintl*-Phase with a Si_3 Chain

Summary. The investigation of the quarternary systems *EA*/Mg/Li/Si (*EA* = Ca, Sr, Ba) led to a new *Zintl* compound $Sr_{12}Mg_{17.8}Li_{2.2}Si_{20}$ with an interesting allylic Si_3 unit. The electronic configuration was analyzed with the help of quantum chemical calculations.

Keywords. Crystal structures; Silicon; *Zintl* phase; Silicon chain, Electron localization function.

Einleitung

In der letzten Decade sind zahlreiche Untersuchungen zu ternären Siliciden in den Systemen *EA*/Mg/Si (*EA* = Ca, Sr, Ba) durchgeführt worden, bei denen eine große Menge neuer stöchiometrisch gut definierter *Zintl*-Phasen mit zum Teil neuartigen oligomeren Anionen des Siliciums gefunden wurde [1–7]. Mit der zunehmenden Zahl unterschiedlicher Siliciumeinheiten in den neuen binären und ternären Alkali- und Erdalkalimetallsiliciden wurden mehr und mehr *Zintl*-Anionen charakterisiert, die jedoch alle wie auch die entspechenden Phasen der Homologen von Silicium eine gerade Zahl von Si-Atomen enthalten. Vielfach sind deren Bindungsverhältnisse nicht mehr zufriedenstellend im Rahmen der einfachen oder auch erweiterten (8-*N*)-Regel zu verstehen; zugleich tritt oft ein Verlust der für *Zintl*-Phasen charakteristischen Halbleitereigenschaft auf. Zahlreiche bisher durchgeführte theoretische Untersuchungen haben gezeigt, daß die in Betracht zu ziehenden elektronischen Zustände trotz allem im wesentlichen Siliciumcharakter besitzen und dass es in der Regel für das Verständnis nicht notwendig ist, kationische Zustände mit einzubeziehen. Aufgrund dieser Erkenntnisse war zu vermuten, daß die in Frage kommenden Verbindungen in gewissen Grenzen eine elektronische Phasenbreite aufweisen. Die Strukturen sollten demnach gegenüber kleinen Veränderungen der Valenzelektronenzahl pro Formeleinheit stabil sein.

Die ternären Systeme *EA*/Mg/Si (*EA* = Ca, Sr, Ba) schienen für eine experimentelle Untersuchung einer solchen elektronischen Phasenbreite prädestiniert, da

* Corresponding author

sich hier die Möglichkeit bot, durch eine Dotierung mit Lithium die Valenzelektronenzahl zu verändern. Die Ergebnisse dieser Untersuchungen zeigen jedoch, daß die Phasen weitaus empfindlicher auf eine Lithiumdotierung reagieren als erwartet und mit dem Li/Mg-Verhältnis ein zusätzlicher Parameter erhalten wird, der die strukturelle Bandbreite der ternären Systeme nochmals in interessanter Weise erweitert [3]. Im Rahmen der Untersuchungen im System Sr/Mg/Li/Si haben wir eine neue Phase $Sr_{12}Mg_{17.8}Li_{2.2}Si_{20}$ gefunden, die das erste ungeradzahlige *Zintl*-Anion des Siliciums enthält.

Ergebnisse und Diskussion

Die Kristallstruktur von $Sr_{12}Mg_{17.8}Li_{2.2}Si_{20}$ konnte mit Hilfe einer Röntgenstrukturbestimmung am Einkristall ermittelt werden [8]. Abbildung 1 zeigt eine

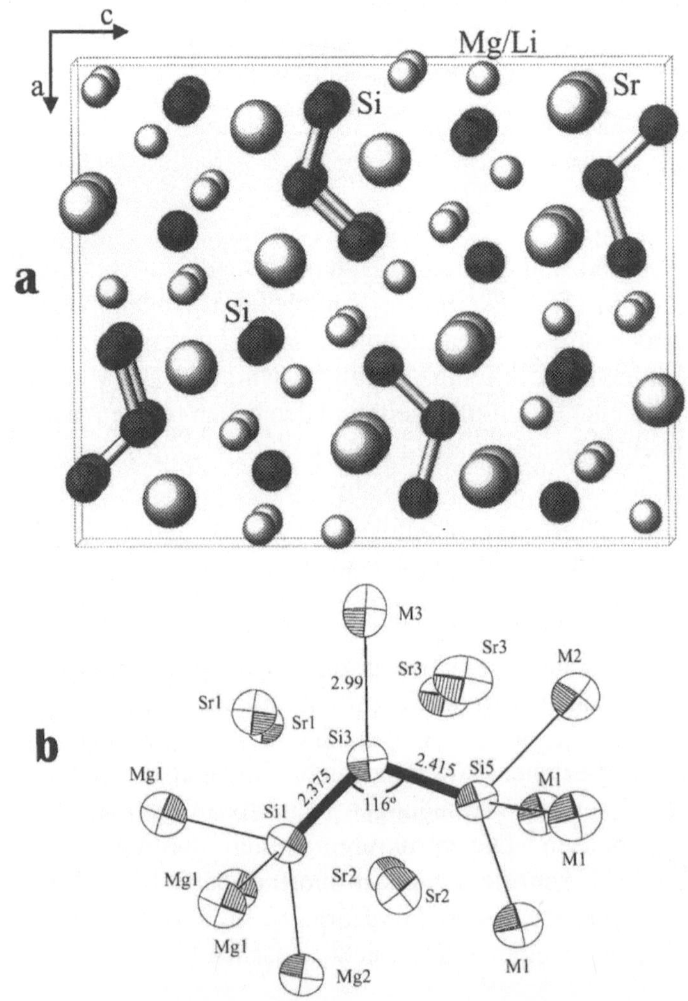

Abb. 1. a) Perspektivische Darstellung der Struktur von $Sr_{12}Mg_{17.8}Li_{2.2}Si_{20}$; b) Koordination der Silicium-Dreierkette in $Sr_{12}Mg_{17.8}Li_{2.2}Si_{20}$

perspektivische Darstellung der Struktur, die wie bei vielen neuen *Zintl*-Phasen einen neuen Strukturtyp definiert. Sie weist neben isolierten Siliciumatomen Si-Dreierketten auf. Sowohl diese Siliciumbaueinheit als auch ganz allgemein ungeradzahlige E_n-Cluster waren bislang im Bereich der *Zintl*-Phasen der vierten Hauptgruppe nicht bekannt. Bei den Pnictiden kennt man die zu Cl_2O isostere Dreierkette in Form von P_3^{5-} [9] oder As_3^{5-} [10] schon seit längerem.

Die Dreierkette kann als einfacher Grundbaustein allgemeiner Ketten $(1b)X\{(2b)X\}_n(1b)X$ aus Hauptgruppenelementen X mit $6n+14$ Valenzelektronen verstanden werden. Dabei können durch unterschiedliche Verknüpfungen Ketten realisiert werden, deren Grenzformen die *cis-trans-* bzw. die *all-trans*-Ketten sind. Eine Verknüpfung in *all-cis*-Konformation führt je nach Bindungswinkel zu Ringen verschiedener Größe.

Die Si-Dreierketten schließen einen Winkel von 116° ein und besitzen zwei unterschiedliche Si–Si Bindungslängen von 2.38 und 2.42 Å. Dies deutet auf einen partiellen Doppelbindungscharakter in Richtung einer siliciumanalogen Allyleinheit hin. Hierfür spricht auch die Elektronenbilanz der Verbindung. Führt man einen formalen Ladungsübertrag auf das Siliciumteilgerüst durch, so erhält man $Si_{20}^{61.8-}$. Unter Berücksichtigung einer Ladung von $q=-4$ für die isolierten Siliciumatome verbleiben somit nach $Si_{20}^{61.8-} = (Si^{4-})_8(Si_3^{7.45-})_4$ noch 7.45 negative Ladungen auf der Dreierkette. Für eine einfach gebundene zu Cl_2O bzw. P_3^{5-} isostere Dreierkette wäre hingegen eine formale Ladung von $q=-8$ zu erwarten. Interessanterweise würde diese Ladung gerade dann realisiert, wenn alle mischbesetzten Mg/Li-Lagen ausschließlich mit Magnesiumatomen besetzt wären; eine entsprechende Verbindung wurde bisher jedoch nicht beobachtet.

Die endständigen Siliciumatome sind durch Magnesiumzentren koordiniert, welche zusammen mit der quadratisch-planaren Anordnung der Strontiumatome um die Si–Si Bindungsmitten in typischer Weise ein quadratisches Antiprisma bilden (vgl. Abb. 1b). Das zentrale Si3-Atom weist eine trigonal-prismatische Koordination durch Strontium auf, wie sie in ähnlichen *Zintl*-Phasen stets durch die größeren Kationen ausgebildet wird. Damit hat auch dieses ungewöhnliche kurze Kettenglied bezüglich der Kationenumgebung alle typischen Merkmale, die bereits in längeren Ketten gefunden wurden.

Die Si_3-Kette läßt sich formal durch eine mischbesetzte Li/Mg-Lage im Abstand von 2.9 Å zu der aus $Ba_5Mg_{18}Si_{13}$ [11] bekannten trigonal planaren Viereinheit, hier jedoch in verzerrter Form, ergänzen. Neben der Dreierkette existieren zwei isolierte Siliciumlagen. Die Si4-Position ist näherungsweise von einem einfach überkappten quadratischen Antiprisma aus Sr- und Mg-Atomen umgeben. Die Si2-Lage weist eine ähnliche Koordination auf, die jedoch weit mehr verzerrt ist und eher an ein dreifach überkapptes trigonales Prisma erinnert.

Auf Basis der EHMO-Methode [12] wurden quantenchemische Bandstrukturrechnungen an $Sr_{12}Mg_{17.8}Li_{2.2}Si_{20}$ durchgeführt. Ausgehend von der experimentellen Valenzelektronenzahl von 141.8e$^-$/Formeleinheit erwies sich die erhaltene Bandstruktur (unter Verwendung der Standardparameter für Strontium, Magnesium, Lithium und Silicium [13]) als unabhängig von der jeweiligen Besetzung der *M*-Lage (Magnesium oder Lithium). $Sr_{12}Mg_{17.8}Li_{2.2}Si_{20}$ sollte demnach eine metallische Leitfähigkeit aufweisen (vgl. Abb. 2), die sich weitgehend im Rahmen der Zustände des Siliciumteilgerüstes verstehen läßt.

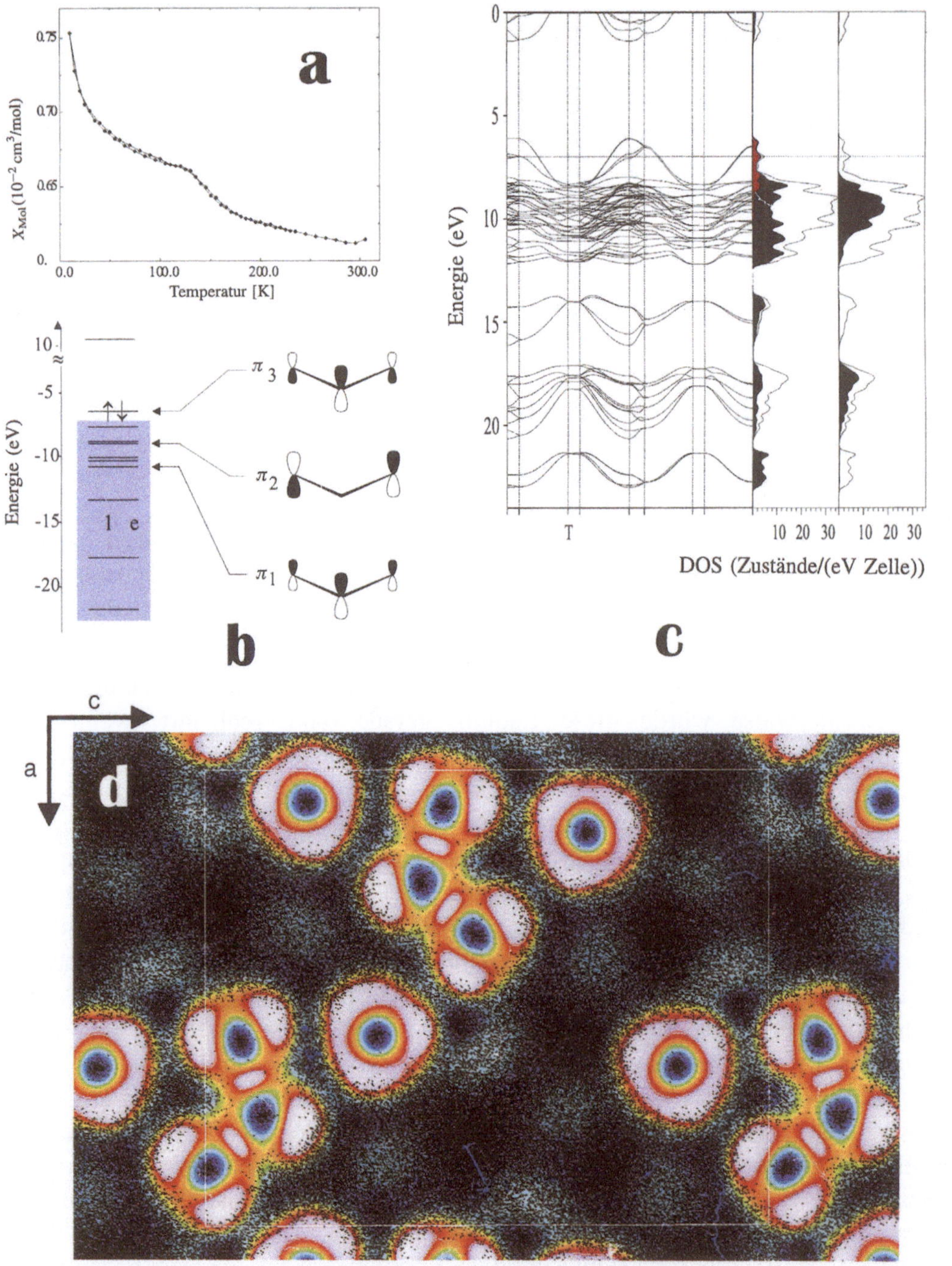

Abb. 2. a) Temperaturabhängige Messungen der Molsuszeptibilität von $Sr_{12}Mg_{17.8}Li_{2.2}Si_{20}$; b) MO-Diagramm der Si_3-Kette aus $Sr_{12}Mg_{17.8}Li_{2.2}Si_{20}$; c) EHMO-Bandstruktur und elektronische Zustandsdichte von $Sr_{12}Mg_{17.8}Li_{2.2}Si_{20}$ mit entsprechenden Projektionen auf die Anteile des Silicium-Teilgerüstes; d) ELF-Verteilung in einer Ebene senkrecht zu [010] durch die Struktur von $Sr_{12}Mg_{17.8}Li_{2.2}Si_{20}$

Eine detaillierte Analyse der EHMO-Bandstruktur zeigt, daß die oben vorgeschlagene formale Ladungsverteilung von $Sr_{12}Mg_{17.8}Li_{2.2}Si_{20} = (Sr^{2+})_{12}(Mg^{2+})_{17.8}(Li^{+})_{2.2}(Si^{4-})_{8}(Si_3^{7.45-})_4$ durchaus begründet ist. Alle von den isolierten Si^{4-}-Einheiten herrührenden Zustände liegen deutlich unterhalb des *Fermi*-Niveaus, während im Bereich des *Fermi*-Niveaus hauptsächlich die π-Zustände der Si_3-Kette zum Tragen kommen.

Betrachtet man zunächst das MO-Diagramm dieser Kette, so entspräche einem einfach gebundenen System mit 20 Valenzelektronen (VE) die in Abb. 2a vorgeschlagene Besetzung; in einer allylischen 18 VE-Einheit wäre hingegen das 10. MO, ein antibindendes π-Orbital, unbesetzt. Es ist gerade dieses MO, das im Bereich des *Fermi*-Niveaus die Bandstruktur bestimmt, wie aus einer Projektion der DOS auf dieses Fragmentorbital ersichtlich ist (vgl. Abb. 2c, rot unterlegt). Die vier von diesem Fragmentorbital abgeleiteten Bänder (es existieren vier Si_3-Einheiten in der Elementarzelle) zeigen wie alle π-Bänder des Systems eine beträchtliche Dispersion von *ca.* 3 eV in [010]-Richtung ($Y \rightarrow \Gamma$, $X \rightarrow S$, $R \rightarrow U$, $Z \rightarrow T$), die, wie bereits an anderer Stelle gezeigt wurde, aus einer Wechselwirkung zwischen den in dieser Richtung ekliptisch gestappelten Si_n-Einheiten resultiert (vgl. Abb. 1a). Da diese Bänder nicht vollständig besetzt sind, entsteht, auf das MO-Bild übertragen, eine ungeradzahlige Besetzung des HOMO ($n = 1.45$), deren Verständnis im Rahmen der Bandstrukturvorstellungen keinerlei Schwierigkeiten bereitet.

Daß eine lokale Interpretation mit isolierten Anionen in erster Näherung weiterhin ihre Berechtigung hat, belegt die auf der Basis der EHMO-Bandstruktur berechnete Elektronenlokalisierungsfunktion ELF [14]. Abbildung 2d zeigt diese für einen zweidimensionalen Schnitt senkrecht zu [010] durch die Struktur von $Sr_{12}Mg_{17.8}Li_{2.2}Si_{20}$. Deutlich zeichnen sich in dieser Ebene die isolierten Si-Atome und Si_3-Fragmente mit Bindungs- und n-Elektronenpaaren durch Bereiche hoher Lokalisierung (orange bis weisse Bereiche) ab.

Experimentelles

Die Darstellung von $Sr_{12}Mg_{17.8}Li_{2.2}Si_{20}$ erfolgt in Niobampullen bei 1350 K aus den Elementen (Aufheizrate 150 K/h, Haltezeit 4 h, Abkühlrate 50 K/h). Man erhält hierbei ein röntgenographisch phasenreines Produkt. Es handelt sich bei $Sr_{12}Mg_{17.8}Li_{2.2}Si_{20}$ um ein sprödes, silbrig glänzendes Material, das mit Luft langsam und mit Wasser unter heftiger Gasentwicklung reagiert. Gemäß DTA-Untersuchungen schmilzt $Sr_{12}Mg_{17.8}Li_{2.2}Si_{20}$ kongruent bei ca. 1150 K.

Die Einkristalluntersuchung erfolgte auf einem mit Flächendetektor ausgestatteten Diffraktometer (IPDS, Fa. STOE) an einem $0.1 \times 0.2 \times 0.2$ mm großen Einkristallbruchstück [8]. Die an Proben von $Sr_{12}Mg_{17.8}Li_{2.2}Si_{20}$ durchgeführten temperaturabhängigen magnetischen Messungen (vgl. Abb. 2b) zeigen bei Raumtemperatur eine paramagnetische Molsuszeptibilität von $\chi_{mol} = 6.1 \cdot 10^{-3}$ cm³/mol (Messung bei $H = 1000$ Oe) und stehen somit im Einklang mit einem erwarteten metallischen Verhalten. Weitgehend unklar ist das beobachtete Temperaturverhalten. Man findet bei abnehmender Temperatur eine leichte Zunahme der Molsuszeptibilität mit einer Stufe bei *ca.* 140 K. Inwieweit diese Abweichungen vom typischen Verhalten eines *Pauli*-Paramagneten signifikant sind, kann beim derzeitigen Kenntnisstand nicht entschieden werden. Es ist nicht auszuschließen, daß diese Effekte durch Verunreinigungen bedingt werden. Andererseits können auch intrinsische Effekte wie z.B. Ordnungsphänomene für dieses Verhalten verantwortlich sein. Messungen der elektrischen Leitfähigkeit, die noch nicht durchgeführt werden konnten, sollten bei 140 K ebenfalls eine

Anomalie aufweisen. Denkbar wäre z.B. eine mit einer Paarbildung verbundene *Peierls*-Verzerrung der eindimensional gestapelten Kettenstücke.

Dank

Diese Arbeit wurde vom *Schweizer Nationalfonds* durch das Projekt Nr. 2000-050675.97 und dem C^4-Projekt der ETH Zürich gefördert.

Literatur

[1] Nesper R, Currao A, Wengert S (1995) Organosilicon Chemistry, Vol 2. Verlag Chemie, Weinheim, p 469 ff

[2] Currao A (1996) Synthese, Struktur und Eigenschaften von Zintlphasen der Erdalkali- und Alkali-Erdalkalimetalle mit Silicium. Dissertation, ETH Zürich

[3] Wengert S (1997) Experimentelle und theoretische Lösungsansätze zu grundlegenden Problemen in Zintlverbindungen, Dissertation, ETH Zürich

[4] Currao A, Curda J, Nesper R (1996) Z Anorg Allg Chem **622**: 85–94

[5] Currao A, Wengert S, Nesper R, Curda J, Hillebrecht H (1996) Z Anorg Allg Chem **622**: 501

[6] Currao A, Nesper R (1998) Angew Chem **110**: 843; Currao A, Nesper R (1998) Angew Chem Int Ed Engl **6**: 841

[7] Nesper R, Currao A, Wengert S (1998) Chem Eur J (im Druck)

[8] Vollmatrix-Least-Squares-Verfeinerung an F^2 [15]. Messung der Einkristallintensitäten: Image Plate Diffractometer (IPDS, Fa. STOE); Mo-K_α-Strahlung. $Sr_{12}Mg_{17.8}Li_{2.2}Si_{20}$: Pnma; $a = 1435.2(3)$, $b = 446.40(10)$, $c = 1843.1(4)$ pm. Sr(1): 0.4139(1), $\frac{1}{4}$, 0.3453(1); Sr(2): 0.2059(1), $\frac{1}{4}$, 0.5217(1); Sr(3): 0.1359(1), $\frac{1}{4}$, 0.3065(1); Mg(1): 0.4479(2), $\frac{1}{4}$, 0.5450(1); Mg(2): 0.1655(2), $\frac{1}{4}$, 0.1297(1); M(1)(88% Mg, 12% Li): −0.0219(2), $\frac{1}{4}$, 0.4388(1); M(2)(72% Mg, 28% Li): 0.0365(2), $\frac{1}{4}$, 0.6881(1); M(3)(88% Mg, 12% Li): 0.2321(2), $\frac{1}{4}$, 0.7265(2); Si(2): 0.3520(1), $\frac{1}{4}$, 0.1722(1); Si(3): 0.7470(1), $\frac{1}{4}$, 0.6118(1); Si(4): 0.4181(1), $\frac{1}{4}$, 0.6888(1); Si(5): 0.9080(1), $\frac{1}{4}$, 0.5735(1); $R_{int} = 0.10$, $R_1 = 0.03$, $wR_2 = 0.08$ (alle 1047 unabhängigen Reflexe). Weitere Einzelheiten zur Kristallstrukturuntersuchung (zu den Kristallstrukturuntersuchungen) können beim Fachinformationszentrum Karlsruhe, D-76344 Eggenstein-Leopoldshafen (Fax: (+49) 7247-808-666; E-mail: crysdata@fiz-karlsruhe.de) unter der Hinterlegungsnummer CSD-408937 angefordert werden

[9] Lin J, Hönle W, von Schnering HG (1992) J Alloys Compd **178**: 455

[10] Hartweg M (1987) PhD Thesis, Universität Stuttgart

[11] Nesper R, Wengert S, Currao A, Zürcher F (1998) Chem Eur J (eingereicht 1998)

[12] Häußermann U, Nesper R, Wengert S, Fässler TF (1993) Program MEHMACC: Modified Extended-Hückel Version Based on the QCPE Program EHMACC [16], ETH Zürich

[13] Hückel-Parameter für Lithium, Magnesium, Strontium und Silicium: Sr: $H_{ii}(5s) = -6.62$, $\zeta(5s) = 1.21$, $H_{ii}(5p) = -3.92$, $\zeta(5p) = 1.21$; Li: $H_{ii}(2s) = -5.4$, $\zeta(2s) = 0.65$, $H_{ii}(2p) = -3.5$, $\zeta(2p) = 0.65$; Mg: $H_{ii}(3s) = -9.0$, $\zeta(3s) = 1.38$, $H_{ii}(3p) = -4.5$, $\zeta(3p) = 1.38$; Si: $H_{ii}(3s) = -17.3$, $\zeta(3s) = 1.10$, $H_{ii}(3p) = -9.2$, $\zeta(3p) = 1.10$

[14] Savin A, Nesper R, Wengert S, Fässler TF (1997) Angew Chem **109**: 1892; (1997) Angew Chem Int Ed Engl **36**: 1808

[15] Sheldrick GM, SHELXS, Göttingen 1985 und SHELXI-96, Göttingen 1996

[16] Wangbo MH, Evain M, Hughbanks T, Kertesz M, Wijeyesekera S, Wilker C, Zheng C, Hoffmann R, Program EHMACC: Extended Hückel Molecular and Crystal Calculations

Received June 13, 1998. Accepted (revised) July 13, 1998

Oligosilylanions and their Reactions with Zirconocene and Hafnocene Dichlorides

Christian Kayser and **Christoph Marschner***

Institut für Anorganische Chemie, Technische Universität Graz, A-8010 Graz, Austria

Summary. New group 4 metallocene oligosilyl compounds (RSi(SiMe$_3$)$_2$$MCp_2$Cl; M = Zr, R = H, Me, SiMe$_3$, SiMe$_2{}^t$Bu, SiMe$_2$CMe$_2$CHMe$_2$, SiMe$_2$Si(SiMe$_3$)$_3$; M = Hf, R = H, SiMe$_3$) have been synthesized by reaction of the respective oligosilyl potassium compounds with zirconocene or hafnocene dichloride.

Keywords. Oligsilylanions; Transition metal silicon complexes.

Oligosilylanionen und ihre Reaktionen mit Zirkonocen- und Hafnocendichlorid

Zusammenfassung. Eine Reihe neuer Übergangsmetall-Siliciumverbindungen der 4. Nebengruppe (RSi(SiMe$_3$)$_2$$M$(C$_5H_5$)$_2$Cl; ($M$ = Zr, R = H, Me, SiMe$_3$, SiMe$_2{}^t$Bu, SiMe$_2$CMe$_2$CHMe$_2$, SiMe$_2$Si (SiMe$_3$)$_3$; M = Hf, R = H, SiMe$_3$) wurde durch Umsetzung der entsprechenden Oligosilylkalium-verbindungen mit Zirconocen- oder Hafnocendichlorid synthetisiert.

Introduction

The chemistry of transition metal silicon compounds [1, 2] has long been limited mainly to monosilyl ligands and the ubiquitous *tris*(trimethylsilyl)silyl group [3]. Based on our interest in the use of transition metal compounds for the functionalization and transformation of polysilanes we are currently exploring the possibilities to obtain transition metal compounds with other polysilyl ligands. Recently, we have reported the synthesis of tertiary, secondary, and primary polysilyl potassium compounds [4]. Herein we present a number of zirconocene and hafnocene complexes with various new oligosilyl ligands.

Results and Discussion

During our studies on polysilylanions we found that potassium *tert*-butoxide does not only react with *tetrakis*(trimethylsilyl)silane, but also with higher and differently substituted polysilanes in a clean and almost quantitative way (Scheme 1) [4].

* Corresponding author

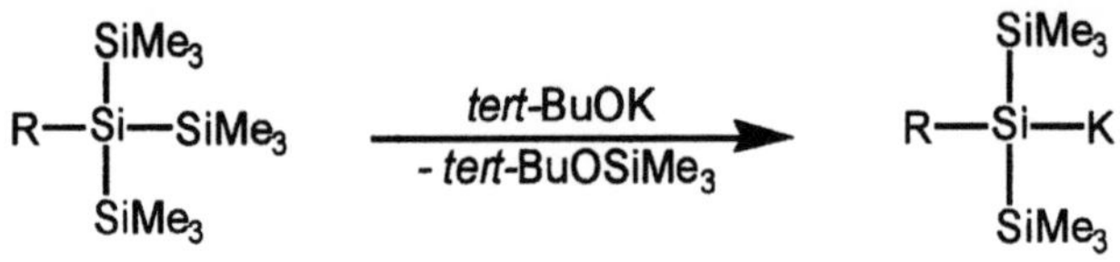

1 : *R* = H	**8**
2 : *R* = Me	**9**
3 : *R* = $SiMe_3$	**10**
4 : *R* = $SiMe_2{}^tBu$	**11**
5 : *R* = $SiMe_2CMe_2CHMe_2$	**12**
6 : *R* = $Si(SiMe_3)_3$	**13**
7 : *R* = $SiMe_2Si(SiMe_3)_3$	**14**

Scheme 1. Synthesis of polysilyl potassium compounds

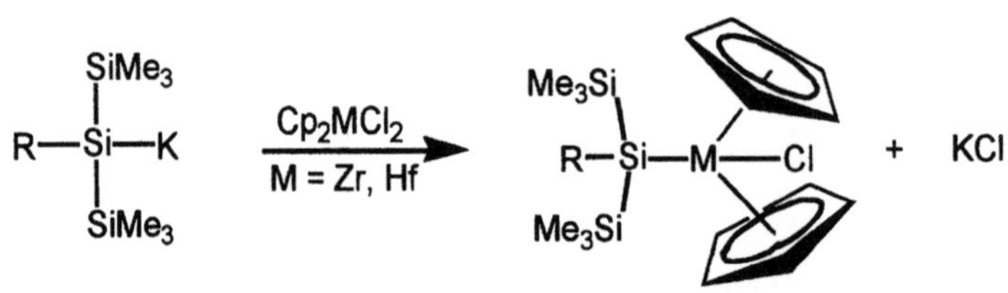

8 : *R* = H	**15, 16** (*M* = Hf)
9 : *R* = Me	**17**
10 : *R* = $SiMe_3$	**18, 19** (*M* = Hf)
11 : *R* = $SiMe_2{}^tBu$	**20**
12 : *R* = $SiMe_2CMe_2CHMe_2$	**21**
14 : *R* = $SiMe_2Si(SiMe_3)_3$	**21**

Scheme 2. Synthesis of polysilyl group 4 metallocenes[a]; *M* = Zr if not stated otherwise

With this set bulky polysilylanions at hand, we decided to investigate its use for the synthesis of silyl complexes of group 4 transition metals. Reactions of silyl potassium compounds were expected to proceed analogously to the reactions with silyl lithium compounds described by *Tilley et al.* some years ago [5, 6] (Scheme 2).

Unfortunately, hydrosilane **1** does not react with potassium *tert*-butoxide in a completely selective manner. The reaction mixture consists of the products **8** and **10**, the latter of which is probably formed by some sort of deprotonation reaction, in a 70:30 ratio. This mixture was used for the reaction with the metallocene chlorides. As a result, mixtures of **15** and **18** (*M* = Zr) as well as **16** and **19** (*M* = Hf) were obtained. Separation by means of recrystallization could not be achieved so far. A hafnocene with the same ligand has been synthesized recently by

Tilley et al. by means of metathesis reaction with 1,1,1,3,3,3-hexamethyltrisilane [7].

The secondary silyl potassium compound **9**, containing only two stabilizing trimethylsilyl groups attached to the silicon atom bearing the negative charge, is expected to be less stable and thus more reactive. Therefore, in contrast to all other silyl potassium compounds used in this study, **9** could not be used as an *in situ* generated solution in *THF*. In order to obtain a reasonably clean reaction with the metal chlorides, toluene had to be used instead of *THF*.

The synthetic problems encountered in the reactions with silyl anions **11–14** reflect the increasing steric influence of the polysilyl groups. Whereas clean reactions occur with the smaller anions **8** and **10** leading to products **15**, **16** and **18**, **19** [5] in high yields, compounds **20–22** are formed in slightly lower yields and are difficult to purify. Besides other side reactions, slow decomposition to the respective hydrosilanes occurs almost from the beginning of the reaction. A particularly sluggish reaction takes place between the metallocene chlorides and **13**, the sterically most demanding *pentakis*(trimethylsilyl)disilanyl potassium compound. Whereas some signals in the ^{29}Si NMR spectrum possibly can be attributed to the expected product, a definite statement on what is actually formed cannot be made at the present state of investigation.

Currently, we are studying the reactivity of compounds **15–22**. We are especially interested in the use of these compounds for the selective functionalization of polysilanes.

Experimental

All reactions were carried out in flame-dried glassware under an inert atmosphere of dry argon or nitrogen. Solvents were distilled prior to use from either sodium or sodium potassium alloy. Starting materials **1**, **3**, **6**, and **7** were prepared as reported previously [4]. In a similar way, compounds **4** and **5** were obtained from the reaction of *tris*(trimethylsilyl)silyl potassium with the respective chlorosilanes. **2** was obtained by the reaction of *tris*(trimethylsilyl)silyl potassium with dimethylsulfate.

^{1}H, ^{13}C, and ^{29}Si NMR spectra were recorded with a Bruker MSL 300 instrument at the indicated frequencies. ^{29}Si NMR spectra were measured in C_6D_6. In all cases the INEPT pulse sequence was employed [8].

General procedure for the synthesis of transition metal silyl complexes:

Starting material and one equivalent of potassium *tert*-butoxide were mixed in a flask and *THF* was added. Almost immediately the solution turned yellow and after some time sometimes orange. Complete formation of the silyl potassium compound was checked by derivatization with ethyl bromide and subsequent GC/MS analysis. The obtained solution was added dropwise over a period of 20 min to a stirred suspension of either Cp_2ZrCl_2 or Cp_2HfCl_2 in pentane at 0°C (in the case of **9**, *THF* was removed prior to addition and replaced by toluene). After complete addition, the reaction mixture was stirred at 0°C for another 60 min after which volatile compounds were removed *in vacuo*. Extraction with 10 ml of pentane, followed by filtration and crystallization on concentrating and cooling, gave the desired products as air sensitive red (M = Zr) or yellow (M = Hf) crystals in 60–80% yields. Yields and selected spectroscopical data (in ppm relative to *TMS* as external standard) are given in Table 1.

Table 1. Selected ^{29}Si NMR data of synthesized polysilyl group 4 metallocenes (δ in ppm, J in Hz)

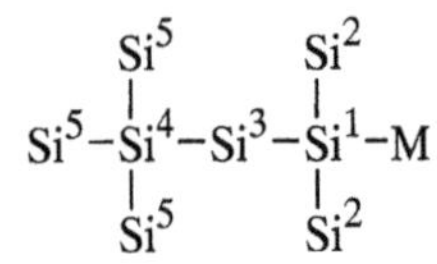

	Si^1	Si^2	Si^3	Si^4	Si^5	Yield (%)
15	−55.6, $J_{Si,H} = 129$	−7.9				80[a]
16	−49.6, $J_{Si,H} = 130$	−6.1				78[a]
17	−29.9	−7.9				68
18	−85.5	−6.1				85
19	−79.7	−5.3				78
20	−85.6	−7.1	10.4			60
21	−83.7	−6.9	13.8			72
22	−63.0,	−6.9	−17.8	−114.1	−9.0	79

[a] Yield calculation based on a 70% pure solution of starting material

Acknowledgements

The authors wish to thank the *Wacker Chemie GmbH, Burghausen*, for the gift of chlorosilanes as starting materials. *Ch. Marschner* is the recipient of an *APART* (Austrian Program for Advanced Research and Technology) scholarship of the *Austrian Academy of Science*, Vienna, which is gratefully acknowledged. The study was carried out within the *Schwerpunktprogramm Silicium-chemie* of the *Fonds zur Förderung der wissenschaftichen Forschung in Österreich.*

References

[1] Tilley TD (1989) Transition-Metal Silyl Derivatives. In: Patai S, Rappoport Z (eds) The Chemistry of Organic Silicon Compounds. Wiley, New York, p 1415
[2] Tilley TD (1991) Transition-Metal Silyl Derivatives. In: Patai S, Rappoport Z (eds) The Silicon-Heteroatom Bond. Wiley, New York, p 309
[3] Gilman H, Holmes JM, Smith CL (1965) Chem Ind (London) 848
[4] Marschner C (1998) Eur J Inorg Chem 221
[5] Campion BK, Falk J, Tilley TD (1987) J Am Chem Soc **109**: 2049
[6] Elsner FH, Tilley TD, Rheingold AL, Geib SJ (1988) J Organomet Chem **358**: 169
[7] Casty GL, Lugmair CG, Radu NS, Tilley TD, Walzer JF, Zagarian D (1997) Organometallics **16**: 8
[8] Morris GA (1980) J Am Chem Soc **102**: 428

Received June 13, 1998. Accepted (revised) September 14, 1998

Transition Metal Silyl Complexes LVIII [1]. Silyl Group Exchange between $(R''_3P)_2Pt(SiR_3)_2$ and $HSiR'_3$

Ulrich Schubert*, **Dominique Kalt**, and **Hilmar Gilges**

Institut für Anorganische Chemie, Technische Universität Wien, A-1060 Wien, Austria

Summary. Reaction of *cis*-$(R_3P)_2Pt(SiPhMe_2)_2$ with an excess of $HSi(OMe)_3$, $HSiPh_3$, HSi_3Ph_7, $HPh_2SiSiMe_3$, or H_2SiPh_2 results in a stepwise exchange of the silyl ligands. Exchange of the first $SiPhMe_2$ ligand is faster than that of the second. The exchange rate depends on the nature of the groups R' in $HSiR'_3$ and decreases in the order $HSi(OMe)_3 \gg H_2SiPh_2 > HPh_2SiSiMe_3 \gg HSi_3Ph_7 > HSiPh_3$. The rate of the exchange reaction of $(R_3P)_2Pt(SiMe_2Ph)_2$ with an excess of $HSi(OMe)_3$ depends on the R_3P ligand and decreases in the order $PPh_2Me > PEt_3 > PPhMe_2$. When the chelated complex $(Ph_3P)(Ph_3Si)\overline{Pt(PPh_2CH_2CH_2SiMe_2})$ is reacted with $HSi(OMe)_3$ or H_2SiPh_2, only exchange of the non-chelated silyl ligand is observed.

Keywords. Silyl complexes; Platinum complexes; Ligand exchange.

Übergangsmetall-Silyl-Komplexe, 58. Mitt. [1]. **Silylgruppenaustausch zwischen $(R''_3P)_2Pt(SiR_3)_2$ und $HSiR'_3$**

Zusammenfassung. Umsetzung von *cis*-$(R_3P)_2Pt(SiPhMe_2)_2$ mit einem Überschuß an $HSi(OMe)_3$, $HSiPh_3$, HSi_3Ph_7, $HPh_2SiSiMe_3$ oder H_2SiPh_2 führt zum stufenweisen Austausch der Silylliganden. Der Austausch des ersten $SiPhMe_2$-Liganden erfolgt rascher als der des zweiten. Die Austauschgeschwindigkeit hängt von der Art der Reste R' in $HSiR'_3$ ab und sinkt in der Reihenfolge $HSi(OMe)_3 \gg H_2SiPh_2 > HPh_2SiSiMe_3 \gg HSi_3Ph_7 > HSiPh_3$. Bei der Reaktion von $(R_3P)_2Pt(SiMe_2Ph)_2$ mit einem Überschuß an $HSi(OMe)_3$ wird der Austausch in Abhängigkeit vom R_3P-Liganden in der Reihenfolge $PPh_2Me > PEt_3 > PPhMe_2$ langsamer. Bei der Reaktion des Chelat-Komplexes $(Ph_3P)(Ph_3Si)\overline{Pt(PPh_2CH_2CH_2CH_2SiMe_2})$ mit $HSi(OMe)_3$ oder H_2SiPh_2 wird nur Austausch des nichtchelatisierten Silylliganden beobachtet.

Introduction

Silyl group exchange by reaction of transition metal silyl complexes with hydrosilanes according to Eq. (1) is an interesting possibility for the preparation of metal complexes with silyl ligands which are otherwise difficult to introduce.

* Corresponding author

Metal complexes with $SiMe_3$ ligands are particularly interesting for preparative purposes because of the high volatility of the leaving silane $HSiMe_3$ [2–6].

$$L_nM\text{–}SiR_3 + HSiR'_3 \rightleftharpoons L_nM\text{–}SiR'_3 + HSiR_3 \quad (1)$$

Exchange reactions in metal catalyzed reactions with hydrosilanes could liberate the organosilicon product and thus regenerate an initial stage of the catalytic cycle. On the other hand, however, they could also lead to the undesirable premature extrusion of the silicon component when the transformation of the silyl group at the metal center involves several steps. An example for the latter case would be the exchange of a growing oligosilyl ligand at the metal center by monomeric hydrosilanes during metal catalyzed dehydrogenative polymerization.

Thermal or photochemical silyl group exchange reactions have not been investigated in detail, although several examples are known [3–15]. Their mechanism is unknown in most cases. It could involve oxidative addition of the entering silane followed by reductive elimination of the leaving silane, but σ-bond metathesis is also possible.

Bis(silyl) complexes are potential intermediates in the conceivable metal catalyzed dehydrogenative silane coupling based on an oxidative addition/reductive elimination mechanism [16]. In the present article we will discuss exchange reactions starting from the *bis*(silyl) complexes *cis*-$(R_3P)_2Pt(SiR_3)_2$ and their chelated analogues $(Ph_3P)\overline{Pt(PPh_2CH_2CH_2Si}Me_2)SiPh_3$. Some exchange reactions involving Pt(II) silyl complexes have already been reported. We have previously prepared the chelated complex (*dppe*)$\overline{PtSiMe_2CH_2CH_2Si}Me_2$ (*dppe* = $Ph_2CH_2CH_2PPh_2$) by thermal reaction of (*dppe*)$Pt(SiMe_3)_2$ with $Me_2HSiCH_2CH_2SiHMe_2$ [3]. $SiMe_3$ group exchange was also observed when *trans*-$(Et_3P)_2Pt(Br)SiMe_3$ was reacted with $HSiMe_2Ph$ or $H_2SiMePh$ [7]. In a detailed study, *Clemmit* and *Glockling* [8, 17] have examined the exchange reactions between (*dppe*)$Pt(EMe_3)_2$ or (*dppe*)$PtCl(EMe_3)$ and $HE'Me_3$ (E, E' = Si, Ge, Sn). They found that a large excess of trimethylsilane was required to partially eliminate $HGeMe_3$, whereas the exchange of $SiMe_3$ by $GeMe_3$ or of $GeMe_3$ by $SnMe_3$ was quantitative with stoichiometric amounts of $HEMe_3$. This reflects the decreasing Pt–E bond strengths in the order E = Sn≫Ge>Si, favouring the formation of the stronger Si–H bond.

Results and Discussion

In this study we intended to simulate conditions of catalytic processes in which an intermediate *bis*(silyl) metal complex would be exposed to a high excess of a hydrosilane. For this reason we employed an excess of the hydrosilane relative to the *bis*(silyl) complexes. The reactions were followed by ^{31}P NMR spectroscopy at room temperature. The different chemical shifts allowed a clear distinction between the starting complexes and the exchange products. Partially exchanged complexes were identified by their typical pattern of two doublets (with ^{195}Pt satellites) due to the chemically inequivalent phosphorus atoms. We observed no side reactions or decomposition.

The exchange reactions of both silyl ligands of $(R_3P)_2Pt(SiPhMe_2)_2$ ($R_3P = Me_2PhP$, $MePh_2P$, Et_3P) with $HSi(OMe)_3$ (Eq. (2)) were complete in less than an hour. The ^{31}P NMR spectra of the reaction mixtures then showed only a new singlet with ^{195}Pt satellites.

(2)

$R_3P = PhMe_2P, Ph_2MeP, Et_3P$

The exchange reactions of *cis*-$(Me_2PhP)_2Pt(SiPhMe_2)_2$ with $HSiPh_3$, HSi_3Ph_7, $HPh_2SiSiMe_3$ and H_2SiPh_2 were much slower under the same conditions, and the rates showed a clear dependence on the nature of the entering silyl group. In each case, the unsymmetrically substituted complex $(Me_2PhP)_2Pt(SiPhMe_2)(SiR_3)$ was observed (Eq. (3)), although the NMR signals could not always be assigned unequivocally due to overlap with the signals of the starting and fully exchanged complexes. The time for the disappearance of the starting complex increased in the order H_2SiPh_2 (~1 h) < $HPh_2SiSiMe_3$ (~3 h) < HSi_3Ph_7 (~1 d) < $HSiPh_3$ (~4 d). Exchange of the second silyl ligand, *i.e.* formation of the symmetrically substituted complexes *cis*-$(Me_2PhP)_2Pt(SiR_3)_2$ (Eq. (4)), was even slower. Formation of these complexes with the more reactive silanes H_2SiPh_2 and $HPh_2SiSiMe_3$ took place within hours. The doubly exchanged products were not observed with HSi_3Ph_7 and $HSiPh_3$ during several days at room temperature. However, when a *THF* solution of $(Me_2PhP)_2Pt(SiPhMe_2)_2$ and $HSiPh_3$ was heated to 60°C for 10 d, the fully exchanged product $(Me_2PhP)_2Pt(SiPh_3)_2$ was formed.

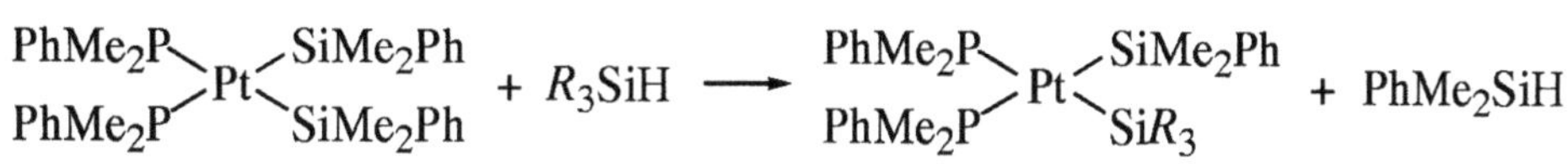

(3)

$R_3SiH = Ph_3SiH, Ph_2SiH_2, Me_3SiSiPh_2H, Ph_7Si_3H$

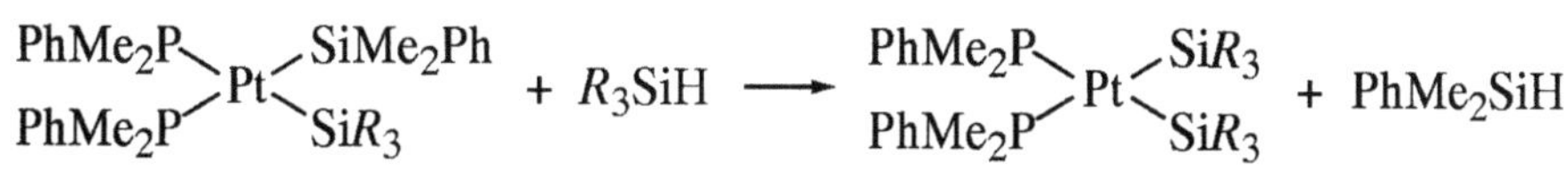

(4)

$R_3SiH = Ph_2SiH_2, Me_3SiSiPh_2H$

Chelation is known to stabilize transition metal complexes towards reductive eliminations. Thus, when the *bis*(silyl) complex $(Ph_3P)(Ph_3Si)Pt(PPh_2CH_2CH_2SiMe_2)$ [18] was treated with $HSi(OMe)_3$ or H_2SiPh_2, only products resulting from the exchange of the non-chelated silyl group were obtained, even with the highly reactive $HSi(OMe)_3$ (Eq. (5)).

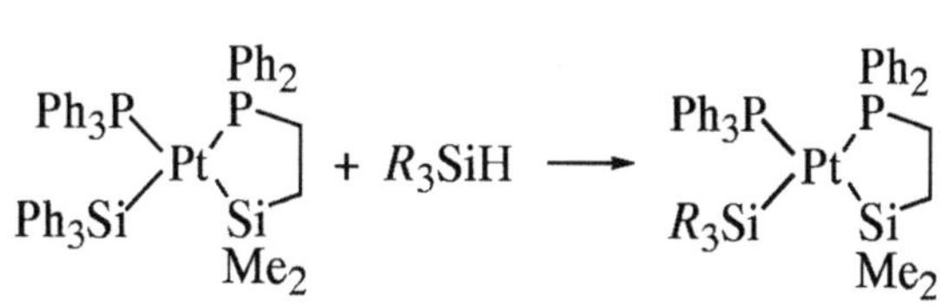

(5)

$R_3SiH = (MeO)_3SiH, Ph_2SiH_2$

Only the reaction of $(R_3P)_2Pt(SiMe_2Ph)_2$ with an excess of $(MeO)_3SiH$ results in the rapid exchange of both silyl ligands (Eq. (2)). Although the reaction is fast compared to the reactions with other silanes, there is a significant influence of the phosphine ligands on the exchange rate. It increases in the order $PPh_2Me > PEt_3 > PPhMe_2$. This trend can be correlated with the cone angle of the phosphine ligands [19, 20]: PPh_2Me (136°) > PEt_3 (132°) > $PPhMe_2$ (122°). Since PEt_3 is the most basic phosphine ligand in this series, the steric influence of the PR_3 ligands appears to play a more important role than their electronic influence. This indicates that (reversible) elimination of a PR_3 ligands is the initial step of the reaction which is easier for bulky substituents. Phosphine dissociation is often the preliminary step in reaction of d^8 systems.

Exchange of the first SiR_3 ligand was always more rapid than that of the second. The exchange reaction rates for the exchange of both the first and second silyl ligand depend on the entering silane and decrease in the order $HSi(OMe)_3 \gg H_2SiPh_2 > HPh_2SiSiMe_3 \gg HSi_3Ph_7 > HSiPh_3$, *i.e.* the reaction rate mainly depends on the electronic properties of the entering silane. Electron withdrawing substituents increase the reaction rate. The slow reaction of HSi_3Ph_7 indicates that the steric properties also play a role. Exchange of the second $SiMe_2Ph$ ligand with HSi_3Ph_7 or $HSiPh_3$ is so slow at room temperature that no significant concentration of the *bis*-exchanged complexes was observed within days. These complexes were only obtained at higher temperatures.

The strength of metal-silicon bonds is influenced by the substituents on silicon. Electronegative substituents such as halogens strengthen and electron donating moieties such as methyl groups weaken the *M*–Si bond. Since the substituents at silicon in the starting *bis*(silyl) complex ($SiMe_2Ph$) are more electron donating than those of the employed hydrosilanes, the formation of the exchange products is thermodynamically favoured in each case. Due to the large excess of the silane, the original silyl groups should therefore exchange completely.

The electronic situation at the metal center is somewhat different for the exchange of the first and the second silyl ligand. The electronic properties of the starting metal complex are determined by the two $SiMe_2Ph$ groups and the spectator ligands. After replacement of the first $SiMe_2Ph$ ligand, the electron density at the metal is decreased because the new silyl ligand is less electron donating than $SiMe_2Ph$. Thus, oxidative addition of the second silyl ligand and/or initial dissoziation of the PR_3 ligand is somewhat disfavoured relative to the first step. The degree of this effect depends on the difference in the electronic properties of the silyl groups involved in the exchange reaction.

This might explain why the exchange of the second silyl ligand is slower than that of the first one. Rapid exchange was only observed with $HSi(OMe)_3$. However, the order of reactivity with regard to the employed hydrosilanes is still the same as in the first step. Thus, the effect that a lower electron density at the silicon atom promotes the addition of hydrosilanes to the metal overrides the effect that the same substituents in a silyl ligand lower the reactivity of the metal center.

Conclusions

Exchange of the silyl ligands upon reaction of Pt(II) *bis*(silyl) complexes with hydrosilanes is slow unless there is a large difference in the electronic properties of the involved silyl groups. Rapid exchange is only observed if the substituents in the entering silyl group are much more electronegative than those of the leaving silane. The exchange rates for a given silyl complex depend both on the properties of the spectator ligands and the substituents at the hydrosilane.

The incorporation of a silyl group in a chelate system impairs their exchange, *i.e.* opening of the chelate ring of $(Ph_3P)Pt(PPh_2CH_2CH_2SiMe_2)SiPh_3$ by the hydrosilane and formation of $(Ph_3P)Pt(PPh_2CH_2CH_2SiMe_2H)(SiPh_3)(SiR'_3)$ was not observed even for the highly reactive $HSi(OMe)_3$. This is another proof for the stabilizing influence of chelation compared to complexes with electronically equivalent non-chelated ligands.

Experimental

All operations were performed in an atmosphere of dry and oxygen-free argon with standard *Schlenk* tube techniques using dried and argon saturated solvents. ^1H and ^{31}P NMR spectra were recorded on a Bruker AC250 spectrometer (^1H: 250.13 MHz, δ rel. to internal *TMS*; ^{31}P: 101.25 MHz, δ rel. to external 85% H_3PO_4). All NMR spectroscopic investigations were carried out in benzene-d_6. The solvent was dried with molecular sieve (4 Å) and stored under argon. The starting complexes and the hydrosilanes were prepared according to the literature [18, 21].

Reactions of platinum(II) complexes with hydrosilanes, general procedure

20 molar equivalents of the hydrosilane were added to a benzene-d_6 solution of *cis*-$(R_3P)_2Pt(SiPhMe_2)_2$ in an NMR tube, or 4 molar equivalents of the hydrosilane to a solution of $(Ph_3P)(Ph_3Si)Pt(PPh_2CH_2CH_2SiMe_2)$. The NMR tube was sealed under argon. ^{31}P NMR spectra were recorded in regular intervals to monitor the progress of the reaction.

Reactions of $(PhMe_2P)_2Pt(SiMe_2Ph)_2$ (^{31}P NMR: −6.48 ppm ($^1J_{PtP}$ = 1514 Hz))

(a) With $(MeO)_3SiH$: after 1 h at room temperature, only $(PhMe_2P)_2Pt[Si(OPh)_3]_2$ was observed; ^{31}P NMR: −37.22 ppm ($^1J_{PtP}$ = 1201 Hz).

(b) With Ph_2SiH_2: when the starting complex had disappeared after 1 h at room temperature, a mixture of $(PhMe_2P)_2Pt(SiMe_2Ph)(SiHPh_2)$ (major product) and $(PhMe_2P)_2Pt(SiHPh_2)_2$ (minor product) was spectroscopically observed. Due to signal overlap, the resonances of $(PhMe_2P)_2$ $Pt(SiMe_2Ph)(SiHPh_2)$ could not be unequivocally assigned. ^{31}P NMR of $(PhMe_2P)_2Pt(SiHPh_2)_2$: −58.58 ppm ($^1J_{PtP}$ = 1135 Hz).

(c) With Ph_7Si_3H: when the starting complex had disappeared after 1 d at room temperature, only $(PhMe_2P)_2Pt(SiMe_2Ph)(Si_3Ph_7)$ was observed. ^{31}P NMR: −2.48 ($^1J_{PtP}$ = 3202, $^2J_{PPtP}$ = 17 Hz), −6.46 ($^1J_{PtP}$ = 2256 Hz) ppm.

(d) With $Me_3SiSiPh_2H$: when the starting complex had disappeared after 3 h at room temperature, a mixture of $(PhMe_2P)_2Pt(SiMe_2Ph)(SiPh_2SiMe_3)$ (major product) and $(PhMe_2P)_2Pt(SiPh_2SiMe_3)_2$ (minor product) was spectroscopically observed. Due to signal overlap, the resonances of $(PhMe_2P)_2Pt(SiMe_2Ph)(SiPh_2SiMe_3)$ could not be unequivocally assigned. ^{31}P NMR of $(PhMe_2P)_2Pt(SiPh_2SiMe_3)_2$: −4.37 ppm ($^1J_{PtP}$ = 1665 Hz).

(e) With Ph_3SiH: when the starting complex had disappeared after 4 d at room temperature, only $(PhMe_2P)_2Pt(SiMe_2Ph)(SiPh_3)$ was observed. ^{31}P NMR: −3.11 ($^1J_{PtP}$ = 1544, $^2J_{PPtP}$ = 18 Hz), −5.81 ($^1J_{PtP}$ = 2272 Hz) ppm. $(PhMe_2P)_2Pt(SiPh_3)_2$ was obtained after heating a *THF* solution to 60°C for 4 d; ^{31}P NMR: −10.11 ppm ($^1J_{PtP}$ = 1487 Hz).

Reaction of $(Ph_2MeP)_2Pt(SiMe_2Ph)_2$ (^{31}P NMR: −0.12 ppm ($^1J_{PtP}$ = 3582 Hz)) *with* $(MeO)_3SiH$

After 30 min at room temperature, only $(Ph_2MeP)_2Pt[Si(OMe)_3]_2$ was observed; ^{31}P NMR: 15.14 ppm ($^1J_{PtP}$ = 2818 Hz).

Reaction of $(Et_3P)_2Pt(SiMe_2Ph)_2$ (^{31}P NMR: 10.49 ppm ($^1J_{PtP}$ = 2560 Hz)) *with* $(MeO)_3SiH$

After 45 min at room temperature, only $(Et_3P)_2Pt[Si(OMe)_3]_2$ was observed; ^{31}P NMR: 19.55 ppm ($^1J_{PtP}$ = 2024 Hz).

Reaction of $(Ph_3P)(Ph_3Si)\overline{Pt(PPh_2CH_2CH_2SiMe_2)}$ (^{31}P NMR: 57.80 ($^1J_{PtP}$ = 1648, $^2J_{PPtP}$ = 17 Hz), 29.30 ($^1J_{PtP}$ = 1762 Hz) ppm)

(a) With $(MeO)_3SiH$: after 15 min at room temperature, only $(Ph_3P)[(MeO)_3Si]\overline{Pt(PPh_2CH_2CH_2SiMe_2)}$ was observed; ^{31}P NMR: 61.81 ($^1J_{PtP}$ = 1651, $^2J_{PPtP}$ = 21 Hz), 33.50 ($^1J_{PtP}$ = 1542 Hz) ppm.

(b) With Ph_2SiH_2: after 15 min at room temperature, only $(Ph_3P)(Ph_2HSi)\overline{Pt(PPh_2CH_2CH_2SiMe_2)}$ was observed; ^{31}P NMR: 61.33 ($^1J_{PtP}$ = 1795, $^2J_{PPtP}$ = 21 Hz), 31.60 ($^1J_{PtP}$ = 1546 Hz) ppm.

Acknowledgements

We thank the *Fonds zur Förderung der wissenschaftlichen Forschung* (FWF), Vienna, for supporting this work (grant No. 57907-CHE)

References

[1] Part 57: Schubert U, Grubert S (1998) Monatsh Chem **129**: 437
[2] Schubert U, Grubert S (1996) Organometallics **15**: 4707
[3] Schubert U, Müller C (1991) J Organomet Chem **418**: C6
[4] Karch R, Schubert U (1997) Inorg Chim Acta **259**: 151
[5] Gilbert S, Knorr M, Mock S, Schubert U (1994) J Organomet Chem **480**: 241
[6] Tobita H, Izumi H, Ohnuki S, Ellerby MC, Kikuchi M, Inomata S, Ogino H (1995) J Am Chem Soc **117**: 7013
[7] Yamashita H, Tanaka M, Goto M (1997) Organometallics **16**: 4696
[8] Clemmit AF, Glockling F (1970) Chem Commun 705
[9] Tobita H, Ueno K, Ogino H (1988) Bull Chem Soc Jpn **61**: 2797
[10] Anderson FR, Wrighton MS (1984) J Am Chem Soc **106**: 995
[11] Hazeldine RN, Malkin LS, Parish RV (1979) J Organomet Chem **182**: 323
[12] Ruiz J, Maitlis PM (1986) J Chem Soc Chem Commun 862
[13] Woo HG, Heyn RH, Tilley TD (1992) J Am Chem Soc **114**: 5698; Woo HG, Walzer FF, Tilley TD (1992) J Am Chem Soc **114**: 7047
[14] Kreutzer KA, Fisher RA, Davis WM, Spaltenstein E, Buchwald SL (1991) Organometallics **10**: 4032
[15] Antinol A, Carillo-Hermosilla F, Castel A (1998) Organometallics **17**: 1523
[16] Schubert U (1996) In: Marciniec B (ed) Education in Advanced Chemistry, vol 2. Wydawnictwo Poznanskie, p 155, and vol 3 in press
[17] Clemmit AF, Glockling F (1971) J Chem Soc A 1164

[18] Gilges H, Kickelbick G, Schubert U (1997) J Organomet Chem **548**: 57
[19] Tolman CA (1977) Chem Rev **77**: 313
[20] Kobayashi T, Hayashi T, Yamashita H, Tanaka M (1988) Chem Lett 1411
[21] Chatt J, Eaborn C, Ibekwe SD, Kapoor PN (1970) J Chem Soc A 1343

Received June 19, 1998. Accepted July 13, 1998

Group IV Metallocene *Bis*(trimethylsilyl)acetylene Complexes as Catalysts in the Dehydrocoupling Polymerization of 1,2-Disubstituted Hydrodisilanes

Florian Lunzer[1], **Christoph Marschner**[1,*], **Berthold Winkler**[2], **Normen Peulecke**[3], **Wolfgang Baumann**[3], and **Uwe Rosenthal**[3]

[1] Institut für Anorganische Chemie, Technische Universität Graz, A-8010 Graz, Austria
[2] Institut für Chemische Technologie Organischer Stoffe, Technische Universität Graz, A-8010 Graz, Austria
[3] Institut für Organische Katalyseforschung an der Universität Rostock e.V., D-18055 Rostock, Germany

Summary. The catalytic activities of five *bis*(trimethylsilyl)acetylene complexes of group IV metallocenes in the dehydrocoupling reaction of various 1,2-disubstituted hydrodisilanes were investigated. The reactions were monitored by GC/MS analysis in the early stages, and the polymers were characterized by NMR spectroscopy and gel permeation chromatography.

Keywords. Alkyne complexes; Dehydropolymerization; Polysilanes; Titanocene; Zirconocene; Catalysis.

Bis(trimethylsilyl)acetylen-Komplexe von Metallocenen der 4 Nebengruppe als Katalysatoren in der dehydrierenden Polymerisation von 1,2-disubstituierten Hydrodisilanen

Zusammenfassung. Die katalytische Aktivität von fünf *Bis*(trimethylsilyl)acetylen-Komplexen von Metallocenen der 4. Nebengruppe bei der dehydrierenden Polymerisation von verschiedenen 1,2-disubstituierten Disilanen wurde untersucht. Zu Beginn (während der Bildung von Oligomeren) wurden die Reaktionen mittels GC/MS verfolgt. Die erhaltenen Polymere wurden mittels NMR-Spektroskopie und Gelpermeationschromatographie charakterisiert.

Introduction

Since the discovery of the dehydropolymerization of hydrosilanes with group IV metallocenes (Eq. (1)) by *Harrod et al.* in 1985 [1], several catalyst precursors have

* Corresponding author

been investigated with respect to this reaction. Most of them are dialkyl- or diarylmetallocenes or precursors thereof [1–4].

$$n\, RR'SiH_2 \xrightarrow[-(n-1)\, H_2]{\text{catalyst}} H(RR'Si)_nH \qquad (1)$$

Whereas silylmetallocene compounds can directly enter the catalytic cycle of σ-bond metathesis [5, 6], the carbon substituted metallocenes are proposed to enter the catalytic cycle *via* the naked metallocene Cp_2M. [3, 7] As metallocene-alkyne complexes are sources for Cp_2M moieties under mild conditions [8, 9], they are likely to exhibit catalytic activities similar to that of the above mentioned catalyst precursors. In a previous report the potential of these complexes in the dehydrocoupling reaction of methylphenylsilane, diphenylsilane, and phenylsilane has been demonstrated [10]. In the present study we would like to disclose our results on the catalytic behavior of the same type of metal complexes with 1,2-disubstituted hydrodisilanes. The latter are known to exhibit a chemistry different from that of phenylsilane [11–13].

Results and Discussion

The reactivities of $Cp_2Ti(\eta^2\text{-}Me_3SiC{\equiv}CSiMe_3)$ (**1**), $Cp_2Zr(THF)(\eta^2\text{-}Me_3SiC{\equiv}CSiMe_3)$ (**2**), $Cp_2Zr(py)(\eta^2\text{-}Me_3SiC{\equiv}CSiMe_3)$ (**3**), *rac*-$(EBTHI)Ti(\eta^2\text{-}Me_3SiC{\equiv}CSiMe_3)$ (**4**, *EBTHI* = ethylene*bis*(tetrahydroindenyl), and *rac*-$(EBTHI)Zr(\eta^2\text{-}Me_3SiC{\equiv}CSiMe_3)$ (**5**) with respect to 1,2-dimethyldisilane, 1-methyl-2-phenyldisilane, and 1,2-diphenyldisilane have been investigated. Catalysts were evaluated according to the GPC data of the produced polymers. These are shown in Table 1 for the polymerizations of 1,2-diphenyldisilane and 1-methyl-2-phenyldisilane. The polymerization experiments with 1,2-dimethyldisilane resulted in the formation of a completely insoluble material for which no GPC trace could be obtained. The early stages of the reactions were analyzed by GC/MS to obtain some insight into the structure of the polymers formed and also to learn more about the mechanism of the reactions. Both structural and mechanistic details reflected in the GC/MS data are outlined below.

^{29}Si and ^1H NMR spectra of the polymers were not found to be particularly useful with respect to assignments of stereoselectivity. They seem to be essentially

Table 1. GPC[a] data for the dehydropolymerization of 1,2-diphenyldisilane and 1-methyl-2-phenyldisilane catalyzed by titanocene and zirconocene alkyne complexes

	Metallocene				
Monomer	**1**	**2**	**3**	**4**	**5**
$H_2PhSiSiPhH_2$	M_n = 900	1.400	1.050	–	530
	M_w = 1.100	1.800	1.400	–	600
$H_2MeSiSiPhH_2$	M_n = 1.400	2.900	1.500	940	1.050
	M_w = 2.600	16.000	2.700	1.400	1.600

[a] Obtained against polystyrene standards (eluent *THF*) of samples subjected to polymerization conditions for 7 days at 40°C

identical with those obtained from reactions catalyzed by Cp_2ZrMe_2 [11, 12] or Cp_2ZrCl_2/2BuLi. This may be caused by the higher molecular weight of the polymers in comparison to polyphenylsilane. On the other hand, in their recent review *Gauvin*, *Harrod*, and *Woo* [4] point out that what was first assigned to be a stereochemically more homogeneous polymer actually was later found to consist largely of cyclics.

Polymerization of 1,2-dimethyldisilane

The results of the polymerization reactions of 1,2-dimethyldisilane catalyzed by metallocene alkyne complexes are in good accordance with the results of *Weinberger* and *Henge* [11] who used Cp_2TiMe_2 and Cp_2ZrMe_2 as catalysts. In the reaction catalyzed by **1**, GC/MS indicates the formation of linear oligomers up to the hexasilane at the beginning of the reaction. Oligomers with even and odd numbers of Si atoms in the chain are formed to the same extent. This indicates that the disilane is cleaved in the first step of the polymer formation reaction. As the reaction proceeds, the concentrations of branched and cyclic oligomers up to the decasilane increase as indicated by GC/MS [14]. After two days of reaction at room temperature, a completely insoluble crosslinked polymethylsilane is obtained.

In the Zr catalyzed reactions (compounds **2** and **3**), more branched silanes than linear ones are formed from the beginning. The isomer patterns of the GC/MS traces obtained from the reaction catalyzed by **2** and **3** are almost identical. Insoluble polymethylsilane was obtained after two days of reaction at room temperature.

In the *Ziegler-Natta* polymerization of propene, ethylene bridged metallocenes such as ethylene*bis*(indenyl) (*EBI*) ziroconocene or ethylene*bis*(tetrahydroindenyl) (*EBTHI*) zirconocene have been found to be of particular use to control tacticity in the polymer. Reports by *Harrod* [15], *Waymouth* [16], and *Corey* [17] seem to indicate a similar effect in the dehydrocoupling polymerization of silanes.

Our experiments with *EBTHI* bridged compounds **4** and **5** gave results different from those with **1–3** in a sense that almost no oligomers with an odd number of Si atoms in the chain were formed. Direct coupling of the disilane predominates over the previously observed cleavage and chain growth by monosilyl observed in the polymerization of disilanes catalyzed by non-bridged catalysts. These findings suggest that the *EBTHI* ligand preferentially effects Si–H activation, whereas the *Cp* catalysts preferentially activate Si–Si bonds.

GC/MS analysis of the reaction mixtures also revealed the presence of 1,2-*bis*(trimethylsilyl)ethene and 1,2-*bis*(trimethylsilyl)ethane, which is quite unusual since 1,2-*bis*(trimethylsilyl)acetylene was observed as the only product deriving from the catalyst precursor in all other reactions studied so far.

Polymerization of 1-methyl-2-phenyldisilane

In the reactions catalyzed by **1**, **2**, and **3**, again the disilane is cleaved in the polymer formation reaction as the GC/MS analysis indicates growth of the oligomers by monosilyl units. From the MS data [14] we find that isomers in which the new silyl group is added to the phenyl substituted end of the disilane are formed to a bigger extent. Therefore, in the isomeric ratios of the trisilanes Me/Ph/Me is favoured over Ph/Me/Me and more Ph/Ph/Me is formed than Ph/Me/Ph. The same is true for the isomeric ratios of the higher oligomers. We attribute this behavior to the preferential activation of the Si–H bond in SiPhH units over those in SiMeH units. Unlike 1,2-dimethyldisilane, 1-methyl-2-phenyldisilane is also cleaved by the *EBTHI* bridged catalysts **4** and **5**. Diphenyldisilane, triphenyltrisilane, and other trisilanes are formed to a bigger extent than the direct coupling products of the starting disilane. **4** and **5** also produce a significant amount of Ph_2SiH_2. The molecular weight data of the polymers are given in Table 1.

Polymerization of 1,2-Diphenyldisilane

In the reactions with all five catalysts, triphenyltrisilane is formed besides the disproportionation products Ph_2SiH_2 and $Ph_2HSiSiPhH_2$. This again indicates polysilane formation by monosilyl units. Due to their molecular weight, higher oligomers than tetrasilane could not be detected by GC/MS.

Conclusion

The common feature of all five alkyne complexes tested as dehydropolymerization catalysts is a lowered reactivity compared to the $Cp_2M\mathrm{Me}_2$ and $Cp_2M\mathrm{Cl}_2$/2BuLi systems (M = Ti, Zr). Whereas the achievable molecular weight of the polymers and the isomer patterns of the low molecular weight oligomers are essentially the same as in the reactions catalyzed by the analogous non-alkyne complexes, the reactions proceed much more slowly and often only at elevated temperatures (40°C). This may indicate that the alkyne ligand is not completely removed from the coordination sphere of the active metal center or still competes with the SiH substrates for the free coordination site. The addition of silane to the complexes induces a color change from black (**3**) or light red (**4**) to yellow for the Zr complexes and from yellow to dark green for Ti compounds. The colors of the reaction solutions are not as intense and appear slightly different from those usually observed in metallocene catalyzed dehydropolymerization reactions. This suggests that either the predominant metal species are different from those in non-alkyne systems or that the catalyst concentration is much lower than expected (*vide infra*).

In an attempt to obtain some information about the active catalytic species present in the system, we tried to react complexes **1** and **3** with 1,2-dimethyldisilane in stoichiometric ratios. Much to our surprise, the expected 1:1 adduct was not formed. The only products detected by NMR spectroscopy were silane oligomers and the metallocene starting materials. Since it has been recently outlined that Si–H activation can occur even in the presence of η^2-alkynyl substituents [18, 19], it appears possible that the actual catalytic species still contains the 1,2-*bis*(trimethylsilyl)acetylene ligand. A different explanation may be that only a very small amount of the complex is converted into the active species.

Experimental

Compounds **1–5** were synthesized according to Refs. [8] and [9]. For 1,2-diphenyldisilane and 1,2-dimethyldisilane, see Refs. [20] and [21]. 1-Methyl-2-phenyldisilane was obtained from *LAH* reduction of the respective triflate compound.

The polymerizations were carried out under an inert atmosphere of argon using *Schlenk* techniques. Solvents were freshly distilled from sodium/potassium alloy. In each case 0.5 ml toluene and 1 g of the monomer were added to the respective metallocene complex (1 mol%). The polymerizations of 1,2-dimethyldisilane were carried out at room temperature, all others at 40°C.

Acknowledgements

Financial support by the *Fonds zur Förderung der wissenschaftlichen Forschung in Österreich* (*SFB Electroactive Materials* at the TU Graz, Projekt No. 904) is gratefully acknowledged. *Wacker Chemie GmbH, Burghausen,* kindly provided various organosilanes as starting materials. *Ch. Marschner* thanks the Austrian Academy of *Sciences* for an *APART* (Austrian Program for Advanced Research and Technology) scholarship.

References

[1] Aitken C, Harrod JF, Samuel E (1985) J Organomet Chem **11**: C279
[2] Nakano T, Nakamura H, Nagai Y (1989) Chem Lett 83
[3] Corey JY, Zhu XH, Bedard TC, Lange DL (1991) Organometallics **10**: 924
[4] For the latest comprehensive review on the dehydrocoupling of hydrosilanes see: Gauvin F, Harrod JF, Woo HG (1998) Adv Organomet Chem **42**: 363
[5] Woo HG, Tilley TD (1989) J Am Chem Soc **111**: 3757
[6] Woo HG, Tilley TD (1989) J Am Chem Soc **11**: 8043
[7] Dioumaev VK, Harrod JF (1997) Organometallics **16**: 1452
[8] Ohff A, Pulst S, Lefeber C, Peulecke N, Arndt P, Burlakov VV, Rosenthal U (1996) Synlett 111
[9] Lefeber C, Baumann W, Tillack A, Kempe R, Görls H, Rosenthal U (1996) Organometallics **15**: 3486
[10] Peulecke N, Thomas D, Baumann W, Fischer C, Rosenthal U (1997) Tetrahedron Lett **38**: 6655
[11] Hengge E, Weinberger M (1992) J Organomet Chem **441**: 397
[12] Hengge E, Weinberger MJ (1992) J Organomet Chem **433**: 21
[13] For some other work on the dehydrocoupling of differently substituted disilanes see: Corey JY, Rooney SM (1996) J Organomet Chem **521**: 75 and references therein
[14] For the analysis of GC/MS data we heavily relied on the method outlined by Hengge and Weinberger in Ref. [11]. By means of careful evaluation of the MS fragmentation patterns and additional ^{29}Si NMR analysis they were able to obtain reliable information about oligomer constitution
[15] Gauvin F, Harrod JF (1990) J Can Chem **68**: 1638
[16] Banovetz JP, Stein K, Waymouth RM (1991) Organometallics **10**: 3430
[17] Shaltout RM, Corey JY (1995) Tetrahedron **51**: 4309
[18] Ohff A, Kosse P, Baumann W, Tillack A, Kempe R, Goerls H, Burlakov VV, Rosenthal U (1995) J Am Chem Soc **117**: 10399
[19] Peulecke N, Ohff A, Kosse P, Tillack A, Spannenberg A, Kempe R, Baumann W, Burlakov VV, Rosenthal U (1998) Chem Eur J **4**: 1852
[20] Aitken C, Barry JP, Gauvin F, Harrod JF, Malek A, Rousseau D (1989) Organometallics **8**: 1732
[21] Wanatabe W, Kabayashi M, Koike Y, Nagashima G, Matsumoto M, Nagay Y (1977) J Organomet Chem **128**: 173

Received June 12, 1998. Accepted (revised) September 14, 1998

Deposition of Amorphous Hydrogenated Silicon (*a*-Si:H): *in situ* Gas Analysis by Time-of-Flight Mass Spectrometry

Gudrun Andrä[1], **Fritz Falk**[1], **Herbert Stafast**[1,2,*], and **Ewa Witkowicz**[1]

[1] Institut fuer Physikalische Hochtechnologie e.V., D-07743 Jena, Germany
[2] Friedrich Schiller University, Faculty of Physics and Astronomy, D-07743 Jena, Germany

Summary. Time-of-flight mass spectra generated by ArF laser photoionization have been obtained in a molecular beam effusing from an *a*-Si:H deposition system with a dc discharge in a gas mixture of SiH_4 and He. The XeF, KrF, and ArF laser wavelength dependence and the kinetics of the SiH_{0-3}^+ and the $Si_2H_{0-2}^+$ signals show that the ions predominantly originate from disilane.

Keywords. Silicon hydrides; Gas phase reactions; Mass spectra.

Abscheidung von amorphem, wasserstoffhaltigem Silicium (*a*-Si:H): *in situ*-Gasanalyse mit Flugzeitmassenspektrometrie

Zusammenfassung. Flugzeitmassenspektren nach Photoionisierung mit einem ArF-Excimerlaser wurden in einem effusiven Molekülstrahl aufgenommen, dessen Ursprung in der *a*-Si:H-Schichtbildungszone einer Gleichstromgasentladung in SiH_4 und He liegt. Die Wellenlängenabhängigkeit bei der Bestrahlung mit XeF-, KrF- und ArF-Lasern sowie die Kinetik der Signale von SiH_{0-3}^+ und $Si_2H_{0-2}^+$ zeigen, daß die Ionen hauptsächlich von Disilan stammen.

Introduction

Thin film deposition of amorphous hydrogenated silicon (*a*-Si:H) by various methods of chemical vapor deposition (CVD) has been known since a quarter of a century and applied to produce technical devices like photovoltaic cells, electrophotographic drums, and thin film transistors [1]. Nevertheless, there are still open questions like those regarding the identity and properties of the film forming species in the gas phase [2] and the origin of mechanical stress in *a*-Si:H thin films [3].

A variety of gas phase species like Si atoms, molecular intermediates SiH_3 and H_2SiSiH_2, as well as stable molecules like Si_2H_6 and Si_3H_8 have been reported to contribute to *a*-Si:H thin film formation. Depending on the details of the

* Corresponding author

deposition process, one or several film forming species appear *a priori* possible. They are, however, hard to identify experimentally, especially due to their very low concentration in comparison with that of the precursor molecules of similar chemical composition. A mass balance in the gas phase during CO_2 laser CVD, for instance, and comparison with the *a*-Si:H film growth led to the conclusion that the film forming species contribute to the silane processing gas in the order of 10^{-5} to 10^{-6} only [4]. To reveal Si_xH_y intermediates in (laser) CVD processes, a molecular beam apparatus was set up to provide sampling in the film forming region and collision free transfer into a time-of-flight (TOF) mass spectrometer with laser photoionization to achieve high selectivity and high efficiency. In the following, our first TOF mass spectra obtained during *a*-Si:H deposition by dc gas discharge are reported.

Results and Discussion

Figure 1 shows a TOF mass spectrum obtained by photoionization of the effusive molecular beam derived from the *a*-Si:H film deposition region. Photoionization was achieved by ArF excimer laser radiation (193 nm), affording the ions SiH_{0-3}^+ (*i.e.* Si^+, SiH^+, SiH_2^+, and SiH_3^+) and $Si_2H_{0-2}^+$ (*i.e.* Si_2^+, Si_2H^+, and $Si_2H_2^+$). With KrF (248 nm) and XeF (351 nm) excimer laser radiation under comparable conditions (laser pulse energy), no ions $Si_xH_y^+$ ($x=1, 2$; $y=0$–3) were obtained.

In a second series of experiments, the temporal evolution of the $Si_xH_y^+$ signals generated by ArF laser radiation was observed while integrating the signals of ions with one ($x=1$) and two ($x=2$) Si atoms, respectively (Fig. 2). No ions are observed without gas discharge (SiH_4/He mixture). Both ion signals rise within a few seconds upon switching on the gas discharge, and after about 1 min they display a constant level (steady state). The steady state signal in both cases amounts to about 80% of the maximum value in the beginning. After switching off

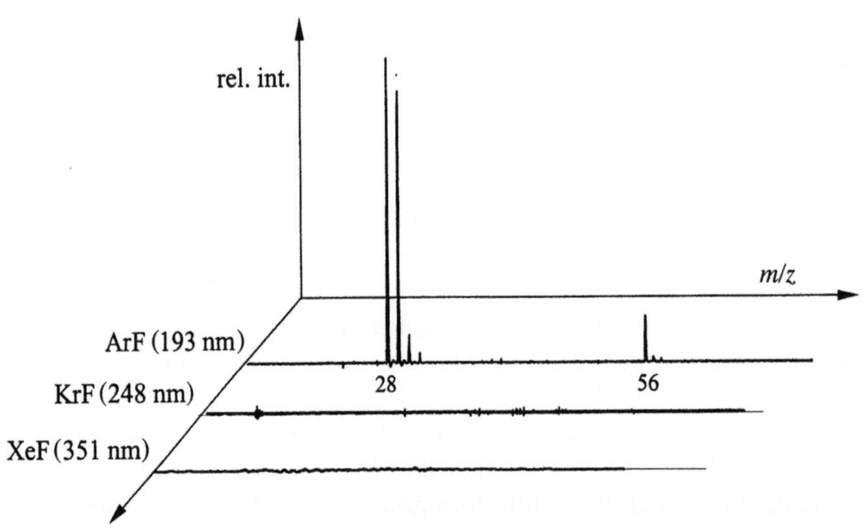

Fig. 1. TOF mass spectrum generated by ArF, KrF, and XeF laser photoionization in a molecular beam effusing from the *a*-Si:H deposition region (dc discharge in SiH_4/He)

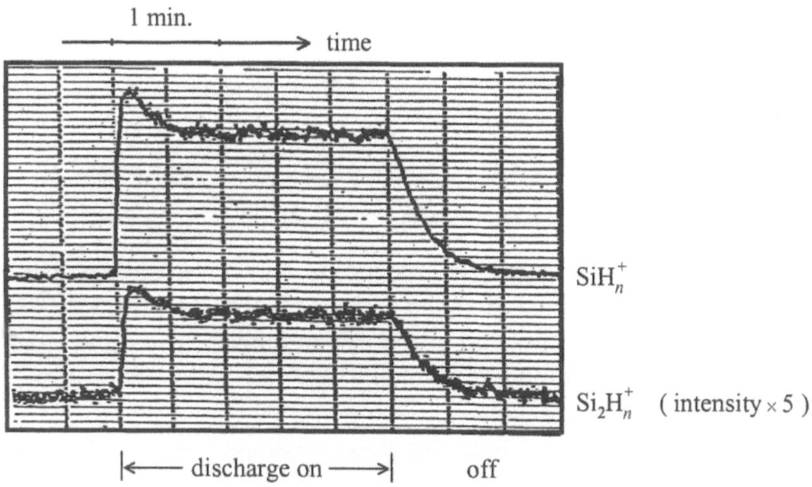

Fig. 2. Temporal evolution of the integrated SiH_{0-3}^+ and $Si_2H_{0-2}^+$ ion signals upon switching on and off the dc gas discharge in SiH_4/He

the discharge, both ion signals decrease with the same time constant of about 30 s, which corresponds to the time for gas exchange in the deposition reactor (flow system).

From the findings presented in Fig. 2 it is concluded that the ion signals, within the error limits, have the same source and originate from a stable species. Ions from short-lived intermediates like SiH_2, SiH_3, or Si_2H_4 should disappear immediately when switching off the discharge. The stable parent of the registered ions is assumed to be disilane (Si_2H_6) in agreement with the laser wavelength dependence of the ion yield. The adiabatic and vertical ionization energies of 10.0 eV and 10.53 eV, respectively, can be derived from the UV photoelectron spectrum of disilane [5, 6]. Consequently, vertical photoionization by XeF (3.53 eV) and KrF laser light (5.00 eV) requires the energy of three photons compared to that of two ArF laser photons (6.42 eV) only. Moreover, XeF or KrF laser radiation require (synchronous) three photon ionization (MPI). Each ArF laser photon, on the other hand, is in resonance with the long wavelength tail of the electronic $4s \leftarrow 2\sigma_{1g}$ transition of disilane ($\sigma = 2 \times 10^{-18}\,cm^2$) [7, 8], leading to (sequential) resonance enhanced MPI (REMPI). Two ArF laser photons can ionize disilane and provide an excess energy of 2.84 eV (273 kJ/mol) in addition to the adiabatic ionization energy. This energy evidently is sufficient to generate the fragment ions observed in the TOF mass spectrum (Fig. 1).

The concentration of trisilane (Si_3H_8) in the deposition reactor is assumed to be negligible: The vacuum UV absorption cross section of trisilane at 193 nm is about five times larger than that of disilane [7], and high ion yields are expected. However, no $Si_3H_y^+$ ions were observed under ArF laser irradiation, possibly because the excess energy in addition to the adiabatic ionization of trisilane is high (3.54 eV, 341 kJ/mol) [5]. On the other hand, the amount of trisilane relative to that of disilane was found to be small under comparable gas discharge conditions [9]. This can be rationalized by the small concentration of Si_2H_6 relative to that of SiH_4

(1–10%, see below), resulting in a small probability for the SiH_2 biradicals to react with Si_2H_6 and to form Si_3H_8.

One can assume a relatively fast equilibration between SiH_4 and Si_2H_6 [10] following the simplified equations

$$SiH_4 \rightarrow SiH_2 + H_2$$
$$SiH_4 + SiH_2 \rightleftharpoons Si_2H_6$$

on a second time scale (steep ion signal rise in Fig. 2). The signal decrease from the maximum to the stationary level then is a measure for the depletion of SiH_4 in the discharge (probe) volume. The small discharge volume of several cm^3 is buffered over some 10 s by the large volume of about 3000 cm^3 of the deposition chamber until stationary conditions are achieved. Hereafter, the stationary Si_2H_6 concentration is in the order of 1–10% of the initial SiH_4 concentration, *i.e.* 10^{-3}–10^{-2} mbar. This corresponds to 10^{-12}–10^{-11} mbar Si_2H_6 in the laser ionization region of the TOF mass spectrometer.

These results represent an important technical step towards detection, identification and characterization of metastable Si_xH_y intermediates of very low concentration in the thin film deposition system. Photoionization mass spectra of Si_2H_n ($n = 2$–5) transient species are known [12]. In our case, however, detection under *a*-Si:H deposition conditions in an effusive molecular beam imposes several experimental difficulties. Further experiments applying tunable narrow-band radiation from a pulsed dye laser are currently in progress.

Experimental

The basic concept of the experimental setup has been described previously [4, 10]. Briefly, the substrate for *a*-Si:H thin film deposition provides a small hole (100 μm diameter) being the source of an effusive molecular beam. This beam passes through a differential pumping stage into the ion source of a mass spectrometer. The gas transfer by the molecular beam is collision-free (no intermediate chemical reactions). Recently, the quadrupole mass spectrometer of the original setup for *in situ* gas analysis has been replaced by a TOF mass spectrometer (reflectron type, Kaesdorf RTF 10) with pulsed laser photoionization. In the above experiments, an excimer laser (Lambda Physik, EMG 201 MSC) was used for photoionization. The TOF spectra were registered on a digital oscilloscope (Tektronix TPS 644B), and the temporal evolution of the ions was recorded by a boxcar integrator (SRS model SR 250). For convenience of operation, *a*-Si:H deposition was performed by a dc gas discharge in SiH_4 (3%)/He (97%) gas mixture at 0.5 mbar total pressure.

Acknowledgements

Funding by the *Deutsche Forschungsgemeinschaft* (DFG) under contract No. II C 10-STA 465/1 is gratefully acknowledged. The authors are grateful to Dr. *J. Bergmann*, Dipl.-Phys. *J. Meinschien*, and Ing. *A. Schumann* for many discussions and technical assistance.

References

[1] Street RA (1991) Hydrogenated Amorphous Silicon. Cambridge University Press

[2] Cf. *e.g.* Veprek S, Veprek-Heijman MGJ (1991) Plasma Chem Plasma Proc **11**: 332 and Dietrich TR, Chiussi S, Marek M, Roth A, Comes FJ (1991) J Phys Chem **95**: 9302

[3] Claasen WAP (1989) Thin Solid Films **168**: 89
[4] Mollekopf G (1993) PhD Thesis, University of Heidelberg, Germany
[5] Bock H, Ensslin W, Feher F, Freund R (1976) J Amer Chem Soc **98**: 668
[6] Stafast H (1988) Appl Phys A **45**: 93
[7] Itoh U, Toyoshima Y, Onuki H (1986) J Chem Phys **85**: 4867
[8] Dietrich TR, Chiussi S, Stafast H, Comes FJ (1989) Appl Phys A **48**: 405
[9] Heintze M, Veprek S (1989) Appl Phys Lett **54**: 1320
[10] Cf. *e.g.* Moffat HK, Jensen KF, Carr RW (1991) J Phys Chem **95**: 145
[11] Golusda E, Lange R, Lühmann K-D, Mollekopf G, Wacker M, Stafast H (1992) Appl Surf Sci **54**: 30
[12] Ruscic B, Berkowitz J (1991) J Chem Phys **95**: 2407, 2416

Received June 12, 1998. Accepted August 5, 1998

Reaktion von atomarem Silicium mit Methanol: Matrixspektroskopische Identifizierung von Methoxysilylen [1]

Günther Maier*, **Hans P. Reisenauer** und **Heiko Egenolf**

Institut für Organische Chemie der Justus-Liebig-Universität Gießen, D-35392 Gießen, Germany

Reaction of Atomic Silicon with Methanol:
Matrix-Spectroscopic Identification of Methoxysilylene

Summary. Codeposition of thermally generated silicon atoms with methanol in an argon matrix at 12 K led to the formation of methoxysilylene (**4**) by insertion of the silicon atom into the O–H bond of the substrate. Its spectroscopic identification was aided by calculated IR spectra and deuteriation experiments. Both the *s-cis* (**4a**) and the *s-trans* conformer (**4b**) of methoxysilylene could be observed and interconverted into each other photochemically.

Keywords. Matrix isolation; *ab initio* Calculations; Photoisomerizations.

Einleitung

Im Jahr 1945 berichtete *Rochow* über die Direktsynthese von Methylchlorsilanen durch die kupferkatalysierte Reaktion von Silicium mit Methylchlorid [2a]. Sie ist von enormer wirtschaftlicher Bedeutung, da Chlormethylsilane die Ausgangsstoffe für die Kunststoffklasse der Silicone wie auch für eine Vielzahl anderer Organosiliciumverbindungen sind. Eine entsprechende Reaktion ist mit Methanol als Substrat möglich, die je nach Reaktionsführung zu Tri- bzw. Tetramethoxysilan führt [2b]. Beide Silane sind ebenfalls wichtige Zwischenprodukte bei der Synthese vieler siliciumorganischer Verbindungen. Die große technische Bedeutung der *Rochow*-Synthesen findet ihren Ausdruck in einer Vielzahl an Studien, die sich mit den mechanistischen Grundlagen der Reaktion befassen [4–6]. Dabei konnten intermediär silylenische Funktionen auf der Siliciumoberfläche nachgewiesen werden, die eine Chemisorption von Methoxyspezies ermöglichen, mit der die beobachtete Reaktivität erklärt werden kann.

Auch für Photoelektroden wird das System Si/Methanol häufig verwendet, da sich Halbleiter-Flüssigkeits-Kontakte durch eine höhere Quantenausbeute im Hinblick auf die zu erzielende Ladungstrennung auszeichnen [7–10].

* Corresponding author

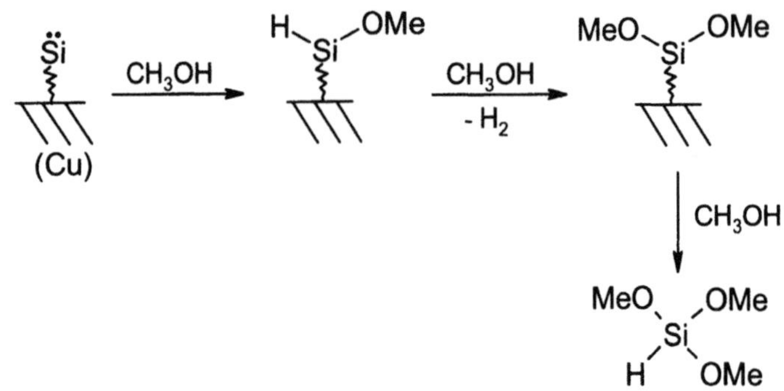

Abb. 1. Mechanismus der Cu-katalysierten Bildung von Trimethoxysilan aus Methanol und Silicium

Experimentelle Informationen über CH_4OSi-Isomere sind jedoch sehr rar. Ein hochsubstituiertes Formylsilan, das nur unter Inertgas haltbar ist, konnte von *Tilley et al.* synthetisiert werden [11]. Mit Methoxysilylen (**4**) [12] bzw. Methylsilanon (**1**) und Hydroxy(methyl)silylen (**2**) [13] werden drei unsubstituierte CH_4OSi-Spezies als Intermediate bei Pyrolyse- bzw. Photolysereaktionen postuliert. Eine zweifelsfreie Identifikation von CH_4OSi-Isomeren stand allerdings noch aus. Einen möglichen Zugang zu solchen Spezies bietet die Reaktion von Siliciumatomen mit Methanol. Die Kenntnis prinzipieller Reaktionsmöglichkeiten atomaren Siliciums [14–17], die vor dem Hintergrund der CVD-Prozesse von Silicium bzw. darauf basierender Materialien auch von großer technischer Bedeutung ist, wird durch den Einsatz einer zusätzlichen Verbindungsklasse (Alkohole) als Substrat erweitert.

Ergebnisse und Diskussion

Die Geometrien sowie IR-Spektren verschiedener Repräsentanten der CH_4OSi-Energiehyperfläche wurde mit der DFT-Methode BLYP/6-31G** [18] untersucht (Abb. 2). Die in einer SCF- bzw. MP3-Studie von *Gordon* und *George* [19] ermittelte Stabilitätsreihenfolge wird dabei, abgesehen von der energetischen Abfolge von Formylsilan (**5**) und 1-Hydroxysilen (**3**), reproduziert. Der hohe Energieinhalt von Formylsilan (**5**) findet seinen Ausdruck darin, daß sich auch dessen substituierte Derivate dem Experiment bisher weitgehend entzogen haben. Mit Methylsilanon (**1**) als globalem Minimum geben die Rechnungen die hohe Oxophilie des Siliciums wieder. Diese wirkt sich auch im (Hydroxymethyl)silylen (**8**) aus, das eine stark verzerrte Geometrie aufweist ($\angle_{SiCO} = 72.9°$, $d_{SiO} = 209.3$ pm), die eine intramolekulare Stabilisierung des silylenischen Zentrums durch die freien Elektronenpaare des Sauerstoffatoms ermöglicht.

Matrixspektren, die nach der Abscheidung von Siliciumatomen und einer Methanol/Argon-Gasmischung (1:80) erhalten werden, zeigen neben den IR-Banden des Substrats und denen der auch in anderen Experimenten [14–17] beobachteten Nebenprodukte eine große Zahl weiterer Absorptionen. Vor allem fallen vier intensive Banden im Bereich um 1900 cm^{-1} auf, die offensichtlich von silylenischen Si–H-Streckschwingungen herrühren. Bei abwechselnder Bestrahlung mit $\lambda = 366$ bzw. 405 nm verschwinden innerhalb weniger Minuten jeweils zwei dieser Banden, während die beiden anderen an Intensität gewinnen. Auch in

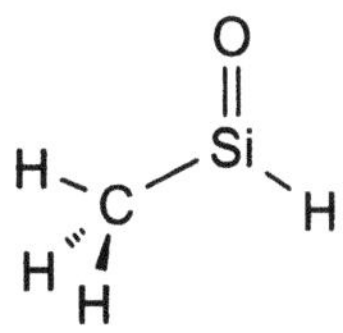

1 C_s (0.0)

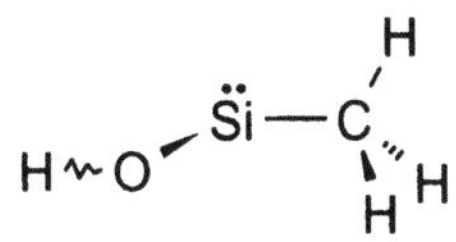

2 C_1

a s-*cis* (3.7)

b s-*trans* (2.9)

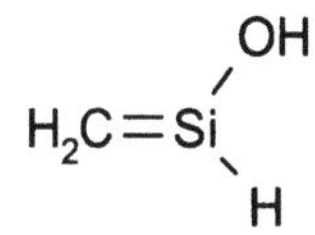

3 C_s (16.8)

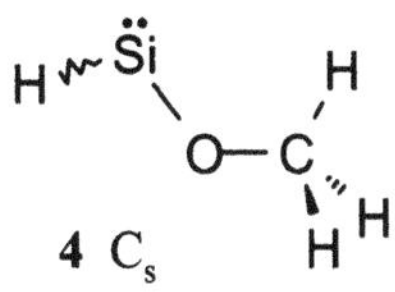

4 C_s

a s-*cis* (20.8)

b s-*trans* (21.1)

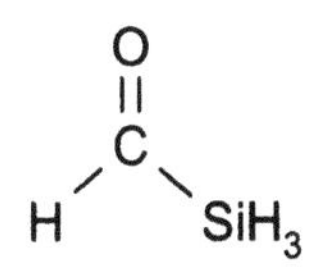

5 C_1 (23.9)

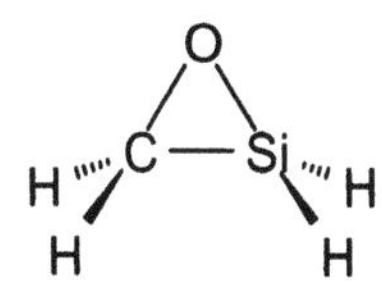

6 C_s (26.1)

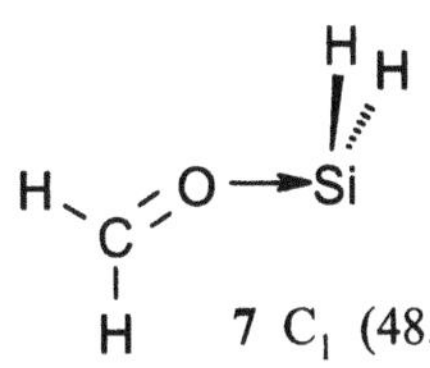

7 C_1 (48.0)

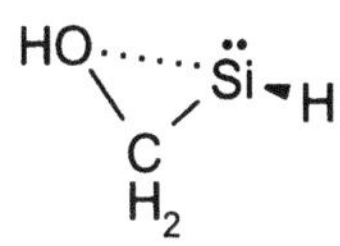

8 C_1 (52.8)

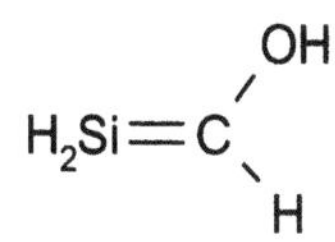

9 C_1 (56.3)

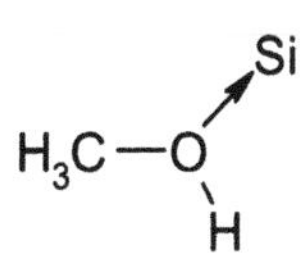

10 C_s

(T: 77.9, S: 96.4)

Si + CH_3OH

11 **12**

(94.3)

Abb. 2. Relative Energien (BLYP/6–31G**, kcal · mol^{-1}) verschiedener CH_4OSi-Spezies; alle Werte sind um die Nullpunktsschwingungsenergie korrigiert

anderen Bereichen des Spektrums fallen Paare von Banden mit sehr ähnlicher Lage und Intensität auf, von denen jeweils eine bei der Belichtung mit einer bestimmten Wellenlänge wächst, während ihr Gegenstück verschwindet. Wechselt man die Wellenlänge von 366 auf 405 nm, so lassen sich die beobachteten Veränderungen vollständig umkehren. Es muß sich also dabei um die reversible Photoisomerisierung zweier strukturell sehr ähnlicher Moleküle handeln. Nach einem Vergleich mit den berechneten Spektren lassen sich die Bandengruppen dem *s-cis-* (**4a**) und dem *s-trans*-Konformer (**4b**) von Methoxysilylen zuordnen (Abb. 3, Tabellen 1, 2).

Die Beobachtung der beiden Methoxysilylene nach der Codeposition von atomarem Silicium mit Methanol zeigt, daß die Siliciumatome bevorzugt in die O–H-Bindung statt in die C–H- bzw. die C–O-Bindung insertieren.

Die Übereinstimmung zwischen den berechneten und experimentellen Spektren der Silylene **4a** und **4b** ist zwar sehr gut, dennoch wurden, auch im

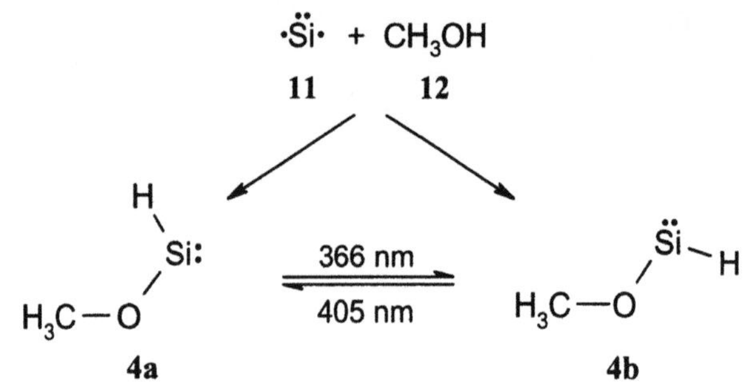

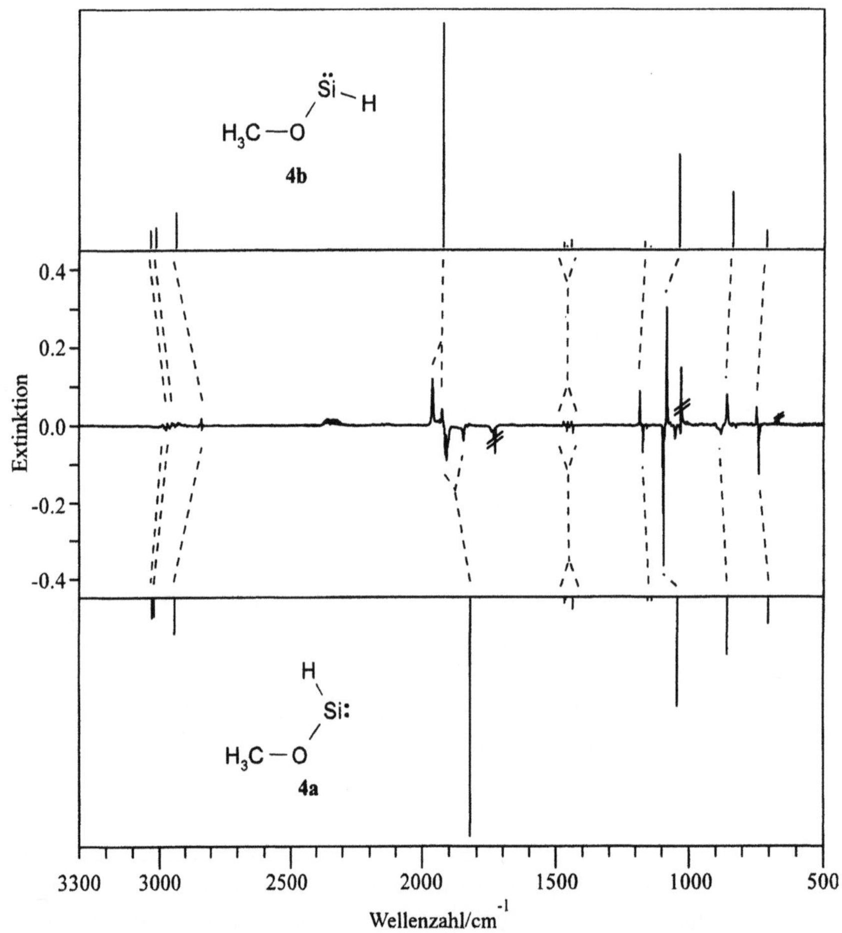

Abb. 3. Oben: berechnetes IR-Spektrum von *s-trans*-Methoxysilylen (**4b**) (BLYP/6–31G**); Mitte: experimentelles IR-Differenzspektrum der Photoreaktion **4a**→**4b** (Ar-Matrix, 12 K, λ = 366 nm, 15 min); von Methanol, CO_2 bzw. SiNN herrührende Banden sind durchgestrichen; unten: berechnetes IR-Spektrum von *s-cis*-Methoxysilylen (**4a**) (BLYP/6–31G**)

Hinblick auf die mögliche Identifizierung weiterer CH_4OSi-Isomerer, Codepositionsversuche auch mit trideuteriertem Methanol (CD_3OH) unternommen. Die Si–H-Banden sind dabei gegenüber den für die Grundkörper **4a** und **4b** beobachteten Werten leicht verschoben, sie können daher den D_3-Isotopomeren der Methoxysilylene (CD_3O–Si–H) zugeordnet werden. Auch die Lagen der übrigen Banden dieser Isotopomere weichen leicht von den für die H_4-Stammverbindungen

Tabelle 1. Experimentelles (Ar-Matrix, 12 K, Wellenzahlen (cm^{-1}), Intensitäten relativ zur stärksten Bande in Klammern), berechnetes (BLYP/6–31G**, C_s-Symmetrie, absolute Intensitäten ($km \cdot mol^{-1}$) in Klammern) und korrigiertes $\tilde{\nu}_{corr.} = \tilde{\nu}_{calc.} \cdot \tilde{\nu}_{exp}$ ($^1H_4$- **4b**)/$\tilde{\nu}_{calc}$ ($^1H_4$-**4b**) IR-Spektrum von *s-trans*-Methoxysilylen (**4b**)

		Mode	Isotop[a]	$\tilde{\nu}_{calc}$	$\tilde{\nu}_{corr}$	$\tilde{\nu}_{exp}$
ν_{15}	a''	CH_3 def/		152.8 (1)		–
		COSi def	D_3	119.0 (1)		–
ν_{10}	a'	COSi def ip		277.1 (4)		–
			D_3	255.5 (3)		–
ν_{14}	a''	SiH def oop/		389.9 (1)		–
		CH_3 def	D_3	380.9 (1)		–
ν_9	a'	SiO str		711.4 (27)	749.3	749.3 (16)
			D_3	681.5 (26)	717.8	716.0 (8)
ν_8	a'	SiH def		837.7 (91)	859.4	859.4 (52)
			D_3	826.4 (68)	847.8	846.4 (15)
ν_7	a'	CO str/		1038.5 (155)	1084.8	1084.8 (84)
		SiH def	D_3	1033.3 (103)	1079.4	1065.6/1077.9[b] (30)
ν_{13}	a''	CH_2 def		1145.7 (<1)		–
			D_3	887.9 (1)		–
ν_6	a'	CH_3 def		1168.5 (8)	1186.2	1186.2 (16)
			D_3	915.4 (24)	929.3	933.0 (9)
ν_5	a'	CH_3 def		1445.4 (11)	1443.9	1443.9 (2)
			D_3	1099.6 (57)	1098.5	1116.8 (33)
ν_{12}	a''	CH_2 def		1461.3 (1)	1457.6	1457.6 (7)
			D_3	1057.5 (<1)	1054.8	–
ν_4	a'	CH_2 def		1472.9 (7)	1473.6	1473.6 (1)
			D_3	1066.7 (7)	1067.2	–
ν_3	a'	SiH str		1924.3 (376)		1929.1/1964.7[b] (100)
			D_3	1924.2 (377)		1930.6/1965.4[b] (100)
ν_2	a'	CH_3 str s		2936.6 (57)		–[c]
			D_3	2105.0 (42)		–[c]
ν_1	a'	CH_3 str as		3014.7 (33)		–[c]
			D_3	2234.0 (18)		–[c]
ν_{11}	a''	CH_2 str as		3034.4 (28)		–[c]
			D_3	2252.0 (17)		–[c]

[a] CD_3O–Si–H; [b] zu möglichen Gründen für die Aufspaltung der Banden vgl. Text; [c] verdeckt von den C–H-Valenzschwingungsbanden von Methanol bzw. dessen Oligomeren

beobachteten Positionen ab (Tabellen 1, 2). Wie beim Grundkörper sind auch hier die Banden der silylenischen Si–H-Streckschwingungen aufgespalten. Die Ursache für das Auftreten von Bandenpaaren ist allerdings unklar. In Frage kommen neben möglichen anharmonischen Resonanzen auch Matrixeffekte, wie sie schon bei anderen H-substituierten Silylenen beobachtet wurden. So findet man beispielsweise auch für die Si–H-Streckschwingung von Methylsilylen [20] zwei Banden, die in Argon um 69 cm^{-1}, in Stickstoff dagegen nur um 12 cm^{-1} gegeneinander verschoben sind. Offensichtlich sind genaue Lage und Anzahl der Banden solcher Schwingungen sehr sensitiv gegen äußere Einflüsse.

Tabelle 2. Experimentelles (Ar-Matrix, 12 K, Wellenzahlen (cm^{-1}), Intensitäten relativ zur stärksten Bande in Klammern), berechnetes (BLYP/6–31G**, C_s-Symmetrie, absolute Intensitäten ($km \cdot mol^{-1}$) in Klammern) und korrigiertes ($\tilde{\nu}_{corr.} = \tilde{\nu}_{calc} \cdot \tilde{\nu}_{exp}$) ($^1H_4$-**4a**)/$\tilde{\nu}_{calc}$ ($^1H_4$-**4a**) IR-Spektrum von *s-cis*-Methoxysilylen (**4a**)

		Mode	Isotop[a]	$\tilde{\nu}_{calc}$	$\tilde{\nu}_{corr}$	$\tilde{\nu}_{exp}$
ν_{15}	a″	CH_3 def/		149.4 (3)		–
		COSi def	D_3	117.5 (2)		–
ν_{10}	a′	COSi def ip		303.0 (5)		–
			D_3	280.0 (3)		–
ν_{14}	a″	SiH def oop/		387.2 (4)		–
		CH_3 def	D_3	378.5 (3)		–
ν_9	a′	SiO str		704.6 (40)	741.8	741.8 (31)
			D_3	674.5 (36)	710.1	708.2 (12)
ν_8	a′	SiH def		858.0 (89)	881.4	881.4 (13)
			D_3	851.8 (67)	875.0	873.5 (4)
ν_7	a′	CO str/		1043.9 (172)	1096.2	1096.2 (85)
		SiH def	D_3	1039.6 (120)	1091.7	1086.8 (20)
ν_{13}	a″	CH_2 def		1139.7 (3)		–
			D_3	881.8 (4)		901.4 (2)
ν_6	a′	CH_3 def		1154.9 (3)	1174.7	1174.7 (11)
			D_3	901.8 (36)	917.3	924.1 (10)
ν_5	a′	CH_3 def		1439.4 (16)	1438.7	1438.7 (3)
			D_3	1089.8 (35)	1089.3	1109.6 (33)
ν_{12}	a″	CH_2 def		1465.1 (2)	1452.6	1452.6 (1)
			D_3	1060.7 (<1)	1051.7	–
ν_4	a′	CH_2 def		1468.3 (7)	1461.3	1461.3 (3)
			D_3	1063.8 (16)	1058.7	–
ν_3	a′	SiH str		1823.8 (384)		1849.6/1910.7[b] (100)
			D_3	1823.1 (385)		1848.3/1913.1[b] (100)
ν_2	a′	CH_3 str s		2943.6 (56)		–[c]
			D_3	2107.2 (40)		–[c]
ν_1	a′	CH_3 str as		3021.6 (27)		–[c]
			D_3	2242.2 (15)		–[c]
ν_{11}	a″	CH_2 str as		3029.4 (30)		–[c]
			D_3	2247.9 (20)		–[c]

[a] CD_3O–Si–H; [b] zu möglichen Gründen für die Aufspaltung der Banden vgl. Text; [c] verdeckt von den C–H-Valenzschwingungsbanden von Methanol bzw. dessen Oligomeren

Die Suche nach weiteren CH_4OSi-Isomeren führte bisher zu keinem eindeutigen Erfolg. Bei der Belichtung einer Matrix, die eines der beiden Methoxysilylenkonformere **4** enthält, mit Licht der Wellenlänge $\lambda = 313$ nm verändern sich die Intensitäten der zugehörigen Absorptionen nicht. Dafür verschwindet eine Gruppe anderer Banden (702.2, 844.5, 972.6, 1083.8, 1183.1, 1381.4, 3619.8 cm^{-1}), deren Herkunft nicht eindeutig festgestellt werden kann. Die recht intensive Bande bei 3619.8 cm^{-1} muß dabei wohl einer O–H-Valenzschwingung zugeschrieben werden. In Frage kommende CH_4OSi-Isomere, die ein solches Strukturelement enthalten, sind beispielsweise der Silicium-Methanol-Komplex (**10**) sowie Hydroxy(methyl)-

silylen (**2**). Die Bildung von Komplex **10** wäre dabei eine Analogie zur Reaktion von atomarem Silicium mit Wasser, bei der eine entsprechende Anlagerung des Si-Atoms an das Sauerstoffatom des Substrats beobachtet wird [21]. Auch die Bildung von Hydroxy(methyl)silylen (**2**) durch Insertion des Siliciumatoms in die C–O-Bindung des Methanols ist aufgrund dessen günstiger energetischer Lage durchaus denkbar.

Andere CH_4OSi-Spezies, die eine OH-Gruppe enthalten, sind Hydroxysilen (**3**) sowie (Hydroxymethyl)silylen (**8**). Das Vorliegen von **3** kann aber mit ziemlicher Sicherheit ausgeschlossen werden, da es sich bei der fraglichen Spezies um ein Primärprodukt der Atomreaktion handelt (die Bande bei $3620\,cm^{-1}$ ist unmittelbar nach der Codeposition zu sehen). Denkbare Primärreaktionen von atomarem Silicium mit Methanol sollten sich jedoch auf die Insertion in σ-Bindungen des Substrats beschränken. Die Bildung von **3** ist aber nicht durch eine solche Reaktion möglich, sondern eher durch einen 1,2-H-shift aus Hydroxy(methyl)silylen (**2**). Solche Isomerisierungen werden jedoch in der Regel photochemisch initiiert [20, 22–24].

Die Bildung von (Hydroxymethyl)silylen (**8**) erscheint ebenfalls unwahrscheinlich, da die hierfür erforderliche Insertion in die C–H-Bindung ungünstiger sein sollte als die Einschiebungen in die C–O- bzw. O–H-Bindung. Als Indiz dafür mag gelten, daß mit den Methoxysilylenen **4** die Insertionsprodukte in die O–H-Bindung in großen Mengen gebildet werden, während die Reaktion von atomarem Silicium mit Methan, die nur unter Insertion in eine C–H-Bindung verlaufen kann, ausbleibt [25].

Die berechneten Schwingungsspektren zusätzlicher CH_4OSi-Isomerer (Tabelle 3) lassen sich allesamt nicht besonders gut mit den oben aufgeführten experimentellen Banden in Einklang bringen. Das berechnete IR-Spektrum des Silicium-Methanol-Komplexes (**10**) hat zwar eine recht große Ähnlichkeit mit der

Tebelle 3. Berechnete IR-Spektren verschiedener CH_4OSi-Isomerer (BLYP/6–31G**, Wellenzahlen (cm^{-1}), absolute Intensität ($km \cdot mol^{-1}$) in Klammern)

2b	6	8	9	10
43.8 (<1)	473.0 (2)	303.5 (36)	277.7 (6)	146.1 (2)
267.7 (4)	558.9 (1)	458.6 (26)	337.2 (107)	178.8 (<1)
505.3 (43)	631.5 (27)	598.9 (27)	377.4 (90)	283.0 (55)
608.4 (42)	695.2 (27)	639.1 (22)	441.9 (37)	495.7 (119)
668.3 (34)	728.1 (92)	692.9 (42)	620.6 (35)	934.6 (99)
726.6 (45)	765.9 (32)	802.9 (38)	748.3 (18)	1037.2 (36)
800.9 (52)	916.6 (32)	874.4 (76)	801.8 (30)	1135.3 (1)
903.5 (165)	1005.9 (85)	1010.5 (38)	912.2 (59)	1310.3 (14)
1243.0 (22)	1075.3 (1)	1137.1 (11)	1168.3 (299)	1435.7 (3)
1418.7 (9)	1091.7 (2)	1238.5 (27)	1209.2 (4)	1460.9 (11)
1428.4 (8)	1447.3 (6)	1443.1 (2)	1432.7 (40)	1470.9 (6)
2937.9 (4)	2194.4 (66)	1959.2 (313)	2175.0 (43)	2973.0 (49)
3009.8 (13)	2215.3 (147)	3026.2 (25)	2207.6 (117)	3050.8 (22)
3046.7 (12)	3004.6 (26)	3115.6 (11)	3019.7 (20)	3099.7 (8)
3683.0 (60)	3082.8 (17)	3653.1 (73)	3633.4 (58)	3649.4 (60)

beobachteten Bandengruppe, jedoch sollte man erwarten, daß dieser Komplex schon bei extrem langwelliger Bestrahlung umlagert, wie dies für den analogen Silicium-Wasser-Komplex beobachtet wird [21].

Von Komplex **10** könnte ebenso eine recht breite Bande bei 979.0 cm^{-1} stammen, die bei der Bestrahlung der nach der Codeposition erhaltenen Matrix mit Licht der Wellenlängen $\lambda > 385$ nm verschwindet und auch nach mehrfachem Wechsel der Wellenlängen nicht wieder beobachtet wird. Diese Bande stimmt ebenfalls recht gut mit der für den Komplex **10** errechneten Lage der O–H-Deformationsschwingung (934.6 cm^{-1}) überein, und die beobachtete Differenz zur entsprechenden Schwingung des freien Methanols liegt mit 54 cm^{-1} in der gleichen Größenordnung wie die nach der DFT-Methode errechnete Differenz (73 cm^{-1}). Allerdings kann dieser Befund – auf der Grundlage dieser einzigen Absorption – nur als Indiz gelten.

Bei der oben erwähnten Belichtung bei 313 nm entstehen neue Banden bei 696.8, 830.3, 925.6, 1074.5, 1191.5, 1259.2, 2186.6, 2204.8 und 3708.0 cm^{-1}. Auch deren Herkunft bleibt unklar. Die Bande bei 3708 cm^{-1} dürfte ebenfalls von einer O–H-Streckschwingung herrühren, während die Absorptionen bei 926, 2187 und 2205 cm^{-1} in typischen Bereichen für die Deformations- bzw. Streckschwingungen von H_2Si-Gruppen liegen. Das einzige CH_4OSi-Isomere, das beide funktionelle Gruppen aufweist, ist Silen **9**, das auch als das Tautomere von Formylsilan (**5**) betrachtet werden kann. Aufgrund ihres recht hohen Energieinhalts ist das Vorliegen dieser Spezies aber nicht unbedingt zu erwarten. Schon eher könnte man mit der Bildung von Oxasiliran (**6**) rechnen, das aus Methoxysilylen (**4**) durch eine Insertion des silylenischen Zentrums in eine C–H-Bindung entstehen könnte. Da die Banden der Silylene **4** jedoch bei der Belichtung nicht abnehmen, kann dieser Reaktionsweg ausgeschlossen werden. Wie schon im Falle der bei der 313 nm–Belichtung verschwindenden Bandengruppe kann auch für die entstehenden Absorptionen keine definitive Zuordnung getroffen werden.

Experimentelles

Die für die Matrixisolation und Photolysen benutzten Apparaturen sowie der verwendete Ofen zur Verdampfung von Silicium wurden bereits beschrieben [15].

Dank

Diese Arbeit wurde von der *Deutschen Forschungsgemeinschaft* und dem *Fonds der Chemischen Industrie* gefördert.

Literatur

[1] 29. Mitteilung über Hetero-π-Systeme; 28. Mitteilung: Lit. [17]
[2] a) Rochow EG (1945) J Am Chem Soc **67**: 963; b) Rochow EG (1948) J Am Chem Soc **70**: 2170
[3] Okamoto M, Osaka M, Yamamoto K, Suzuki E, Ono Y (1993) J Cat **143**: 64 und dort zitierte Lit.
[4] Suzuki E, Ono Y (1990) J Cat **125**: 390
[5] Bateman JE, Horrocks BR, Houlton A (1997) J Chem Soc Faraday Trans **93**: 2427
[6] Okamoto M, Onodera S, Okano T, Suzuki E, Ono Y (1997) J Organomet Chem **531**: 67
[7] Tan MX, Kenyon CN, Krüger O, Lewis NS (1997) J Phys Chem B **101**: 2830

[8] Krüger O, Kenyon CN, Tan MX, Lewis NS (1997) J Phys Chem B **101**: 2840
[9] Kenyon CN, Tan MX, Krüger O, Lewis NS (1997) J Phys Chem B **101**: 2850
[10] Fajardo AM, Lewis NS (1997) J Phys Chem B **101**: 11136
[11] Elsner FH, Woo HG, Tilley TD (1988) J Am Chem Soc **110**: 313
[12] Kwak YW, Lee KK (1997) J Organomet Chem **542**: 219
[13] Pola J, Cuknová D, Ponec R, Stanczyk W (1994) J Organomet Chem **468**: 49
[14] Maier G, Reisenauer HP, Egenolf H (1998) In: Auner N, Weis J (Hrsg) Organosilicon Chemistry III – From Molecules to Materials. VCH, Weinheim, S 31
[15] Maier G, Reisenauer HP, Meudt A, Egenolf H (1997) Chem Ber/Recueil **130**: 1043
[16] Maier G, Reisenauer HP, Egenolf H, Glatthaar J (1998) Eur J Org Chem 1307
[17] Maier G, Reisenauer HP, Egenolf H (1998) Eur J Org Chem 1313
[18] GAUSSIAN 94 (1995) Revision B. I, Frisch MJ, Trucks GW, Schlegel HB, Gill PMW, Johnson BG, Robb MA, Cheeseman JR, Keith T, Petersson GA, Montgomery JA, Raghavachari K, Al-Laham MA, Zakrzewski VG, Ortiz JV, Foresman JB, Cioslowski J, Stefanov BB, Nanayakkara A, Challacombe M, Peng CY, Ayala PY, Chen W, Wong MW, Andres JL, Replogle ES, Gomperts R, Martin RL, Fox DJ, Binkley JS, Defrees DJ, Baker J, Stewart JP, Head-Gordon M, Gonzales C, Pople JA, Gaussian, Inc., Pittsburgh, PA
[19] Gordon MS, George C (1984) J Am Chem Soc **106**: 609
[20] Maier G, Mihm G, Reisenauer HP, Littmann D (1984) Chem Ber **117**: 2369
[21] Ismail ZK, Fredin L, Hauge RH, Margrave JL (1982) J Chem Phys **77**: 1626
[22] Maier G, Mihm G, Reisenauer HP (1981) Angew Chem **93**: 615; (1981) Angew Chem Int Ed Engl **20**: 597
[23] Reisenauer HP, Mihm G, Maier G (1982) Angew Chem **94**: 864; (1982) Angew Chem Suppl 1785; (1982) Angew Chem Int Ed Engl **21**: 854
[24] Maier G, Reisenauer HP, Jung J, Pacl H, Egenolf H (1998) Eur J Org Chem 1297
[25] Egenolf H (1998) Dissertation, Justus-Liebig-Universität Gießen

Received June 13, 1998. Accepted July 24, 1998

Zum Spreitverhalten von Mischungen definierter Trisiloxantenside auf niederenergetischen Oberflächen

Roland Wagner*, **Yuling Wu**, and **Hans v. Berlepsch**

Max-Planck-Institut für Kolloid- und Grenzflächenforschung, D-12489 Berlin, Germany

The Spreading Behaviour of Mixtures of Defined Trisiloxane Surfactants on Low Energy Surfaces

Summary. The spreading behaviour of mixtures of defined trisiloxane surfactants of the general structure $((CH_3)_3SiO)_2CH_3Si(CH_2)_3(OCH_2CH_2)_nOCH_3$, $n = 3$–9, has been investigated. The initial spreading rates of the equimolar binary mixtures $n = 3+9$, $n = 4+8$, and $n = 5+7$ are related to their phase transition temperatures from the lamellar phase (L_α) into the two phase state (2Φ) and considerably exceed those of the corresponding single components.

Keywords. Spreading behaviour; Superspreading; Trisiloxane surfactant.

Einleitung

Wäßrige Lösungen ausgewählter Siloxantenside haben die Fähigkeit, niederenergetische Oberflächen (Festkörperoberflächenspannungen $\gamma_{sv} < 30$ mN/m) schnell zu benetzen [1,2]. Bei den zum Einsatz kommenden „Superspreitern" handelt es sich um Trisiloxanderivate, in denen die permethylierte Siloxaneinheit über einen kurzen C_3-Spacer mit einer Oligoethylenglycolkette verknüpft ist. Für typische Vertreter dieser Tensidgruppe (Silwet L77) ist eine breite Verteilung der Länge der Oligoethylenglycoleinheit charakteristisch. In den komplexen Mischungen werden Spezies mit 3 bis 12 Glycoleinheiten gefunden [3].

Es ist bekannt, daß die Spreitgeschwindigkeit wässriger Lösungen nichtionogener Tenside stark von der Oberflächenenergie des zu benetzenden Festkörpers und der mittleren Kettenlänge der Oligoethylenglycoleinheit abhängt [4]. Es wurde weiterhin gefunden, daß die äußerst komplexen Mischungen zweier kettenlängenverteilter Tenside nicht das Verhalten der entsprechend monomodal verteilten Produkte simulieren [5]. Ein eindeutiger Zusammenhang zwischen Spreitgeschwindigkeiten, Oberflächenenergien und der Art der Kettenlängenverteilung konnte allerdings nicht hergestellt werden. Eine Abhängigkeit der Spreitgeschwindigkeit von der Temperatur wurde trotz damit einhergehender drastischer Änderung des Phasenzustandes ebenfalls nicht beobachtet [5].

* Corresponding author

Zur Klärung der auf Kettenlängenverteilungen beruhenden Fragen haben wir Tenside der Struktur $((CH_3)_3SiO)_2CH_3Si(CH_2)_3(OCH_2CH_2)_nOCH_3$ mit $n = 3$–9 (*EO3-EO9*) definiert synthetisiert [6]. Es konnte gezeigt werden, daß auf einer hydrophoben und unpolaren Oberfläche in Abhängigkeit von der Temperatur eindeutige Maxima der Spreitgeschwindigkeit existieren, welche streng mit dem Phasenverhalten korreliert sind [7, 8]. Die höchsten Spreitgeschwindigkeiten werden stets für Tenside beobachtet, deren Phasenumwandlungstemperatur $L_\alpha \rightarrow 2\Phi$ wenig unterhalb der Spreittemperatur liegt. Schnelles Spreiten auf niederenergetischen Oberflächen ist somit ein Phänomen zweiphasiger, fein dispergierter Systeme. Dieses vom Verhalten polydisperser Systeme abweichende Resultat für Einzelkomponenten legt Spreitversuche mit gezielt hergestellten Mischungen nahe. Speziell soll über die Ergebnisse der Substitution des bei Raumtemperatur schnell spreitenden Hexaethylenglycolderivates durch zwei- und dreikomponentige Systeme berichtet werden.

Ergebnisse und Diskussion

Spreitverhalten der Einzelkomponenten EO3–EO9

Die in Abb. 1 dargestellten Anfangsspreitgeschwindigkeiten belegen, daß bei 23°C das Hexaethylenglycolderivat *EO6* der überlegene Spreiter ist. Es folgen das hydrophilere Heptaethylenglycolprodukt *EO7* und das hydrophobere Pentaethylenglycolderivat *EO5*. Die kurzkettigen Verbindungen *EO4* und *EO3* spreiten nur noch sehr langsam bzw. nicht mehr. Diese Reihung der Tenside wird unter Einbeziehung der Phasenumwandlungstemperatur T_c aus der lamellaren Phase (L_α) in ein zweiphasiges System (2Φ, tensidreiche Phase + Wasserphase) verständlich (Tab. 1.)

Für eine 1%ige (w/w) wäßrige Lösung des Derivates *EO6* beträgt T_c 19.5°C. Bei 23°C liegt das System somit zweiphasig vor. Ein fein dispergierter Zustand

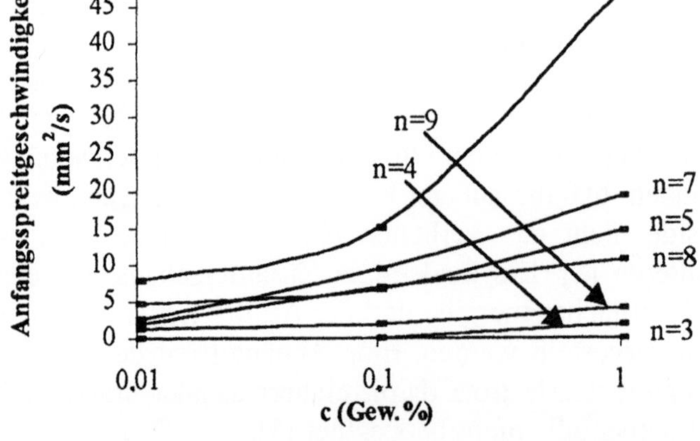

Abb. 1. Konzentrationsabhängigkeit der Anfangsspreitgeschwindigkeit für definierte Einzelkomponenten ($T = 23$°C)

Tablle 1. Phasenumwandlungstemperaturen $L_\alpha \rightarrow 2\Phi$ Einzelkomponenten, binäre und ternäre Tensidmischungen

Tensid	*EO5*	*EO6*	*EO7*	*EO8*	*EO9*	*EO3+9*	*EO4+8*	*EO5+7*	*EO4+6+8*
T_c(°C)	6.5	19.5	31.5	36.5	32.0	<0	15.4	19.0	17.2

kann jedoch über eine längere Zeit stabilisiert werden, da die Triebkraft für die makroskopische Phasenseparation klein ist. Für die hydrophobere Verbindung *EO5* beträgt T_c 6.5°C, weshalb eine schnellere makroskopische Separation zu erwarten ist. Unter identischen Randbedingungen wurden mit dynamischer Lichtstreuung für die Verbindung *EO5* größere Teilchen festgestellt [7].

Die wäßrigen Lösungen der Verbindungen *EO7* bis *EO9* liegen unter den experimentellen Bedingungen einphasig vor. Die im Vergleich zum Derivat *EO6* geringeren Spreitgeschwindigkeiten belegen, daß der Transport von Tensidmolekülen an die Grenzflächen flüssig/gasförmig und fest/flüssig aus der lamellaren Phase heraus weniger effektiv verläuft. Bemerkenswert ist aber, daß der Abfall der Spreitgeschwindigkeit weniger drastisch als bei den kurzkettigen Verbindungen verläuft.

Spreitverhalten equimolarer Mischungen

Um eindeutige Aussagen bezüglich des Einflusses der für kommerzielle Produkte typischen Kettenlängenverteilungen auf das Spreitverhalten zu gewinnen, wurden Lösungen von Einzeltensiden mit equimolaren binären und ternären Tensidmischungen verglichen (Abb. 2). Letztere weisen bei gleichem Gesamttensid-, Trisiloxan- und Ethylenglycolgehalt lediglich eine Längenverteilung der Oligoethylenglycolkette auf. Da für das Hexaethylenglycolderivat *EO6* die höchste

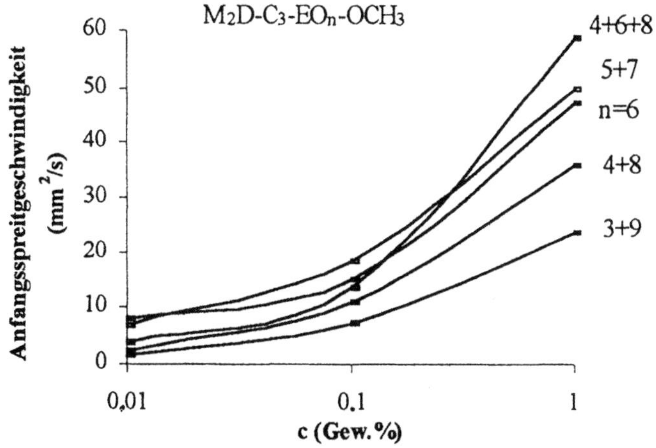

Abb. 2. Konzentrationsabhängigkeit der Anfangsspreitgeschwindigkeit für binäre und ternäre Tensidmischungen ($T = 23$°C)

Spreitgeschwindigkeit einer Einzelkomponente gefunden wurde, können die größten Effekte für auf dieses Derivat zielende Mischungen erwartet werden.

Es ist erkennbar, daß für hohe Tensidkonzentrationen die Einzelkomponente *EO6* deutlich von der ternären Mischung *EO4+6+8* übertroffen wird, obwohl die Moleküle des besten Einzelspreiters nur noch zu einem Drittel enthalten sind.

Aufschlußreich ist die Abfolge der binären Mischungen. Während *EO5+7* mit *EO6* vergleichbar ist, sinken die Spreitgeschwindigkeiten der Mischungen *EO4+8* und *EO+9*, auf hohem Niveau verbleibend, ab. Offensichtlich wirkt sich eine zunehmende Differenz der Kettenlängen negativ auf die Spreitfähigkeit aus.

Daß es sich beim Übergang zu Mischungen um einen qualitativen Umschlag handelt, wird aus einem Vergleich der Spreitgeschwindigkeiten der Mischungen mit denen der jeweiligen Einzelkomponenten deutlich. Die Mischung *EO3+9* (1% (w/w)) spreitet um den Faktor 6 schneller als die schnellste enthaltene Komponente *EO9*. Für die Mischungen *EO4+8* und *EO5+7* werden entsprechende Steigerungen um die Faktoren 3.5 bzw. 2.5 beobachtet.

Die Überlagerung individueller durch kollektive Eigenschaften wird ebenfalls aus der Entwicklung der Phasenübergangstemperaturen T_c erkennbar. Die für Einzelkomponenten gefundene Korrelation zwischen Spreitgeschwindigkeit und T_c ist nicht länger universell gültig. Zwar wird für die binären Mischungen *EO3+9*, *EO4+8* und *EO5+7* eine eindeutige Abhängigkeit der Spreitgeschwindigkeit von T_c gefunden, jedoch spreitet die binäre Mischung *EO3+9* ($T_c < 0$°C) schneller als die Einzelverbindung *EO5* ($T_c = 6.5$°C). Das Vorhandensein einer Hierarchie deutet sich beim Vergleich der ternären Mischung *EO4+6+8* ($T_c = 17.2$°C) mit der binären Mischung *EO5+7* ($T_c = 19$°C) und der Einzelkomponente *EO6* ($T_c = 19.5$°C) an. Für Mischungen identischen mittleren Glycolgehaltes scheint das Zusammenwirken einer Vielzahl von Komponenten von Vorteil zu sein. Die Toleranz gegenüber Abweichungen von T_c wird immer stärker ausgeprägt. Die Ursache für diese erhöhte Toleranz ist noch nicht eindeutig erklärbar, da nur wenige systematische Ergebnisse zum Phasenverhalten definierter Tensidmischsysteme vorliegen [9, 10]. Es ist vorstellbar, daß die langkettigen, hydrophilen Verbindungen als Lösungsvermittler für die sonst praktisch inaktiven kurzkettigen Moleküle wirken und damit eine Erhöhung der Spreitgeschwindigkeit ermöglichen.

Experimentelles

Die Synthese und chemische Charakterisierung der definierten Oligoethylenglycolderivate der allgemeinen Struktur $((CH_3)_3SiO)_2CH_3Si(CH_2)_3(OCH_2CH_2)_nOCH_3(n = 3\text{–}9)$ ist an anderer Stelle beschrieben worden [6].

Alle Spreitversuche wurden bei 23°C and 29% relativer Luftfeuchte durchgeführt. Hierzu wurden Tensidlösungen in Konzentrationsbereich von 0.01 bis 1% (w/w) hergestellt, in Ultraschallbad 2 min behandelt und davon anschließend 10 μl-Tropfen auf die Oberfläche hydrophober, trimethylsilylierter Siliciumwafer aufgesetzt. Die Spreitversuche wurden mit einem Videosystem aufgezeichnet und anschließend die Tropfenflächen als Funktion der Zeit ausgewertet. Aus dem Anstieg der Tropfenflächen-Zeit-Kurven kann die Anfangsspreitgeschwindigkeit berechnet werden.

Die Festkörperoberflächenspannung γ_{sv} der Wafer beträgt 23.5 mN/m und wurde aus Randwinkelmessungen gegen Alkane bestimmt [6, 11, 12]. Der Randwinkel gegen Wasser beträgt 91°.

Dank

Das Projekt *Polyhydroxylierte Siliciumverbindungen* wird von der *Deutschen Forschungsgemeinschaft* unterstützt (Reg. No. WA 1043/1–2).

Literatur

[1] Zhu S, Miller WG, Scriven LE, Davis HT (1994) Colloids Surfaces **90**: 63
[2] Murphy DS, Policello GA, Goddard ED, Stevens PJG (1993) In: Devisetty BN, Chasin DG, Berger PD (eds) Pesticide Formulations and Application Systems, vol 12, ASTM STP 1146. Philadelphia, p 45
[3] Knoche M, Tamura H, Bukovac MJ (1991) J Agric Food Chem **39**: 202
[4] Stoebe T, Lin Z, Hill RM, Ward MD, Davis HT (1997) Langmuir **13**: 7270
[5] Stoebe T, Lin Z, Hill RM, Ward MD, Davis HT (1996) Langmuir **12**: 337
[6] Wagner R, Wu Y, Czichocki G, von Berlepsch H, Weiland B, Rexin F, Perepelittchenko L (1998) Appl Organomet Chem (in press)
[7] Wagner, R, Wu, Y, Czichocki G, v. Berlepsch H, Rexin Perepelittchenko L (1998) Appl Organomet Chem (submitted)
[8] Wagner, R, Wu, Y, Czichocki G, v. Berlepsch H, Weiland B, Rexin F, Perepelittchenko L (1998) In: Auner N, Weis J (eds) Organosilicon Chemistry, vol IV. VCH, Weinheim (in press)
[9] Jonströmer M, Strey R (1992) J Phys Chem **96**: 5993
[10] Strey R, Schomäcker R, Roux D, Nallet F, Olson U (1990) J Chem Faraday Trans **86**: 2253
[11] Driediger O, Neumann AW, Sell PJ (1965) Kolloid-Zeitschr u Z Polym **201**: 101
[12] van Oss CJ, Chaudhury MJ, Good RJ (1988) Chem Rev **88**: 927

Received June 15, 1998. Accepted (revised) September 7, 1998

Springer-Verlag
und Umwelt

Als internationaler wissenschaftlicher Verlag sind wir uns unserer besonderen Verpflichtung der Umwelt gegenüber bewußt und beziehen umweltorientierte Grundsätze in Unternehmensentscheidungen mit ein.

Von unseren Geschäftspartnern (Druckereien, Papierfabriken, Verpackungsherstellern usw.) verlangen wir, daß sie sowohl beim Herstellungsprozeß selbst als auch beim Einsatz der zur Verwendung kommenden Materialien ökologische Gesichtspunkte berücksichtigen.

Das für dieses Buch verwendete Papier ist aus chlorfrei hergestelltem Zellstoff gefertigt und im pH-Wert neutral.

MIX
Papier aus verantwortungsvollen Quellen
Paper from responsible sources
FSC® C105338

If you have any concerns about our products,
you can contact us on
ProductSafety@springernature.com

In case Publisher is established outside the EU,
the EU authorized representative is:
Springer Nature Customer Service Center GmbH
Europaplatz 3, 69115 Heidelberg, Germany

Printed by Libri Plureos GmbH
in Hamburg, Germany